LEARNING AND KNOWLEDGE FOR THE NETWORK SOCIETY

International Series on Technology Policy and Innovation

IC^2 Institute, The University of Texas at Austin, Texas
and
The Center for Innovation, Technology and Policy Research,
Instituto Superior Técnico, Lisbon, Portugal

General Editors:

Manuel V. Heitor, Center for Innovation, Technology and Policy
Research, Instituto Superior Técnico, Lisbon, Portugal

David V. Gibson, IC^2 Institute, The University of Texas
at Austin, Texas

Pedro Conceição, Center for Innovation, Technology and Policy
Research, Instituto Superior Técnico, Lisbon, Portugal

LEARNING AND KNOWLEDGE FOR THE NETWORK SOCIETY

Edited by
David V. Gibson, Manuel V. Heitor, and Alejandro Ibarra-Yunez

Purdue University Press
West Lafayette, Indiana

Library of Congress Cataloging-in-Publication Data

Learning and knowledge for the network society /
edited by David V. Gibson, Manuel V. Heitor,
and Alejandro Ibarra-Yunez.
p. cm. -- (International series on technology
policy and innovation)
Includes bibliographical references and index.
ISBN 1-55753-356-3 (casebound)
1. Technological innovations--Economic
aspects. 2. Technology and state.
3. Knowledge management. 4. Organizational
learning. 5. Information society--Management.
I. Gibson, David V. II. Heitor, M. V. (Manuel
V.), 1957- III. Ibarra Yunez, Alejandro. IV. Title.
V. Series.

HD45.L34 2005
338'.064--dc22

2004019837

Contents

Foreword   ix
     Ramiro Wahrhaftig

Acknowledgments   xi

1.   Introduction: Technological Change in the Network Society: Governance,   1
     Inclusiveness and Development
     Pedro Conceição and Manuel V. Heitor

**PART I: BUILDING THE LEARNING ECONOMY: TRENDS AND
PERSPECTIVES FOR GOVERNANCE**

Introductory Note   21
     Alejandro Ibarra-Yunez

2.   Innovation Policy and Knowledge Management in the Learning Economy   25
     Bengt-Åke Lundvall

3.   Innovation Policies in the Knowledge Era: A South American Perspective   57
     Helena Maria Martins Lastres and José Eduardo Cassiolato

4.   Development Policy and the Economy: Wither the State?   73
     Robert H. Wilson

5.   Gateway Airports, Speed and the Rise of Aerotropolis   99
     John D. Kasarda

6.   Institutions and Knowledge Networks: The Chinese Experience   109
     Leslie Young

7.   Regulatory Shields and Firms' Conduct: Application to Telecommunications   121
     Companies in Chile and Mexico
     Alejandro Ibarra-Yunez

PART II: NETWORKING FOR REGIONAL ECONOMIC DEVELOPMENT: LOOKING FOR INCLUSIVENESS

Introductory Note   137
    Manuel V. Heitor

8. Getting the Tail to Wag: Enabling Innovation in Small/Medium-sized   141
    Enterprises
    John Robert Bessant

9. Productivity and Regional Density   165
    Rui Miguel L. N. Baptista

10. From Technology Policy for Regions to Regional Technology Policy: 185
    Towards a New Policy Strategy in the EU
    Michael Guth

11. Regional Innovation Policy at the Community Level: Evidence from   201
    the RITTS Programme to Promote Regional Innovation Systems
    Fabienne Corvers

12. Information and Communication Technologies and Economic Development   227
    of Peripherally Located Regions: Experiences in the Netherlands
    Marina van Geenhuizen

13. Network Building Between Research Institutions and Small & Medium   245
    Enterprises: Dynamics of Innovation Network Building and
    Implications for a Policy Option
    Junmo Kim

14. Innovation Clusters in Latin America   261
    Isabel Bortagaray and Scott Tiffin

15. Globalization and Industrial Restructuring in Mexico: The Electronics and   307
    Automobile Industries
    Cristina Casanueva Reguart

16. A Network of Knowledge-Intensive Clusters for Regional Development: 341
    The Paraná W-Class Program
    Carlos Quandt and Luiz Márcio Spinosa

17. The Nature of the Networks of Innovation and Technological Information   357
    Diffusion in a Region in the Initial Stages of Industrial Development
    Décio Estevão do Nascimento

18. Digital Cities and the Network Society: Towards a Knowledge-Based View   375
    of the Territory?
    Manuel V. Heitor and José Luiz Moutinho

PART III: LANGUAGE, DEVELOPMENT AND POLICY

Introductory Note   405
     Lawrence S. Graham

19. The Universal Network Language: An Electronic Esperanto?   409
    No, a Language for Computers
     Tarcisio Della Senta and Mambillikalathil G. K. Menon

20. The Color of Voice, The Resonance of Language: Precedents of the   423
    Digital Divide
     Mercedes Lynn de Uriarte

21. Foreign Language Materials for Business Portuguese: The Role of   443
    Technology in the Development of Foreign Language Curricula
     Orlando R. Kelm

22. Portuguese Language without Frontiers   453
     Regina L. P. Dell'Isola

Index   457

About the Contributors   461

Foreword

This book is the result of a process, initiated in the International Conference on Technology Policy and Innovation held in Curitiba in June 2000, which has involved a network of researchers from different parts of the world working in emerging topics worldwide. Curitiba, the capital of Paraná State, in the south of Brazil, had the honor to host this instigating and all-including event, which is becoming one of the most important forums of discussions and proposals of technological policies and models of innovation. For many reasons, I am very grateful for the task to preface this book. First, for being for such a long time close to the friends of Lisbon and Austin that had the excellent initiative to promote this series of conferences; second, for the fact of being one of the first to defend Curitiba as the site of the conference; also, for the positive impact that this accomplishment meant for Paraná and Brazil; for the many friends we were able to make and for the remembrances we keep from them. As responsible for the technological policies of Paraná, and trusting that the promotion of innovation is one of its main components, I must confess that the conference was felt by me as a real accomplishment.

The conference was held at a special moment when Curitiba and Paraná consolidated its option for innovation as a factor of wealth creation, and for technology as a tool of promoting welfare and the reduction of social inequalities. The joint activities with the impressive number of researchers, professors, government authorities and entrepreneurs who work in the promotion of technology and innovation, from all over the world, was short but intense. It provided us the certainty that the processes of regional development are more easily understood and managed with the knowledge of models, the structuring of policies and the study of successful experiences.

As a coincidence, this conference was a point of an important inflection in the curve of growth of the series of International Conferences on Technology Policy and Innovation, which magnified its width, subjects and people. As one of the first promoters of the event, and as a member of the organizing committee, I feel myself honored to have collaborated in its success.

The three great subjects of this book can be seen as chapters of an agenda for development, adequate for the present situation, of challenges and hopes in the new information society. The regulating and inductive role of the government, especially important in emerging countries, is approached in Part I, in excellent and varied points of view. Part II deals with the question of economic development, including the analysis of major determinants of regional innovation systems. Finally, the seldom-remembered relevance of language as a bridge—or a barrier—to the learning process in global societies is argued in a selection of chapters covering the essential part of the subject.

This sample is very representative of what the conference meant and of what it proposes. It has become one of the main forms of the decision-making process and discussions in the area of technological policy and innovation. Matching the academy and the enterprise sector, contrasting regulation and initiative, matching coherence and renewing, the conference was meant to have a bright future assured by the effort and the ability of its promoters and collaborators.

I am convinced that the people interested in technological policies and innovation, in regional development and economic progress with social justice, will find in the chapters that follow enough material for up-to-date reflections of the best quality.

Ramiro Wahrhaftig

Acknowledgments

The editors of *Learning and Knowledge for the Network Society* thank the organizers and sponsors of the 4th International Conference on Science and Technology Policy and Innovation (ICTPI). The sponsoring academic institutions were the Instituto Superior Técnico (IST), Lisbon, Portugal, through the Center for Innovation, Technology and Policy Research, IN+; the University of Texas at Austin, including the IC^2 Institute; and the Pontifical Catholic University of Paraná (PUC-PR). Thanks are also due to the Government of Paraná and the Brazilian Ministry of Science and Technology.

We are also grateful to the conference's International Organizing Committee and Program Committee for planning the conference and for reviewing and selecting papers to be presented. And we are grateful to those who so effectively managed the four-day event, in particular the staff at the Pontifícia Universidade Católica do Paraná (Catholic University of Paraná), Brazil.

The editors thank the contributors to this volume for sharing their insights and regional perspectives on a range of important topics regarding science and technology policy and innovation. Finally, the editors are especially grateful for the dedicated and excellent publication effort including Miguel Silveiro, Publications Coordinator at the Center for Innovation, Technology, and Policy Research, IST, Lisbon, Portugal; Katie Chase, copyeditor; and the staff of Purdue University Press for successfully bringing this volume to publication.

1

INTRODUCTION:
Technological Change in the Network Society: Governance, Inclusiveness and Development

Pedro Conceição and Manuel V. Heitor

This book draws on recent conceptual approaches to economic growth, in which the accumulation of knowledge is the fundamental driving force behind growth. This fact is reflected in the trend in the most developed regions towards an increasing investment in advanced technology, research and development, education, and culture. At the same time, analysis has shown that firms' competencies are characterized by stability and inertia and that the phenomena of increasing returns and path-dependence affect the nature of the innovation processes and the dynamics of firms in any regional context. On the other hand, the specific characteristics of local systems of innovation are expected to play a significant role in shaping the organization of innovative activity. As a result, concepts such as learning ability, creativity and sustained flexibility have gained greater importance as guiding principles for the conduct of individuals, institutions, nations and regions. It is thus legitimate to question the traditional way of viewing the role that contemporary institutions play in the process of economic development and to argue for the need to promote systems of innovation and competence building based on learning and knowledge networks.

Under the broad designation of "learning and knowledge for the network society" this book brings together a range of experts discussing technology, policy and management in a context much influenced by a dynamic of change and a necessary balance between the creation and diffusion of knowledge. While the idea of inclusive development developed in a previous book (Conceição et al., 2002) entails a process of shared prosperity across the globe following local specific conditions, it is crucial to understand the dynamics of the process of knowledge accumulation, which drives a learning society. Thus, this book includes a set of extended contributions that are largely grounded on empirical experiences of different regional and national contexts.

To set the overall context for the book, we argue in this introductory chapter that value-based networks have the potential to make both public administration and markets more effective, which helps promoting learning trajectories for the inclusive development of society, but require effective public investments in intangible structures fostering competence building. The analysis builds on the concept of social capital, as a relational infrastructure for collective action, in a context much influenced by a dynamic of change and a necessary balance between the creation and diffusion of knowledge.

BACKGROUND

While much attention has been devoted to information and communication technologies, a more fundamental change at the start of the new millennium is the increasing importance of innovation for economic prosperity and the emergence of a learning society. In fact, this book clearly shows that innovation should be understood as a broad social and economic activity: it should transcend any specific technology, even if revolutionary, and should be tied to attitudes and behaviours oriented towards the exploitation of change by adding value.

The book also builds on the idea of inclusive learning and argues that it is crucial to understand the features of knowledge-induced growth in rich countries, as well as the challenges and opportunities for late-industrialized and less favoured regions. In this context, Conceição, Heitor and Veloso (2003) emphasized the relative importance of infrastructures and incentives, but considered the increasingly important role of institutions towards the development of social capital. This is because learning societies will increasingly rely on "distributed knowledge bases", as a systematically coherent set of knowledge, maintained across an economically and/or socially integrated set of agents and institutions.

This conclusion is based on the broad innovation framework of Smith (2000), who suggests that the knowledge bases of mature and traditional industries are cognitively deep and complex, as well as institutionally distributed. Thus, rather than relying exclusively on "high-technology" sectors, there is a need to integrate policies relating to education, science and technology, and social and economic development, so that there is a diversification of actions to support the creation and diffusion of distributed knowledge bases. This is particularly applicable to catching-up countries and regions, with the practical consequence that growth will not be based just on the creation of new sectors, but on the internal transformation of sectors which already exist, namely by exploiting their distributed knowledge bases through adequate institutions.

This leads us to conceptualise "learning" and the process of knowledge accumulation, as a framework to understand the new demands for being innovative.

FRAMING THE CONCEPTUAL UNDERSTANDING OF KNOWLEDGE

Many contributions in recent years have confirmed the perception that the creation and dissemination of knowledge are fundamental factors for the promotion of economic growth, although the scarcity of empirical data on intangible economic factors makes it extremely difficult to demonstrate the growing importance of knowledge. In fact, economic growth has traditionally been explained as being the result of increases in the labour and capital factors and technological change. However, in the light of recent empirical analysis, it is necessary to rethink how these three factors influence the process of economic development.

With regard to the contribution of the labour factor, the facts show that a quantitative increase in population is not sufficient, since developed economies produce ever more intangible factors, creating employment mainly in the service sector, in which educational and professional qualifications are required. It is thus essential for growth and job creation to develop human capital, providing access to more and better skills, particularly through education.

With regard to the contribution of capital, it can be seen that the accumulation of intangible assets is gaining in relative importance compared to physical capital. The importance of knowledge is accordingly seen not only in its contribution to technological change, a fact that has led to a rethinking of traditional ways of explaining growth. The new economic growth theories, which are not analysed here, bring together many of these ideas, putting forward the message that the accumulation of knowledge, which we will identify with learning, is the most important factor in explaining economic development.

Our inspiration to frame the process of knowledge accumulation comes from the contribution of Lundvall and Johnson (1994), who challenge the commonplace by introducing the simple, but powerful, idea of learning. Lundvall and Johnson suggest that a "learning economy", rather than a "knowledge economy", describes better the way in which knowledge contributes to development. The fundamental difference between the two expressions is associated with the fact that the former considers a dynamic perspective. According to Lundvall and Johnson, some types of knowledge do indeed become more important, but there is also knowledge that becomes less important. There is both knowledge creation and knowledge destruction. By forcing us to look at the process, rather than at the mere accumulation of knowledge, Lundvall and Johnson add a dimension that makes the discussion more complex and more uncertain, but also more interesting and intellectually fertile. The richness associated with the concept of the learning economy is further demonstrated in the volume edited by Archibugi and Lundvall (2001).

We attempt to extend the concern associated with the process and with its dynamic character even further (see also Conceição, Heitor and Lundvall, 2003). Thus, the title of the book entails a dependence on enhancing the processes of producing and exchanging knowledge and information. This enhancement relies on the build-up of learning and knowledge networks, which must follow local specific conditions to adapt, engage and mobilize local actors and agents. The papers that follow in this book discuss critical aspects associated with the process of building such networks. In this introductory chapter, we will start the analysis by focusing on two crucial elements of "inclusive learning" for development: innovation, on the one hand, and competence building, on the other. Then, based on related literature, we will argue that institutions do matter!

Innovation is the key process that characterizes a knowledge economy understood from a dynamic perspective. Lundvall and Johnson's learning economy is about new knowledge replacing old knowledge. This dynamics is very close to Schumpeter's concept of "creative destruction", which is a standard description of the innovation process. Innovation is associated with creativity, with the generation of new ideas, but also with initiative and risk-taking. Innovation entails bringing new ideas to fruition in the marketplace, satisfying demands or creating new needs, in a process that improves overall welfare.

Competence is the foundation from which innovation emerges, and which allows many innovations to be enjoyed. In other words, it contributes both to the "generation" of innovations (on the supply side of the knowledge economy) and to the "utilization" of innovations (on the consumptions side of the knowledge economy). Competence is also fuelled by innovation itself. Competence is associated with skills and capacities, both individual and collective. When we consider competence, we focus on a "higher order of skills" (e.g., Conceição and Heitor, 1999). These generic skills include higher levels of education (who can ever be against more education?) but also capacities that are more generic, such as creativity, risk-taking, and initiative.

By choosing the themes of innovation and competence building as drivers towards "inclusive learning" we are not considering that these are the exclusive elements. There are clearly other issues of major importance, namely those associated with macroeconomic conditions, but we do not intend to be comprehensive. Our aim is to look for insights through the contributions collected in this book.

Learning as Knowledge Accumulation

The paragraphs above show that, from our perspective, learning is under-stood, broadly, as knowledge accumulation. There are different levels of "learn-ing entities", from individuals, to organizations, to whole economies. A first important step in our discussion is the clarification of our conceptual understanding of terms such as "knowledge" and "learning", often loosely used with dramatically different meanings. The work by Johnson et al. (2002), following the work of Cowan et al. (2000), provides further evidence for the need to clarify these concepts. This conceptual clarification of our understanding of learning as knowledge accumulation is the objective of this section.

We find it useful, as developed in more detail in Conceição and Heitor (1999), to follow Nelson and Romer's (1996) differentiation between ideas and skills, or software and wetware, to use these authors' nomenclature. The conceptual difference between software and wetware lies in the level of codification. While ideas correspond to knowledge that can be articulated in words, symbols, or other means of expression, skills cannot be formalized, but always remain in tacit form. Under this taxonomy, knowledge may be divided into two worlds (Johnson et al., 2002): the world of codified ideas (software) and the world of non-codified skills (wetware).

The difference in the level of codification has implications in terms of the "economic properties" of the two types of knowledge that we consider. The most important implication is associated with the differences in the rivalry associated with the consumption of each type of knowledge. Since the knowledge underlying software is codified, it is easily articulated and reproduced by simple, inexpensive means. Consequently, rivalry in the consumption of software is low. By contrast, the transmission of skills (wetware) is complex, expensive, and slow. Skills result from a combination of factors, ranging from their largely innate quality, through individual experience, to formal training. Thus, rivalry is comparatively higher in the consumption of wetware.

The differences in rivalry between software and wetware have important implications for knowledge production. Dasgupta and David (1994) suggest that there are basically two alternatives for the production of software. The first consists of intervention by the state in the production of ideas, by means of direct production, or by subsidizing production, such as funding of university R&D. The second alternative consists of granting property rights for the creation of ideas, that is by defining regulations for intellectual property specific instruments that include patents, registered trade marks and copyright (see Conceição and Heitor, 2001; Conceição et al., 1998, for a more comprehensive analysis). Therefore, the production of ideas requires more complex institutional mechanisms than those provided by the market. As for skills, the market provides a large proportion of the incentives needed for their production, at least when these are analyzed in isolation, although with important limitations (see, again, Conceição and Heitor, 2001).

We bring our own understanding to the process of knowledge accumulation when the interaction between software and wetware is explored. The idea of interaction between ideas (software) and skills (wetware) is what, in our understanding, defines learning. Analysis of the interaction between ideas and skills leads us to explore the learning processes associated with the generation of each type of knowledge in a more integrated and dynamic way, beyond the mere accumulation of ideas and skills, each in isolation. Our view is yet another perspective on the ongoing debate between the complex and multifaceted interaction between different types of knowledge. Recent manifestations of this debate include Johnson et al. (2002), in which they contest the implicit assumption of Cowan et al. (2000) that codification always represents progress.

Indeed, according to Freeman and Soete (1997), ideas and skills are no more than two sides of the same coin, two essential aspects of the accumulation of knowledge. New ideas spur the development of the skills required to use those new ideas. The bridge from the production of ideas to the usage of ideas is established by producing new skills. Increased use of an idea, which requires its diffusion, will lead to a constellation of other ideas, aimed at improving and extending the initial idea, which will lead to the need for further skills and so on, in a self-reinforcing cycle that leads to the accumulation of knowledge. The accumulation of knowledge results from the production, usage, and diffusion of both software and wetware, in an interactive learning process that leads to knowledge accumulation, as initially proposed by Conceição and Heitor (1999). The close and complex interdependence between ideas and skills that lead to overall knowledge accumulation depends on two types of learning processes. First, learning by codifying (Foray and Lundvall, 1996), associated with the production of ideas, through the codification of knowledge. Second, learning by interpreting (OECD, 1997), related with the production of skills, through the usage, or more broadly, the interpretation of ideas.

Conceição and Heitor (1999) show how this conceptual understanding can be used to analyze broad historical interactions between knowledge and development (such as in the evolution of China and Europe; Landes, 1998) as well as the adoption and diffusion of specific technological innovations (such as standards of videotape recorders). The model also acknowledges the indivisibility of ideas, as discussed by David (1993) (once created, an idea remains at least potentially accessible everywhere, and there is no need to rediscover it).

This conceptual understanding of the learning processes can also be used to draw implications in terms of the complex relations associated with the building-up of innovation systems (e.g., Christensen, 2002), again as proposed by Conceição and Heitor (1999). In this introductory chapter we develop, next, the importance of stimulating innovation (generation of ideas) and the parallel importance of developing competencies.

The Importance of Stimulating Innovation

The section above made explicit the way in which we understand learning as knowledge accumulation, which is a result of a complex set of learning processes where there is considerable interdependence between the accumulation of ideas and of skills. We now turn for the analysis of innovation as the concept that best fits with the idea of the knowledge economy understood from a dynamic perspective.

It is by now well understood that the early conceptualisations of innovation as a linear process were clearly insufficient to describe the complexity and contingency of the innovative effort of people, firms and countries (e.g., Nelson and Romer, 1996; Kline and Rosenberg, 1986; Dosi, 1988; Nelson and Winter, 1982). Still, what is surprising is the extent to which the linear perspective still informs much of today's public perceptions about innovation, as well as policy design and implementation. The reliance on simple and direct indicators such as expenditure of R&D by the private sector, and the obsession in some circles associated with improving these types of indicators, reflects the dominance of the linear perspective.

We do not question the importance of these and other indicators, but it should have also become clear by now that they provide an incomplete description of the innovation process and are tied to the linear perspective (see, for the continuation of the linear perspective, Guellec and Pottelsberghe, 2000). Romer (1990; 1993) recognizes the importance of what he calls appreciative theories of growth and innovation in helping more formal approaches to better describe the richness of the innovation process, but somehow the link has been hard to accomplish.

The link between the complexity of the innovation process and the special economic characteristics of knowledge, and of conceptualisations of the learning process such as the one advanced above in this introductory chapter, could be a bridge. In fact, Romer (1990; 1993) constructs his theory of endogenous growth drawing on the non-rival nature of ideas. Dasgupta and David (1994) advance new ideas about the economics of science building also on the same principles associated with the special characteristics of knowledge. Thus, the conceptual understanding of learning advanced above could serve more than just being an interesting modelling tool, allowing the development of new conceptual approaches. It could also become a useful guide for policy, especially in light of the still predominant domination of the linear model. In a series of papers, Conceição and Heitor (2001) and Conceição et al. (1998) have explored the implications of this conceptual model to advance policies associated with innovation (that is, the generation of ideas, or software). We turn, next, to the other side of our conceptual model of learning: the importance of wetware.

The Relevance of Competence Building

Competence is the foundation on which innovation is generated and diffused. Competence is associated with individual skills, but also with collective capacities. It is also on competence that a learning society can be constructed and sustained. Some suggest that technological change is (or has become) skill-biased (Autor et al., 1997). Empirical work supporting the skill-biased technological change conjecture includes studies such as Krueger (1993). Thus, for some, the connection between innovation and competences is primarily understood as being related with this hypothesis.

However, the skill-biased technological change hypothesis is far from being uncontroversial. From a conceptual point of view, critics note that the treatment of technological change rarely goes beyond asserting that new technologies, and especially computers, are responsible for a steady increase in the demand for skills. Technology is conceptualized as in the linear models of innovation. Criticisms based on empirical analysis include DiNardo and Pischke (1997) and the realization that there is a mismatch in the timing of the increase in inequality and the spread in the diffusion of computers, and the fact that the increased adoption of information technology has not noticeably contributed to increased productivity (see Galbraith, 1998, for a comprehensive review). Alternatives to the skill-biased technological change include the perspective advanced by Bresnahan (1999), who proposes an organizational complementarity between information and communication technologies (ICTs) and highly skilled workers.

But the relationship between competences and innovation is not only seen through the skill biased technological change perspective. And competence building also entails much more than formal skills. For example, Dore (1976) differentiates "education" from "schooling", which refers to "mere qualification-earning", leading to an "educational inflation" spiral. Several other authors, (e.g., Bourdieu and Passeron, 1970; Boudon, 1973; Jencks, 1972; Bowles and Gintis, 1976), are similarly skeptical about a direct relationship between increases in the level of education and economic performance. The differences between the economists of human capital and these other authors, who come primarily from sociology, remain until today. In fact, some of the critiques have important parallels with economic perspectives, such as Bourdieu and Passeron's theory of the social filter, whereby schools work as filters to preserve and maintain social and educational differences, and the "inheritance of inequality" perspective of Meade (1964).

However, if one is ready to accept the existence of a labor market where wages reward, at least partially, productivity and skill, Katz and Murphy (1992) provide strong evidence that supply and demand go a long way in explaining the patterns in the evolution of inequality. Most of the recent studies on inequality focus on a single-country longitudinal analysis of the evolution of the dispersion of income. Examples of the same methodology applied to other single country studies include Schmitt (1995) for the United Kingdom, and Edin and Holmlund (1995) for Sweden.

This discussion clearly highlights the link between competence (skills, education), and innovation (technological change) towards inclusive learning. The connection between education, skills and competence, on the one hand, and the learning society, on the other, must consider the manifold interconnections between competence and the learning society and links them with the broader context of the anxieties and concerns, hopes and expectations that we live with today.

An important issue is to know what it takes to be part of the learning society. We may not know exactly what the learning society is, but we do know that there are requirements to be part of it. We need, in particular, to build competence, of which skills are a part. However, for some cases, the need for new skills is not associated with technological change, but with an organizational change, and the new skills provided are not particularly intensive in specialized knowledge. It is important to stress this point because the discussion can easily be drawn into the skill-biased technological change discussion.

Naturally, technological change does indeed play a role in increasing the demand for "a higher order of skills", but there are other elements of change driving this demand. What is hardly questionable is that those that do not possess the skills nor the ability or possibility to acquire them become excluded.

THE NECESSARY CONDITIONS FOR KNOWLEDGE AND LEARNING

Besides the holistic view of systems of innovation and competence building, we need to frame the empirical evidence on the bases of a unified conceptual framework of analysis, for which we may consider the incentive structure of "the market" which is determined by competition. Competition in product and factor markets provides signals to economic actors about the potential returns among alternative options, thus determining their investment patterns. Endogenous growth theories, because they are based on the existence of dynamic externalities and imperfect markets, require a careful understanding of the structure of competition. On the one hand, because of the nature of knowledge, investment of private agents often fails to acknowledge spillover effects, or may not be able to anticipate the full extent to which there is further learning potential in a new technology. On the other hand, incentives to invest in new knowledge depend on the existence of some degree of monopolistic rents. These rents may not exist in latecomer countries exposed to international competition, if they are solely adopting foreign technology.

As a result, private investment levels (which result from the incentive structure provided by the market to economic agents) in activities with learning or spillover potential tend to be lower than the social optimum, and may even generate what is known in the literature as "low-level equilibrium traps". This happens when private but not social returns from productivity-enhancing investments—that is, accounting for spillovers—are below those of nonproductivity-enhancing investments, causing stagnation in growth. This situation may be overcome by inducing decision-makers to include the spillover effects in their accounting processes, or by creating monopolistic markets that generate above-normal returns.

In principle, these shortcomings of the market mechanism call for some sort of government intervention—a major factor affecting the firms' incentive structures. Governments are concerned with making sure that societal costs and benefits are endogenized in the decisions of private firms. In a learning environment this may mean subsidizing research activities, investing in education, protecting infant industries, promoting exports, or even disciplining firms. But government intervention must balance the potential distortions on competition that may come from intervention with the needs to "correct market failures": artificial restraints on competition can also divert profits to activities other than building technological capabilities. In relatively closed regimes with strong pressure to substitute imported for local goods, there may be little incentive for firms to improve, since they can capture the local market regardless of their own productivity.

In the neoclassical view, infrastructure is related with the existing amount of labor, capital, and natural resources. The new theories bring to stage other important factor inputs, in particular human capital, and R&D expertise embodied in firms, universities, and laboratories. Thus, infrastructure will encompass, in addition to labor and capital, what we call technology infrastructure, or technostructure. Considering a distinction between labor and capital on one hand, and technostructure on the other, enables a separate analysis of the roles played by each of these aspects in the development path of a particular industry or region.

The examples discussed in the book show how the interaction between sets of incentives and the technostruture of a particular region, industry, or nation fosters and hampers the patterns of knowledge accumulation and the development process. Nevertheless, it will also be clear that, although incentives and infrastructure greatly inform our understanding of the behavior of firms, government policies, and industrial trajectories, they do not tell the whole story about the differences across countries and regions. That is because both incentives and infrastructure do not operate in a vacuum, being shaped by and shaping the particular context where they operate. In other words, for a market system to function well, the country or region must have embedded a set of social capabilities that allow it to function according to the theoretical principles of allocative efficiency and Pareto optimum social welfare.

Integrated Learning and Social Capital

If one considers innovation as a broad social and economic activity, two key questions need to be considered. First, the understanding of conditions for integrated learning processes. This has led us to build in the previous paragraphs on Lundvall and Johnson's learning economy and to discuss the learning society in terms of innovation and competence building. Further, the ability to learn seems to be the main driver of long-term growth, but learning can occur at different levels. Individual people, firms and organizations, and countries all are dependent of learning for development. There are also different ways through which people, firms, and countries can learn. Learning can be an unintended consequence of experience and augmentation of scale. On the contrary, formalized and intentional learning methods such as education, training or R&D is often the result of a utility maximization rational decision from the point of view of the firms. The new growth theories attempt to formalize the way in which learning mechanisms can impact on economic growth.

Second, the relevance of considering distributed knowledge bases across economically and/or socially integrated set of agents and institutions, which leads us to the concept of social capital. In the broadest sense, social capital is associated with the "social capabilities" that allow a country or region to move forward in the process of development. In a more sophisticated treatment, Coleman (1988) states that social capital is "a variety of different entities, with two elements in common: they all consist of some aspect of social infrastructure, and they facilitate certain actions of actors—whether personal or corporate actors—within the structure."

None of the case studies presented in this book provides single and definitive answers to the problem of achieving learning societies. But it is our aim to argue that social capital is key, and that infrastructure (in the broad sense described above) and institutions are the elements out of which social capital is born. Different types of institutions can be effective, as long as they enable collective learning and collective innovation. As in every situation where institutions are important, history matters. Path dependence and increasing returns lead to self-reinforcing cycles, whereby events, often sporadic and serendipitous, define current patterns of development. But the good news is that if we understand the dynamics of institutional change and evolution (that is, of "collective learning"), we can also create conditions for future development.

The Institutional Framework and the Changing Role of the State

The OECD has called for our attention that from the diffusion of information technology and the growth of the knowledge economy to the globalisation of markets and radical managerial innovations, the factors driving and being driven by social change are both wide-ranging and deep. It is a tide of pervasive transformation that is simultaneously washing away and reshaping the social foundations provided by cultural traditions, social symbols and institutions of authority and security. From the family and school to the firm and parliamentary fora, long-standing social reference points are being called into question, reformed and reinvented. Exploring the challenges posed by this transition to new, more dynamic social foundations are critical to promote innovation for most late-industrialized and developing regions.

In this context, Petit and Soete (2000) provide insight into the impact of globalization and technical change on social cohesion and exclusion in the European Union. The most important relates to the fields of the regulatory system (where it is argued the European policy makers take the lead in setting up appropriate frameworks in emerging science-based industries), science and technology policy (where user-learning could be more central), territorial policy (where the notion of knowledge capital could be much more central in the Structural Funds), and labour market policy (where a twin strategy of targeting small sectors with relatively large spillovers, together with boosting jobs in areas such as personal services is proposed).

Again, we believe that we learn by comparing the regulatory framework among OECD countries, mainly because in the past two decades an increasing number of countries have been reforming their regulatory environments in both the labor and product markets. It should be noted that regulation is essentially aimed at improving the functioning of market economies, by establishing the "rules of game" in areas such as market competition, business conduct, labor market, consumer protection, public safety and health, and the environment. In this context, many national reforms have been driven by comparisons with he policies implemented and the results obtained by other countries. In addition, cross-country comparisons allow identifying and analyzing to what extent regulatory arrangements and their economic implications are country-specific or can apply more generally.

Figure 1.1 shows sample results collected in OECD countries making use of formal economic (i.e., constraints and incentive mechanisms concerning market access, the use of inputs, output choices, pricing, and incremental trade and investment) and administrative regulations (i.e., interface between government agencies and economic agents) that affect product markets, but ignore other important regulatory areas, such as environmental, health and safety regulations. Also, provisions concerning financial markets and land-use, which are likely to affect entrepreneurship, are not considered. The analysis do not assess the overall quality of regulations, focusing exclusively on the relative friendliness of regulations to market mechanisms in terms of the impact on the intensity of product market competition. Although it is clear that a market-oriented and administrative regulatory environment is only a necessary condition for enhancing product market competition, the analysis is particularly important to extract lessons for southern European countries, namely in terms of apparent relation establish among product market regulations and employment protection. In fact, based on a simple average of the summary indicators for regular and temporary contracts through factor analysis, the Mediterranean countries, and Portugal in particular, appear with the tightest regulations.

Figure 1.1
Product Market Regulation and Employment Protection Legislation in the OECD

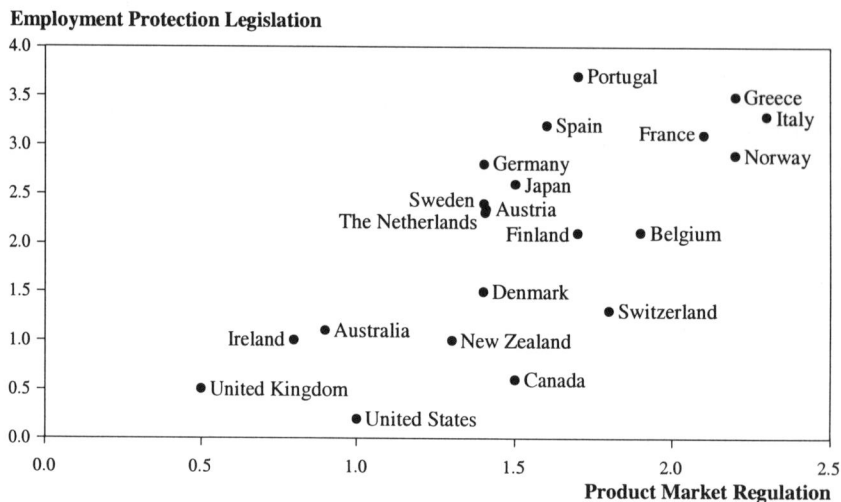

Source: Adapted from Nicoletti, Scarpetta, and Boylaud (2000).

The evidence from the results is that restrictive product market regulations are matched by analogous employment protection legislation restrictions to generate a tight overall regulatory environment for firms in their product market as well as in the allocation of labor inputs. In addition, the analysis suggests the possible existence of compounded effects on labor market outcomes, making regulatory reform in one market less effective than simultaneous reform in many markets. Making use of selected summary indicators for product market regulations (state control; barriers to entrepreneurship; and barriers to international trade and investment) and employment protection legislation (regular and temporary employment), Figure 1.1 identifies three clusters of countries, as follows: (1) the United States, the United Kingdom, and common-law countries characterized by a relatively liberal approach; (2) continental European countries with relatively restrictive product market regulations; and (3) Mediterranean countries characterized by a tight overall regulatory framework. This raises fundamental issues for European policies, namely in terms of the economic effects as product market regulations and employment protections interact.

The question that thus arises is how far the impact of de-regulation depends upon the broad socio-economic context and overall institutional framework? For example, the strong regulatory framework of Norway, together with the expected high levels of social capital of the Nordic countries, at least as measured by the levels of "thrust", clearly result in a context which differ from that found in Mediterranean countries. Certainly "unemployment protection" for the former may represent a risk incentive, so that regulatory frameworks are not directly comparable. Anyway, there are a number of implications for innovation, but in general the analysis calls for a renewed attention for de-regulation, which should definitely be accompanied by the development of new competencies and complementary actions at the levels of knowledge creation and diffusion.

It is also clear that the question of regulation must be considered within a more complex and ambiguous tendency that is emerging at the outset of the 21st century—that is, the perception that there is a changing role for the state. This is a controversial area, since it involves ideology and issues associated with personal beliefs on the effectiveness and fairness of social and political systems. In very broad terms, the changing role of the state can be characterized by an increased detachment from holding economic assets and from shying away from determining the direct allocation of economic resources.

In general, our argument is that the ways innovative capacity and new competencies, namely in conventional engineering, economics and management, may positively influence the development of a country and/or region depend on the institutional framework, which is currently particularly determined by regulation policies and the process of market liberalization.

INTRODUCING THE BOOK

Following the discussion above, the material included in this book is presented in three main parts. We start in Part I by discussing the economic context leading to learning societies, in a way that gives special relevance to governance issues. Six papers from experts from different parts of the world are included, as a sample of the diversified economic arrangements we live on.

Part II addresses the problem of regions as platforms of policy articulation, particularly when it involves the adaptation of competition or innovation policies to regional economic development concerns by societies and economies of divergent interests. We set this in order to define the "regional question" through related issues of scale and complexity. We find that the "regional question" thereby contributes to the renewed focus of regional innovation systems (RIS), by embracing the growing attention to skilled labor supply as central to this debate, rather than the mere positive externality of agglomeration economies that it is often attributed.

In these terms, although there is an emerging set of literature on technological innovation and industrial economics looking at the distinctive features and institutional characteristics of specific regions (e.g., Wolfe and Gertler, 1999; Gambardella and Malerba, 1999, focusing on European regions), there have been few attempts to build analytical frameworks to improve understanding and to allow the development of well-sustained technology policies for less favored zones and late industrialized regions. In fact, the neoclassical approaches in industrial economics have emphasized the analysis of the microeconomic behavior of firms and built theories specialized in the American, and Anglo-Saxon systems and related market dynamics. On the other hand, evolutionary economics have attempted to improve our understanding of learning processes and the role of institutions in economic development, but have not specialized on the specific historical context of regions, namely those characterized by late industrialization (e.g., Cooke and Morgan, 1998). Building on the evolutionary approaches and in system theory, the concept of "national system of innovation" (e.g., Lundvall, 1992) has led to numerous studies of individual countries, but there is still a long way to go in order to assess the specificity of transition economies and late industrialized regions and countries. Among the various aspects raised, it should be noted that the sectoral specificity in the organization of innovative activities, on one hand, and the specific characteristics of local systems of innovation, on the other hand, are expected to play a significant role in shaping the organization of innovative activity in any regional context. The prevalence of one effect over another depends on history and competitiveness of firms and their degree of internationalization.

The book concludes by addressing in Part III the critical issue of how language and culture faces globalization and relates to technology policy and innovations linked to the use of computer-based technologies. To achieve these objectives, we have considered four complementary papers encompassing the diversity to be found in approaching the issue of language toward learning societies.

REFERENCES

Archibugi, D., and Lundvall, B.-Å. (eds.) (2001), *The Globalizing Learning Economy*. New York: Oxford University Press.

Autor, D., Katz, L., and Krueger, A. (1997), *Computing Inequality: Have Computers Changed the Labor Market?*, NBER Working Paper 5956.

Boudon, R. (1973), *L'Inegalité des Chances*. Paris: Armand Collin.

Bourdieu P., and Passeron, J.-C. (1970), *La Réproduction: Éléments pour une Théorie du Système d'Enseignement*. Paris: Éditions du Minuit.

Bowles, B., and Gintis, H. (1976), *Schooling in Capitalist America*. London: Routledge.

Bresnahan, T. (1999). "Computerisation and Wage Dispersion: An Analytical Reinterpretation," *Economic Journal*, 109(456), pp. 390-415.

Christensen, J. F. (2002), "Corporate Strategy and the Management of Innovation and Technology", *Industrial and Corporate Change*, 11(2), pp. 263-288.

Coleman, J. (1988), "Social Capital in the Creation of Human Capital," *American Journal of Sociology*, 94, pp. s95-s120.

Conceição, P., Gibson, D., Heitor, M. V., Sirilli, G., and Veloso, F. (eds.) (2002), *Knowledge for Inclusive Development*. Westport, CT: Quorum Books.

Conceição, P., and Heitor, M. V. (1999), "On the Role of the University in the Knowledge Economy," *Science and Public Policy*, 26(1), pp. 37-51.

Conceição, P., and Heitor, M. V. (2001), "Universities in the Learning Economy: Balancing Institutional Integrity with Organizational Diversity," in Archibugi, D. and Lundvall, B.-Å. (eds.), *The Globalizing Learning Economy*. New York: Oxford University Press, pp. 83-107.

Conceição, P., Heitor, M. V., and Lundvall, B.-Å. (eds.) (2003), *Innovation, Competence Building, and Social Cohesion in Europe—Towards a Learning Society*. London: Edward Elgar.

Conceição, P., Heitor, M. V., and Oliveira, P. (1998), "Expectations for the University in the Knowledge-Based Economy," *Technological Forecasting and Social Change*, 58(3), pp. 203-214.

Conceição, P., Heitor, M. V., and Veloso, F. (2003), "Infrastructures, Incentives and Institutions: Fostering Distributed Knowledge Bases for the Learning Society, *Technological Forecasting and Social Change*, 70(7), pp. 583-617.

Cooke, P., and Morgan, K. (1998), *The Associational Economy: Firms, Regions, and Innovation*. New York: Oxford University Press.

Cowan, R., David, P. A., and Foray, D. (2000), "The Explicit Economics of Knowledge Codification and Tacitness", *Industrial and Corporate Change*, 9, pp. 211-253.

Dasgupta, P., and David, P. (1994), "Toward a New Economics of Science," *Research Policy*, 23, pp. 487-521.

David, P. (1993), "Knowledge, Property, and the System Dynamics of Technological Change", in Summers, L. H. and Shah, S. (eds.), *Proceedings of the World Bank Annual Conference on Development Economics 1992*, Supplement to The World Bank Economic Review.

DiNardo, J., and Pischke, J. (1997), "The Returns to Computer Use Revisited: Have Pencils Changed the Wage Structure Too?," *Quarterly Journal of Economics*, 112(1), pp. 291-303.

Dore, R. (1976), *The Diploma Disease: Education, Qualification, and Development*. Berkeley, CA: University of California Press.

Dosi, G. (1988), "Sources, Procedures and Microeconomic Effects of Innovation," *Journal of Economic Literature*, 26(3), pp. 1120-1171.

Edin, P., and Holmlund, B. (1995), "The Swedish Wage Structure: The Rise and Fall of Solidarity Wage Policy?," in Freeman, R. and Katz, L. (eds.), *Differences and Changes in Wage Structures*. Chicago, IL: The University of Chicago Press.

Foray, D., and Lundvall, B.-Å. (1996), "The Knowledge-Based Economy: From the Economics of Knowledge to the Learning Economy," in Foray, D. and Lundvall, B.-Å. (eds.), *Employment and Growth in the Knowledge-based Economy*. Paris: OECD.

Freeman, C., and Soete, L. (1997), *The Economics of Industrial Innovation* (third edition). Cambridge, MA: MIT Press.

Galbraith, J. K. (1998), *Created Unequal: The Crisis in American Pay*. New York: The Free Press.

Gambardella, A., and Malerba, F. (1999), *The Organization of Economic Innovation in Europe*. New York: Cambridge University Press.

Guellec, D., and Pottelsberghe, B. V. (2000), "The Impact of Public R&D Expenditure on Business R&D," *OECD STI Working Papers*. Paris: OECD.

Jencks, C. (1972), *Inequality*. New York: Basic Books.

Johnson, B., Lorenz, E., and Lundvall, B.-Å. (2002), "Why All This About Codified and Tacit Knowledge?", *Industrial and Corporate Change*, 11(2), pp. 245-262.

Katz, L., and Murphy, K. (1992), "Changes in Relative Wages, 1963-1987: Supply and Demand Factors," *Quarterly Journal of Economics*, 107(1), pp. 35-78.

Kline, S. J., and Rosenberg, N. (1986), "An Overview of Innovation", in Landau, R. and Rosenberg, N. (eds.), *The Positive Sum Strategy: Harnessing Technology for Economic Growth*. Washington, DC: National Academy Press.

Krueger, A. (1993), "How Computers Have Changed the Wage Structure? Evidence from Micro Data," *Quarterly Journal of Economics*, 108(1), pp. 33-60.

Landes, D. (1998), *The Wealth and Poverty of Nations: Why Some are so Rich and Some so Poor*. New York: W. W. Norton & Company.

Lundvall, B.-Å. (1992), *National System of Innovation—Towards a Theory of Innovation and Interactive Learning*. London: Printer.

Lundvall, B.-Å., and Johnson, B. (1994), "The Learning Economy," *Journal of Industry Studies*, 1/2, pp. 23-42.

Meade, J. E. (1964), *Efficiency, Equality and the Ownership of Property*. London: Allen and Unwin.

Nelson, R. R., and Romer, P. (1996), "Science, Economic Growth, and Public Policy," in Smith, B. L. R. and Barfield, C. E. (eds.), *Technology, R&D, and the Economy*. Washington, DC: Brookings Institution Press.

Nelson, R. R., and Winter, S. G. (1982), *An Evolutionary Theory of Economic Change*. Cambridge, MA: The Belknap Press of Harvard University Press.

Nicoletti, G., Scarpetta, S., and Boylaud, O. (2000). "Summary Indicators of Product Market Regulation with an Extension to Employment Protection Legislation", *OECD Economic Dept. Working Paper*, 226, ECO/WKP(99)18.

OECD (1997), *Technology and Industrial Performance*. Paris: OECD.

Petit, P., and Soete, L. (2000), *Technology and the Future of European Employment*. Cheltenham: Edward Elgar.

Romer, P. (1990), "Endogenous Technological Growth", *Journal of Political Economy*, 98(5), pp. s71-s102.

Romer, P. (1993), "Idea Gaps and Object Gaps in Economic Development", *Journal of Monetary Economics*, 32, pp. 543-573.

Schmitt, J. (1995), "The Changing Structure of Male Earnings in Britain 1974-1988," in Freeman, R. and Katz, L. (eds.), *Differences and Changes in Wage Structures*. Chicago, IL: The University of Chicago Press.

Smith, K. (2000), "What is the 'Knowledge Economy'? Knowledge-Intensive Industries and Distributed Knowledge Bases", Paper presented to DRUID Summer Conference on The Learning Economy—Firms, Regions and Nation Specific Institutions. Rebild/Denmark, June 15-17.

Wolfe, D., and Gertler, M. (1999). *Innovation and Social Learning*. London: Macmillan/ New York: St. Martin's Press.

PART I:
BUILDING THE LEARNING ECONOMY: TRENDS AND PERSPECTIVES FOR GOVERNANCE

Introductory Note

Alejandro Ibarra-Yunez

The knowledge society has been a term coined as a paradigm of modern and competitive economies, subject to a set of new rules by key actors. Yet policy approaches and even intellectual inquiry have been inconclusive in relevant aspects: (a) How is innovation made an integral part of human aspirations? (b) What are key roles of governments, financial institutions, firms and social groups? (c) How would lagging parts of society move from nonknowledge activities towards knowledge economic ones? and (d) How does knowledge imply and create marketable value?

Knowledge is a human characteristic that becomes a value similar to psychological motivation and group dynamics for improvement and quality of life. In analyzing knowledge, learning-by-doing, innovation, and competition, economics, political science, social psychology and anthropology seem to be at odds with each other regarding the subject of analysis, let alone theories and methodologies. And each discipline seems to be expectant of what the other presents about new approaches. Social science and policy, however, cannot bear such a luxury. Economics, with all its strengths and weaknesses, is the underlying framework in Part I by a group of academics in the field, from different parts of the world.

In chapter 2 a detailed and deep discussion of origins of knowledge—epistemological, technical or instrumental, and contextual—Lundvall presents a new framework to encompass not only information, but also skill and collective tacit knowledge, and even social cohesion and norms. The author's approach tries to balance the intrinsic character of learning-by-doing, with the roles played by policy makers mainly with applications to Europe.

Chapter 3, by Martins Lastres and Cassiolato, emphasizes the use of information technologies in the quest to foster knowledge-based economies. An underlying motivation of these authors seems to be the pressing need of emerging economies to tackle policy and market problems in lagging sectors and social groups in the dynamics of change towards competition, such as the Brazilian experience.

Chapter 4, by Wilson, spells out the changing role of state expenses on R&D, and the new way subsidies to private innovators have moved during the past decade. The chapter also addresses property protection, infrastructure and education policies, and how they enter into asymmetric priorities across developed and developing economies. An important lesson derived by Wilson is between the European purposeful policy towards a knowledge society, and the U.S. market model that has given low priority to international cooperation even within the Western Hemisphere.

Kasarda follows with a case analysis of the information and "aerotropolis" in the United States in chapter 5. This author relates the development of gateway airports and hubs, to historical experiences of competitive infrastructure for knowledge.

Young's chapter 6 compares the emerging economic powers of mainland China with those declining ones in Hong Kong's capital "Growth Enterprise Market." The author stresses the need for capital markets for innovation and governance structures, and presents the hypothesis that established rules for mercantile success are a necessary but not sufficient condition for economic and social improvement in the technology era.

In chapter 7 Ibarra-Yunez presents an economic analysis of market performance and economic effort by telecommunications companies in Chile and Mexico that were subject to different regulatory shields by their respective governments. The chapter emphasizes that governance structures for innovation, efficiency, and capital formation are affected by policies that pursue objectives that might deviate from modernization objectives pressed by what is called the new economy.

A new body of economics literature, that of governance structures, has been increasingly addressing important parts of the knowledge economies. Some countries have well-established institutional settings, while others, mainly in emerging economies, have lagged behind. A diverse set of rules, observance of the rule of law, undeveloped processes of dispute settlements, insufficient funds and incompatible objectives by players in the technology-based competition affect the way companies are structured and behave in their various markets. Additionally, some countries, pressed by the need to complement scarce domestic savings and resources by foreign investment and capital, face the trade-off between attracting capital and firms with promotion of local capitalists, or conditioning foreign firms to performance measures related to innovation, its transmission and adaptation, the environment, the promotion of joint-venturing or even labor development and training programs, as Lundvall points out. The referred literature of economics addresses in a related but key part, what is called the problem of the incentive mechanism, where players (internally within firms, across them and even organizations in relation to governments), decide for specific conducts. Lundvall, Martins Lastres and Cassiolato, and Ibarra-Yunez make this point explicit.

The economics approach to learning-by-doing is contrasted and criticized by others who pursue a more systemic and wider framework. However, the so-called holistic framework is not generally accepted as scientific by economists because it is not prone to analytical techniques used by economics, or even the subject of inquiry. While for economics the substantive element of knowledge is productivity in a static sense, and advance in a dynamic one, for other disciplines it is behavior, cooperation, social cohesion or competition.

Not only optimal decision-making is important in the topic of knowledge and learning, but also in the "world of mind and body," as Lundvall points out. Also, communication of information through networks is an important element in learning-by-doing. This point is apparent in Kasarda's chapter of speed and aerotropolis. The Lundvall, Kasarda and, to a lesser extent, Martins Lastres and Cassiolato chapters, however, show different points of view as to what is the underlying framework for technology, innovation and learning. While Kasarda's proposition is that competition generates creativity, knowledge acquisition and use through innovation in the market, Lundvall's proposition aims at a much broader idea that calls for policy. The former emphasizes the need for property protection, adequate infrastructure, and modern market-clearing mechanisms related to economies intensive in information technologies. The latter calls for social, not only economic policies, some that include education and collective learning. Wilson, on his part, spells out various policy instruments, while stressing the need to coordinate national with subnational policy objectives and instruments.

Although these chapters differ in their focus, in the case of Young, incentive mechanisms in capital markets are key elements of technology policy. In addition, deviations from market viability can be caused by pyramidal governance structures and inadequate management of incentives. In the end, Part I remains with the question of whether cooperation in policy design and implementation is an integral part of successful learning societies or, alternatively, are competitive economic structures to be promoted by technology policy, which brings us back to the incentive mechanism problem.

In any case, knowledge, learning, innovating and competition need social and economic environments that are transparent, stable and clear. Some economies face a mature set of rules, while others suffer from insufficient and inadequate sets. This has become a point stressed by multilateral agencies. And technology policy calls for coordination of various government agencies, ranging from the treasury, to economic promotion, labor ministries, environment, infrastructure and education. Much is needed. European experiences seem to be at the forefront of addressing diversity and valuing knowledge and the learning society. The objective of equalization of opportunities in less developed regions has also provoked public policy in financing and upgrading through market clearance mechanisms, but also by the use of training programs and collective learning. Market-driven experiences in the United States and other countries are also noteworthy. The developing economies, on their part, have a pressing need for policy prioritization and then action. The reader should gain from the chapters that follow.

2

Innovation Policy and Knowledge Management in the Learning Economy

Bengt-Åke Lundvall

INTRODUCTION

The following is based upon a hypothesis that our societies are in the midst of a transformation of the same importance as the industrial revolution of more than two centuries ago. The core element in the emerging new mode of production is knowledge and the most important process is learning. But our knowledge about how knowledge is created, transferred and used remains partial, superficial and partitioned between scientific disciplines, giving different interpretations and meanings to the basic concepts of knowledge and learning. The indicators used to measure knowledge and learning are correspondingly weak. It is fair to say that we have not yet reached a stage where we can systematically apply knowledge to the production of knowledge. As we shall see, this is true for learning taking place within firms as well as for learning in its broader societal context. According to Karl Marx, the real breakthrough of the industrial revolution took place when machinery was used to produce machinery. It is tempting to see an analogy where the full-scale transformation into a learning economy will have to await the systematic application of knowledge to the production of knowledge.[1]

The most important dividing line in the analysis of knowledge and learning goes between knowledge as information and knowledge as tacit skills. Information is knowledge that has been reduced into bits and then can be transmitted between localities and agents through telecommunication networks. Tacit knowledge is constituted by skills and routines embodied in people and embedded in organisations. It can normally be transferred only by experience-based learning in the form of apprenticeships or network relationships. Learning tacit skills involves using all senses. Often it involves interaction with others more skilled in the area. Information, on the other hand, is accessed rather than learnt. Learning skills changes people's capability to change the world, while increased access to information affects what they know about the world.

It is a major task for economists to contribute to the understanding of knowledge and learning in the context of economic development and cooperation. In doing so they must take into account that learning shapes the life of citizens in many other respects as well. This is true, for instance, for political and social citizenship. General skills related to a mother tongue, foreign languages, mathematics and information technology become increasingly important prerequisites for taking active part in civil society and in local, national and global politics. Just to cope with the challenges of everyday life is becoming increasingly demanding in these respects. More irregular life careers of individuals and frequent changes in their relative position in local and national communities increase the need for understanding culture and for fostering insights and values that make change understandable and bearable. So, while understanding the role of knowledge and learning in relation to the economy is fundamental, it is equally fundamental to take into account knowledge and learning in the broader societal and cultural context.

At the same time concepts such as social capital indicate that the borderlines between learning for the economic sphere and learning in relation to the social and cultural sphere tend to become perforated. Learning is a socially embedded and context-dependent process. The social context and the basic norms and values of citizens, including their willingness to find collective solutions, has a major effect on economic development in an economy where learning increasingly has become the key to competitiveness. Social capital is an important factor contributing to the production of intellectual capital and capabilities in firms. To find ways to operate that secure the reproduction of social capital in forms that support learning processes has become a major responsibility for those in charge of management strategies and public policies.

In this chapter the focus is upon how knowledge management strategies and innovation policies should be designed in the new context of the learning economy. First, we analyse different concepts of knowledge and learning in economics and show how they lead to quite different practical implications. Second we define the learning economy and point to the need to find ways to overcome the extreme specialisation in management practises and in innovation policy in order to avoid that a speed up of narrow learning goes hand in hand with diminishing wisdom. Third we discuss how benchmarking and learning by comparing may be used to support management and policy learning. Finally we confront our analysis of the learning economy with current debates on the new economy.

KNOWLEDGE AND ECONOMIC DEVELOPMENT—ON THE NEED FOR AN INTERDISCIPLINARY APPROACH

Economists have always been aware of the importance of knowledge for economic development. This is illustrated by the statement below of Adam Smith where he links learning and knowledge creation to the extension of the division of labour. The first part of the quote refers to learning processes in production (learning-by-doing and learning-by-using) resulting in new technology. The second part is full of foresight in emphasising the importance for economic development of R&D-investments and of the increasing division of labour in the production of knowledge as a separate activity (learning-by-searching).

> A great part of the machines made use of in those manufactures in which labour is most subdivided, were originally the inventions of common workmen, who, being each of them employed in some very simple operation, naturally turned their thoughts towards finding out easier and readier methods of performing it. ...
> All the improvements in machinery, however, have by no means been the inventions of those who had occasion to use the machines. Many ... have been made by the makers of the machines, when to make them became the business of a peculiar trade: and some by ... those who are called philosophers, or men of speculation, whose trade is not to do anything but to observe everything: and who, upon that account are often capable of combining together the powers of the most distant and dissimilar objects ... Like every other employment ... it is subdivided into a number of different branches, each of which affords occupation to a peculiar tribe or class of philosophers; and this subdivision of employment in philosophy, as well as in every other business, improves dexterity and saves time. (Smith, 1776, p. 8)

Modern economics is aware of the importance of knowledge and learning more than ever. New growth theory and new trade theory assume a strong link between the increase in the stock of knowledge and the rate of growth in productivity. Learning is treated as a fundamental process in the analysis of market transactions by the Austrian economists. Recent decades have witnessed an explosive growth in institutional economics and economics of innovation. In these new fields of economics, the role of knowledge and learning for economic development is pivotal. New theories of the firm focus on the building of capabilities and competencies. In the management literature the concept of learning organisations has become central for both theoretical developments and practitioners.

But the understanding of knowledge and learning remains narrow in almost all of these contributions. At the core of standard economics it is still assumed that rational agents make choices on the basis of a given amount of information. The only kind of learning allowed for is one where agents get access to new information sets. As we shall see below, this corresponds to a world of the mind where knowledge is identical with information to the neglect of competence building and tacit knowledge.

The more recent developments outside standard economics have been less bounded in these respects and there are many new insights from the research on institutional and technical change. Institutional economics, evolutionary economics, socioeconomic research and economics of innovation have typically been developed in close interaction with historical and empirical research programmes. This is why we now know much more than before about, for instance, how innovation takes place in different parts of the economy. These new developments in economics are compatible with a world of the mind and body and with tacit knowledge as a strategic element in the economy. But economists are still at an early stage of opening up this black box. It is only recently that systematic attempts have been made to specify who learns what and how learning takes place in the context of economic development. It also remains to open up for the wider world of the mind, body and wisdom where normative dimensions of human life and the formation of wisdom are taken into account. The basic distinction between codified knowledge, experience based knowledge and wisdom has roots back to Aristotele:

> Knowledge has been at the centre of analytical interest from the very beginning of civilisation. One of the most interesting taxonomies was developed by Aristotele when he made the distinction between Episteme, Techne and Phronesis. These three categories refer to respectively:
>
> - *Episteme*: knowledge that is universal and theoretical: "know-why".
> - *Techne*: knowledge that is instrumental, context-specific and practise-related: "know-how".
> - *Phronesis*: Knowledge that is normative, experience-based, context-specific and related to common sense: "local wisdom".

It is obvious that economists need to collaborate with scholars from other disciplines in order to integrate knowledge and learning into their models. They are confronted with new issues and their traditional tool box is not sufficient. Scholars in philosophy, psychology, education, anthropology and other disciplines have illuminated different aspects of these activities much more in depth. The increasing division of labour in the production of knowledge, so strongly hailed by Adam Smith, has had as one of its negative consequences a lack of a more holistic understanding of the complex process of knowledge creation and learning.

Below we will point to a need for an integrated perspective on competence building and innovation at the level of management and public policy. Integrated perspectives at these levels are a necessary (but not a sufficient) condition for wise action.

Economic Perspectives on Knowledge

It may be argued that, in a sense, all economic theory is about information and knowledge. At the core of economic theory since Adam Smith has been coordination problems where individual agents, independently of each other, make choices on the basis of information offered by the market. A very important distinction between different economic models and theories are the assumptions made about what agents know and about to what degree they learn anything from what they do. This separates neoclassical economics from the Austrian economics, the former taking fully informed agents as the reference case while the latter emphasise ignorance as the starting point for learning (von Hayek). It also separates those who assume hyperrationality (including rational expectations) and rationality from those assuming bounded rationality (Herbert A. Simon).

An important dividing line goes between those who try to build consistent general and deductive theory of the type characterising natural science, and especially Newtonian physics, and those who believe that such theories need to be combined with context-specific and historical analysis. At the one end, we find theories that are axiomatic, precise and intensive users of mathematics, but have little to say about economic change and development. At the other end, we find theories that explicitly focus on change and development. These operate with dialectical concepts (cf. Georgescu Roegen), look for context specificity and are open to integrate elements from other disciplines in social science.[2]

For noneconomists these distinctions may seem to be of limited interest. Here, I will argue that they are interesting because they are based on two interpretations of knowledge and learning that are present in a less-clear-cut form in much policy debates. Specifically I will argue that the neoclassical tradition operates with a world of mind while the others open up for a world of mind and body.[3]

In the World of Mind Knowledge Equals Information

Let us assume a world where people are reduced to computers. They carry around in their heads algorithms (software programmes) that make it possible for them to pursue calculations on the basis of information input. In this world, learning has two dimensions. With the algorithms given, the choices they make will reflect the kind and amount of information they get access to. In this case learning is identical to access to more or more accurate information and it results in a change in the assumptions about the state of the world. This is the world in which most models of standard economics operate.

Taking into account the most recent neoclassical models where knowledge creation and technology is endogenous—new growth theory—you can add to this model the assumption that new algorithms may be found in a process similar to the search for scarce minerals. This search will typically demand investments by private individuals. When the new algorithms (software programmes) have been found they are privately owned and protected by instruments of intellectual property rights.

In the world of the mind, everything is reversible; there is no cumulativeness and no path dependency. Nobody is better off when it comes to take on a new algorithm. There are no problems to unlearn old algorithms before taking on new ones. Normally firms are treated as individuals in all important respects. While the individual algorithm is to maximise utility, the firm algorithm is to maximise profit. It is characteristic for the world of the mind that there is no collective knowledge and that collective outcomes can always be deduced from the choices made by individuals (the dictum of methodological individualism).

It is interesting to note that some of the economists who have gone deepest in their analysis of this world now tend to point to a need to go outside it and enter the world of mind and body. In the context of an activity by the Organisation for Economic Cooperation and Development (OECD) on knowledge management at Stanford University, Kenneth Arrow, Hal Varian and Paul Romer all, in different ways, expressed that this world is too narrowly defined when it comes to capture learning and knowledge-production. Kenneth Arrow pointed to the extreme importance for the formation and development of Silicon Valley of face-to-face interaction. Hal Varian made a distinction between the learning of codified knowledge that could be learnt through the use of computers and relational knowledge where the personal intervention of teachers is crucial. Finally, Paul Romer pointed to the importance of developing an analysis of the interaction and communication between users and producers of knowledge. All three, thus, gave examples of different avenues that take us into another world where the communication of information gets intertwined with social interaction.[4]

Innovation Policy and Management Strategies in the World of Mind

In this world the most important challenge for innovation policy would be to design incentives to produce and access information. There is a basic dilemma between the need to protect intellectual property rights and to get an optimal diffusion of information. If anything, the bias would be toward strong protection since knowledge is regarded as information that can be easily spread. Only in extreme cases where knowledge production is resulting in solutions to fundamental human and societal needs (such as a general cure for cancer or a new missile system) a weakening of legal property rights may be considered and be combined with government support to R&D. It is recognised that government has a legitimate role in supporting basic research and general training, because it is assumed that these activities result in generic knowledge that is difficult to protect and/or that is characterised by a social rate of return that is higher than the private.

Developing economywide information and communication technology-based networks and the technical competence to access and use such networks becomes another major issue, since access to information is identical to access to knowledge.

In the education and training system and in professional life, formal skills related to reading, writing and solving mathematical problems tend to become the ones given the highest status. Big differences in incomes between those with much and those with little formal training are regarded as important incentives for individuals to engage in formal education.

Management strategies will also focus on developing the right incentives for individuals to search for information. Schemes of reward (including for instance transforming key experts into shareholders) must be worked out that strike the right balance between keeping well-informed people with the organisation over time and hiring new ones.

Information sharing through systematic writing down and codifying experiences into common databases would be given much attention. Information technology would be regarded as a major factor in promoting information sharing.

The World of Mind and Body

In this world, the emphasis is on what people do rather than on what they believe about the state of the world; it is on know-how rather than on know-that types of knowledge. The focus is on more or less skilful behaviour in different situations. The characteristics of human behaviour that have to do with how different types of knowledge is learnt and how learning affects behaviour are specified. One fundamental difference is that it is recognised that important elements of knowledge remain in a tacit form either because of their intrinsic nature or because of the fact that the effort needed to make them explicit is prohibitive. It follows that knowledge is not easy to transfer and that its tacitness and the limited absorptive capacity at the receiving end are more important barriers for knowledge diffusion than legally specified property rights. This reflects that critical skills can only be learnt through doing things and through direct interaction with others (cf. the importance of face-to-face contact and hands-on experimenting in connection with the creation of new ideas) and when they have become internalised they are embedded in the whole body, not in a mind that is separated from the body.

Here, we also have to give up methodological individualism. When individuals interact, they form relationships that are durable and that contain collective capabilities. Modern organisation theory recognises that corporations invest substantial resources in building a culture in order to coordinate the activities of their employees. Shared routines and problem-solving approaches are layered into the organisation and one of the major reasons for establishing firms as organisations may be the development of a common language that is not the property of any specific individual but belongs to the organisation as such (cf. Arrow on Limits of Organisations, 1974). In the present phase of economic development, networks of organisations tend to become carriers of shared competencies. Many of these knowledge elements will remain implicit and one cannot find them written down in any document. They represent *collective tacit knowledge*.

In the world of mind and body, the most fundamental economic problem is how to organise society so that it becomes efficient in the production of new knowledge and in learning. This involves the need to specify a broader set of institutional and organisational characteristics (cf. the analysis of innovation systems) than those referred to above. Also, it is necessary to take into account that most of the learning that matters involves social interaction and is socially embedded. Finally the institutional and organisational characteristics stimulating knowledge-creation and learning will differ depending on the field of knowledge that is addressed.

Innovation Policy and Management Strategies in the World of Mind and Body

The policy problems that came out most strongly in the analysis of the mind world now tend to appear in a rather different light. While the major problem in the mind world is to protect others from getting access to private information, the major problem in the mind and body world is how to organise learning processes where, not least, experience-based knowledge is efficiently shared with others.

To create framework conditions for firms that support learning becomes important. In this context well-functioning markets for skills, knowledge, finance and labour are markets that support learning and the efficient use of competence rather than just efficient allocation mechanisms. But it also becomes important for governments to stimulate the formation of learning organisations and learning networks where they do not evolve spontaneously. Inequality becomes an issue related to the distribution of learning capabilities between regions and people.

At the level of the firm, management needs to focus on human resource development, on building learning organisations and on positioning themselves in networks in such a way that their competence base evolves along with the new demands from new technologies and customers. To strike the right balance between knowledge production and competence building in-house and buying it from the outside becomes one major strategic question. To define what is core competence needs to be done in a dynamic context where the *competence to learn* along certain trajectories while being open to enter new ones is the most important.

In this world management will accept that computers and Internet are important *drivers* but they will not focus on them as being the most important aspect of the *response*. Information and communication technology (ICT) will primarily be regarded as useful in supporting learning processes and in linking people together in interactive learning. The costs of building internal information databanks that everybody can draw upon need to be compared to the benefits from establishing strong interpersonal relationships through, for instance, a shared enterprise culture and job circulation schemes.

Finally, the division of tasks between the public sector and the firms becomes blurred. Especially when it comes to develop human resources, there is a strong need for cooperation and interaction. But it is also true that private firms will have a strong interest in the efficiency of the public sector, since framework conditions become crucial for their performance. And the public sector needs to take a much stronger interest in the dynamic efficiency of private firms. In spite of being exposed to competition, the resistance to change—for instance, in the direction of building learning organisations—is strong also at the top level of many private organisations.

The World of Mind, Body and Wisdom

While the second world widens the perspective as compared to the first, it is still narrow by its instrumental perspective on behaviour. The focus is on skilful action in relation to individual and often very narrowly defined partial objectives. In a society where the division of labour is extreme, the pursuit of this kind of knowledge is not incompatible with trajectories that are both ecologically and socially unsustainable. There is no direct link from a growth of knowledge in terms of information, or in terms of tacit know-how, to the growth of wisdom. The extreme division of labour, while promoting specialist learning, may endanger the cohesion of modern societies. Where the perspective on problems becomes partial and limited, unsustainable trajectories may be the outcome of the competition between specific interests.

There are important institutions, not least those of political democracy, that have as their major task to counterbalance partial perspectives. If there are no restraints embodied in individuals and embedded in their mutual relationships when it comes to how far they are prepared to pursue their own interests, such institutions will founder. This is why processes of socialisation where social norms and the understanding of wider interests imprinted in people are fundamental for the cohesion of society. These are also processes of learning that take place in the family, in school and in civic affairs but also in the business sector. The way the economy is organised in terms of incentives and career paths may undermine or support such processes of social learning. The resulting knowledge is more similar to tacit know-how than to information. It is imprinted and implicit in shared norms and it is not possible to buy this kind of knowledge in the market.[5]

It is obvious that the real world is a mixed one where access to information, tacit skills and ethical values are all important. Innovation policy and knowledge management needs to take all three kinds of knowledge into account. In the next section we will argue that this can be done satisfactorily only if the extreme specialisation in management and policy making is complemented by new kinds of integration across functions and sectoral responsibilities.

INNOVATION POLICY AND MANAGEMENT STRATEGIES IN THE LEARNING ECONOMY

Structural analysis of industrial developments by the OECD secretariat shows that the sectors most intensive in their use of knowledge inputs such as R&D and skilled labour are the ones that grow most rapidly. At the same time, the skill profile is upgraded in almost all sectors in the economy. The most rapidly growing sectors in terms of employment and value added are, in most OECD countries, knowledge-intensive business services.

These observations have led more and more analysts to characterise the new economy as knowledge-intensive or knowledge-based and there is little doubt that there has been a relative shift in the demand for labour toward more skilled workers. Even so, this term may be misleading because it does not fully capture the dynamics in what is going on. The acceleration in the rate of change implies that knowledge and skills are exposed to a moral depreciation that is more rapid than before. Therefore, the increase in the stock of knowledge might be less dramatic than it looks at first sight. The alternative hypothesis put forward here is that we are moving into a *learning economy* where the success of individuals, firms, regions and countries will, more than anything else, reflect their capability to learn. The speed-up of change reflects the rapid diffusion of information technology, the widening of the global marketplace, with the inclusion of new strong competitors, deregulation and less stability in market demand.

Defining learning in the context of the learning economy. In the present context we define learning as a process of enhancing competence and skills that make learning individuals more successful in pursuing their own goals or the goals of the organisation they belong to. But it will also involve a change in the context of meaning and purpose for individuals and affect their preexisting knowledge. This corresponds to what is meant by learning in everyday language and in what experts on learning understand with the concept. It is also the kind of learning most crucial for economic success. But it differs from most definitions of learning in standard economic theory where it is either synonymous with information acquisition or treated as a black-box phenomenon assumed to be reflected in productivity growth.

The information technology revolution stimulates codification but it also increases the relative importance of tacit knowledge. The radical development of information and communication technologies has an equivocal impact on the codification of tacit knowledge. On the one hand it gives stronger incentives and more effective procedures for codification. On the other hand the very growth in the amount of information made accessible to economic agents increases the demand for skills in selecting and using information intelligently. The major impact of the information technology revolution is that it speeds up the process of change in the economy. For this and other reasons, experience-based learning tends to become even more important than before.

Protecting and widening the knowledge base of the firm. The competitive advantage of a firm is rooted in a set of competences not easily imitated by potential competitors. Elements of collective tacit knowledge are at the very core of the competitiveness of the firm. The tacit knowledge embodied in individuals can easily disappear in fluid labour markets while codified knowledge may be more easily copied by outsiders.

Intellectual property rights such as patents represent one way to limit the access of competitors to the core competences of the firm. They play different roles in different sectors but on average the move toward a learning economy tends to make them less adequate—at least as the core of knowledge management strategies. As the speed of change accelerates, it becomes more important for the firm to get access to new sources of knowledge (through recruitment, internal learning and networking) than to hinder others to get access to its own competences. Patents are increasingly used as chips for knowledge trading between competing and collaborating firms rather than as instruments of secrecy.

Learning is a social process based on trust and social capital. Know-how is typically learnt in something similar to apprenticeship-relationships where the apprentice follows the master and relies upon him as his trustworthy authority (Polanyi, 1958/1978, p. 53 et passim). Know-who is learnt in social practise and some of it is learnt in specialised education environments. Communities of engineers and experts are kept together by reunions of alumni and by professional societies giving the participant access to know-how trading with professional colleagues (Carter, 1989). It also develops in day-to-day dealings with customers, subcontractors and independent institutes.

The learning economy needs a lot of trust in order to be successful. And as Kenneth Arrow has pointed out, "trust cannot be bought: and if it could be bought it would have no value whatsoever" (Arrow, 1971). The fundamental role of trust raises strong doubts about how to interpret the standard assumption in economic theory that the most efficient economy is one where individuals act as economic men who *calculate* the outcomes of all alternatives in order to select the one which is best for themselves. *In the learning economy the importance of the ethical dimension and social capital increases enormously.* Little can be learnt and information cannot be used effectively in a society where there is little trust.

The most important inherent contradiction of the learning economy has to do with polarisation and social exclusion. The most immediate benefits of intensified competition and accelerated change and learning are growing productivity, lower prices and a higher level of consumption. Another primary benefit is that the employees of innovative and flexible organisations may earn a premium or at least avoid bankruptcy and unemployment.

But there is also a clear and strong tendency toward polarisation in the learning economy. The distribution of the benefits and costs of economic development has become more uneven during the last decade, with the low-skilled of the labour force as the major losers (OECD, 1994, p. 22 et passim). Within Europe the catching-up of the poorest regions has slowed down in this period (Fagerberg et al., 1997). On a global scale inequality between rich and poor countries has increased (World Bank, 2000).

In the learning economy there is a growing tension between the process that excludes a growing proportion of the labour force and the growing need for broad participation in the change process. It is not obvious that, in the long run, a learning economy can prosper in a climate of extreme social polarisation. This is why there is a growing need at all levels of society to combine elements of the old new deal with a new new deal that puts the emphasis on a more even distribution of skills and competences and especially of the capability to learn (Lundvall, 1996).

Another, even broader, problem is that the speed-up of change puts a pressure on traditional communities. It contributes to the weakening of traditional family relationships, local communities and stable workplaces. This is important since the production of intellectual capital (learning) is strongly dependent on social capital. Social capital—the social capability of citizens and workers to collaborate without too much friction—is not easily reestablished if once devaluated. How new forms of social capital can be created and accumulated is a major issue in the learning economy.

The basic hypothesis—a speedup of the rate of change and learning—calls for reintegrating narrow perspectives and strategies. This new context of accelerating change calls for new strategies at the level of the firm, the region and the nation state. It also requires a rethinking of industrial relations and the role of trade unions. While knowledge production and policy making through decades have been characterised by growing specialisation in knowledge production and in the economy at large as well as by more narrow fields of responsibilities for policy makers, the learning economy calls for lateral thinking and for a reintegration of separate perspectives and strategies. In this section, I will focus on the need for integrating organisational change, labour market dynamics and innovation at the level of the firm and on the implications for policy strategies. To begin with, I present the major concepts and elements of the learning economy.

Building learning organisations and integrating strategies of competence building at the level of the firm. Our research shows that there is a strong synergy between the introduction of *new forms of organisation* and the performance and innovative capacity of the firm (Lundvall, 1999; Lundvall and Nielsen, 1999). Establishing the firm as a learning organisation characterised by decentralised responsibility, team work, circulation of employees between departments and investment in training has a positive impact on a series of performance variables. Flexible firms are characterised by higher productivity, by higher rates of growth and stability in terms of employment and they are more innovative in terms of new products. Our research also shows that success in terms of innovation is even greater when such a strategy is combined with active networking in relation to customers, suppliers and knowledge institutions.

But we also find that, so far, there is only a small minority of all firms (10-30% depending on the strictness of the definition) that have introduced the major traits of the learning organisation. There is an enormous unexploited reserve of economic competitiveness, especially in manufacturing and business service firms. In some sectors such as construction, agriculture and transport the efficacy of building learning organisation can be fully exploited only after a period of de- and reregulation. (Actually we found that, on average, construction firms got worse off when introducing the characteristics of the learning organisation.)

Our conclusion is that a new kind of *integrated competence-building strategy* is needed and that such a strategy should take into account how to combine the three different major sources of competence building: Internal competence building, hiring and firing and network positioning (see Figure 2.1).

Figure 2.1
Knowledge Management in the Learning Organisation

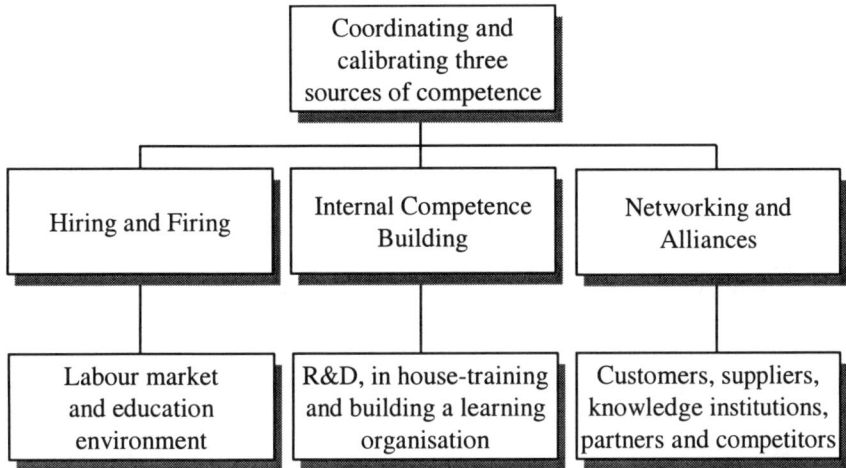

```
                    ┌─────────────────────┐
                    │   Coordinating and  │
                    │   calibrating three │
                    │ sources of competence│
                    └─────────────────────┘
                              │
        ┌─────────────────────┼─────────────────────┐
        │                     │                     │
┌───────────────┐   ┌───────────────────┐   ┌───────────────┐
│               │   │ Internal Competence│   │               │
│ Hiring and Firing│ │     Building      │   │ Networking and│
│               │   │                   │   │   Alliances   │
└───────────────┘   └───────────────────┘   └───────────────┘
        │                     │                     │
┌───────────────┐   ┌───────────────────┐   ┌───────────────────────┐
│ Labour market │   │ R&D, in house-training│ │ Customers, suppliers, │
│ and education │   │ and building a learning│ │ knowledge institutions,│
│ environment   │   │   organisation    │   │ partners and competitors│
└───────────────┘   └───────────────────┘   └───────────────────────┘
```

Firms differ in how strongly they emphasise each of these elements both between and within national innovation systems. Japanese firms have emphasised internal competence building while most hi-tech firms in Silicon Valley depend on learning through high interfirm mobility of employees within the industrial district. Hewlett Packard is one U.S. firm that has given strong emphasis to internal competence building, but it is now moving toward a compromise strategy with more openness to hiring experienced employees from other firms. In Denmark the institutional setup of the training system and the labour market support networking firms and high mobility in the labour market, making it attractive for firms to locate in industrial districts.

There is no single optimal strategy in this respect even if the relative success of information-technology (IT)-based firms in the United States and the weakening of the Japanese firms might be interpreted as an indication that high interfirm mobility of labour is an advantage in the learning economy context. Under all circumstances, management needs to be aware of its priorities in this respect and the different mechanisms need to be attuned to each other so that the firm can become an efficient competence-creating system.

In this context it is important to take into account that labour markets and education systems still have strong national characteristics. Strategies covering multinational operations need to take into account such differences—it is not possible to have one single global knowledge management strategy that neglects local and national specificities.

Industrial Relations and the Role of Trade Unions in the Learning Economy

When Danish managers were asked about what factors stimulated or hampered the movement toward learning organisations, many of them referred to shop stewards (*tillidsmænd*) as a positive factor and only a small minority mentioned them as raising barriers to organisational change. This indicates that trade unions at the central and local levels may be a positive factor when firms need to cope with the new challenges of the learning economy. The relative strength of organised labour in Europe may be regarded as a positive factor in global competition—at least potentially.

Giving workers and their representatives the right incentives to participate positively in building learning organisations may be a question of creating a minimum of security in processes of restructuring—in Denmark the unemployment support level, and its duration, has done so (in spite of high interfirm mobility, Danish workers express fewer worries of increased insecurity in their job situations than do workers in other European countries with much less labour mobility).

The fact that access to learning capability is what constitutes success among the members of trade unions should affect the priorities of the trade union movement. When demanding shorter working hours they could combine such demands with requiring real access to skill upgrading for their members. Agreements between business and labour on the development of new forms of work organisation and skill development become more and more important for both parties.

There is a risk that old priorities lead to short-termism on both sides. Obtaining nominal wage increases for union members whose skill position is stagnating may be highly counterproductive to the long-term interests of those involved. On the business side, routine lamenting on tax levels and government regulation might get in the way of long-term considerations regarding the transformation of training systems and labour-market institutions. Organisations on both sides may need to take on the task to convince their members that the advent of the learning economy involves a new game to be played according to new rules.

A special new responsibility which affects both sides is to cope with the growing tendency toward social exclusion and not least the exclusion of workers of foreign origin. There is a need at the central level of trade unions to focus on the upgrading of the learning capability of those segments of workers who have narrow skills and to find ways to shelter those segments of the workforce (older unskilled workers) that cannot take part in the learning race. In general, trade unions need to be prepared to develop new kinds of solidarity that focus on redistribution of learning capabilities.

Management also has a responsibility for this problem. Our research shows a strong Mattheus-syndrome in the human management strategy of most firms: It is primarily those with extensive training that are offered even more training within firms. It is tempting for firms to focus skill upgrading on those who are rapid learners and leave the rest to public training programmes. In light of a growing scarcity of new entrants into the labour market and the need for broad participation of employees in learning organisations, this might need to change in the future. Under all circumstances, coordinated efforts between business and labour to reduce social exclusion are necessary to make the remaining tasks of governments manageable.

Also in the field of industrial relations there is a need for reintegrating functions and responsibilities. Traditional interests in terms of pay, working time and job security must be linked to and assessed in relation to competence building and the distribution of learning capabilities. Again, having traditions of concertation between government, business and labour may be turned into a comparative advantage if there is a willingness on all sides to take up these challenges.

The Challenge of the Learning Economy and the Need for a New Type of Policy Coordination

As pointed out in the introduction, there is a growing consensus on the need to focus on long-term competence building in firms and in society as a whole. At the same time, the prevailing institutional setup and global competition tend to give predominance to short-term financial objectives in policy making. At the institutional level this is reflected in the fact that ministries of finance have become the only agency taking on a responsibility for coordinating the many specialised area policies. Area-specific ministries tend to identify with their own customers and take little interest in global objectives of society.

The concept of the learning economy has its roots in an analysis of globalisation, technical innovation and industrial dynamics (Lundvall and Johnson, 1994; Lundvall and Borras, 1998). But the concept also implies a new perspective on a broad set of policies including social policy, labour market policy, education policy, industrial policy, energy policy, environmental policy and science and technology policy. Specifically, the concept calls for new development strategies with coordination across these policy areas.

Social and distributional policies need to focus more strongly on the distribution and redistribution of learning capabilities. It becomes increasingly costly and difficult to redistribute welfare, ex post, in a society with an uneven distribution of competence. Therefore there is a need for stronger emphasis on a new new deal where weak learners (regions as well as individuals) are given privileged access to competence upgrading.

The effectiveness of *labour market institutions and policy* has so far been judged mainly from a static allocation perspective. There is a need to shift the perspective and to focus on how far the labour market supports competence building at the individual level and at the level of firms. This implies for instance that some dimensions of flexibility and mobility are more productive than others and that there may be third roads aside from Anglo-Saxon maximum flexibility and Mediterranean contractual job security (cf. for a Danish model characterised by a unique combination of relative income security, high participation and mobility rates).

Education and training policy needs to build institutions that promote simultaneously general and specific competences, learning capability and life-long learning. This points toward a new pedagogy that combines individual learning plans with collective problem-oriented styles of learning. A real commitment among employers, employees and policy makers to life-long learning with a strong interaction between schools and practise-based learning is necessary.

Industrial policy needs to adjust to competition policy and policies aiming at developing learning organisations and competence-building networks. Intensified competition may stimulate superficial change rather than competence building if not combined with organisational change and new forms of interfirm collaboration. *Energy and environment policies* need to take into account their impact on competence building in the economy.

Science and technology policy needs to support incremental innovation and the upgrading of competence in traditional industries as well as the formation and growth of high-technology industries. For instance, the reallocation of academically trained workers toward small and medium-sized firms is a key also to the formation of networks with universities and other knowledge institutions.

These area-specific policies need to be brought together and tuned into a common strategy. In the learning economy it is highly problematic to leave policy coordination exclusively to ministries of finance and to central banks—their visions of the world are necessarily biased toward the monetary dimension of the economy and thereby toward the short term.

It could be decided to establish a new types of institutions and *High Level Councils on Innovation and Competence Building* at regional and national levels (in Finland the prime minister is chairman of a National Council of Science and Technology). Such new institutions could have as their strategic responsibility to develop a common vision for how to cope with the challenges of the learning economy. The basis of such a vision must be a better understanding of the national and regional systems of competence building and innovation. In the framework of such an understanding, international benchmarking and policy learning at the European level becomes meaningful. The European High Level Council could be in charge of the design of the main lines of a *Framework Programme on Innovation and Competence Building* where science is treated as only one among several sources for competence building.

BENCHMARKING AND POLICY LEARNING

Introduction

Sometimes certain management and policy-related concepts seem to get a life of their own and suddenly they become widely diffused without any obvious explanation. Today, in the public as well as in the private sector, it has become high fashion to recommend the use of benchmarking—that is, the systematic comparison of organisations with a best-practice organisation. In spite of the popularity of the concept there are few attempts to explain why and under what circumstances it could be useful. The purpose of this section is to critically assess some standard uses of benchmarking and to suggest that a systemic approach may overcome some of the traditional weaknesses and actually make something similar to benchmarking in its narrow sense a valid method of international policy learning. The focus will be on the more useful aspect of benchmarking and a process of learning-by-comparing.

There are many examples of naïve benchmarking built upon overly simplistic ideas about the real world. There are also examples of abuses of benchmarking aimed at advancing hidden ideological objectives shared by technocrats. At the same time, we will argue, the increased use of benchmarking in the public sector reflects that the current era of extreme uncertainty and rapid change undermines the belief in more precise and complete logical procedures. For example, the use of full-blown mathematical econometric models may be too rigid. In the learning economy intelligent (as opposed to naïve) benchmarking and international comparative analysis may be used to enhance the quality of policy making.

What Is Benchmarking?

Benchmarking always involves some sort of systematic comparison of one organisation's outcomes or processes with either some other organisation or some standard. It is usually a comparison with an institution that is regarded as superior in performance or with a best practice standard.

Bogan and English (1994) are responsible for one of the early definitions which gets cited frequently in management literature:

- Benchmarking is the systematic observation of organizational routines and the comparison of performance with superior units at the levels of resource use and efficiency and effectiveness (inputs and outputs).
- Benchmarking is the search for industry best practices that lead to superior performance.

It is thus basically about comparing one company's performance with another that is considered one of the best, if not the best, in the field. A learning process takes place whereby the benchmarking company can adjust its behaviour based on observation of the benchmarked company and thus improve its efficiency. Crucially, data are generated (i.e., benchmarks) that can be compared between organisations.

Business Benchmarking in the Learning Economy

The basic idea behind benchmarking is less nontrivial than it appears at first sight. For instance, it stands in sharp contrast to standard microeconomics where it is assumed that all firms are equally competent and have equal access to technologies and modes of organisation. The popularity of the model may actually be seen as a demonstration of the fact that practitioners operate differently than it is assumed in models of standard economics.

It may also be seen as a reaction to increasingly turbulent competition where change is accelerating while, at the same time, new technologies become more complex and markets more volatile. In this learning economy (Lundvall and Johnson, 1994) it is the capability to learn that determines the outcome of competition. This is why management is so eager to find ways to speed up the transformation toward more efficient methods and ways. More sophisticated, logical and comprehensive methods such as operational research and linear programming may be too cumbersome and too slow when it comes to cope with a fluid, rapidly changing and uncertain reality. In such a context a more intuitive and interactive procedure may be more efficient. Benchmarking may be seen as a process aiming at establishing consensus on the basis of incomplete, implicit and intuitive models of reality.

But there is, of course, some risk in this movement toward such a postmodern management technique. Benchmarking may, since it is often based on comparing numbers, appear to be a more exact technique than it is. It may be abused by those in power to impose controversial changes on unwilling employees and citizens.

Diversity and Benchmarking

There are many problems with simplistic versions of benchmarking when applied in the private sector. The very idea that there is one single best practice that can be referred to as the benchmark is valid only under specific, quite exceptional conditions. Normally there will be *several competing good practices* that coexist. This corresponds to the insight that competing technologies have coexisted for long historical periods.[6]

A second problem has to do with *the lemming effect*. Stardom among management styles and firms is changing rapidly and some of the firms characterised as parading excellence one year may be treated with disdain the next. Just copying the management style of the current stars may bring a whole population of firms onto the wrong track.

Too much copying may also be problematic because it *reduces diversity*. Even if there were a rather clear best practice to copy from, the copying process may undermine the dynamic capabilities of the industry. One most important insight of economic historians that has been confirmed by the research within evolutionary economics is that diversity in a population of firms is a key to economic evolution, innovation and economic growth.[7]

Benchmarking in Systemic Contexts

The last critical point is the obvious one: that the *context*—defined in its economic, technical, geographic, historical and cultural dimensions—has a great influence on what is a best practice. The literature on national competitive advantage (Porter, 1990), national business systems (Whitley, 1996) and national systems of innovation (Lundvall, 1992; Nelson, 1993) has as one common message that national institutional differences are systemic and reflect the specialisation of the system. Since national systems differ in terms of institutions related to competition, the formation of human skills and labour markets, best-practice governance at the firm level is not the same in, for instance, the United States, Japan and Denmark.

Given these critical points one might ask, why does management in advanced firms bother about benchmarking at all? It seems to create more problems than it solves. There are at least two reasons why benchmarking may be more useful than it appears to be. The most fundamental reason is that *in a world where things change very rapidly, any kind of initiative that gets the members of an organisation to focus and reflect on performance and on what factors may help to make it better is useful.* Benchmarking obviously serves this general purpose.[8]

Second, experienced and competent managers would, of course, avoid promoting the kind of naïve benchmarking sketched above. They would normally take into account the *context* and *adapt* the best practice accordingly. This is not to rule out that naïve benchmarking actually takes place in the private sector. In periods when a new management technique is introduced, consultancy firms may ride on waves of enthusiasm and deliver standard solutions of limited usefulness.

Benchmarking National Systems of Innovation

Empirical research shows that there are systemic differences between how competence building and innovation are pursued in national economies. At the same time it is true that a system's approach to innovation presents policy makers with a more complex agenda. For instance, it does not support simplistic strategies that have as their exclusive aim to increase the national R&D budget. The quality of demand and user competence becomes as important as supply factors. And what matters most is the relationship between the different elements of the system. Having excellent universities does not help a lot if they do not interact with competent users of academic research.

In spite of this fact, the innovation system approach has diffused rapidly to national and international policy making bodies. OECD, the European Commission and UNCTAD have integrated the concept in their policy analyses. A number of national studies have inspired new, more integrated policy strategies aiming at stimulating innovation and competitiveness.

One of the academic analyses that has gone quite far in giving a quantitative characterisation of different national systems of innovation without explicitly referring to benchmarking is Amable, Barré and Boyer (1997). It is interesting to note that they end up grouping 12 different national systems in four different categories with different systemic characteristics. Rather than showing that the Nordic model (Sweden, Norway and Finland) is better or worse than the market-oriented model (United States and United Kingdom) when it comes to promote innovation, their data show that each group of countries has its own specific strengths and weaknesses. What would appear to be a best practice in one group would normally not be so in the other.

Intelligent Benchmarking

It might be useful to consider how benchmarking innovation systems could be turned into a procedure that takes into account some of the criticism developed above. Is it possible to use international benchmarking as a tool for policy learning in Europe? It seems that what has been said leads to the following primary considerations:

- Benchmarking should not be seen primarily as a technical procedure focused on comparing quantitative data. Its rationale is that it focuses attention on the efficiency of a system, stimulates reflection and thereby supports learning among those involved in the process. Benchmarking in the narrow sense can only play a subordinate role in processes of open coordination. A more general process of learning by comparing that specifically takes into account systemic differences has a much greater potential when it comes to stimulate policy learning.
- The setup of benchmarking national systems should be designed in such a way that it stimulates international policy learning in order to produce good and better practices rather than striving for best practice. It will involve comparing and sharing national experiences. In some cases this might include a certain convergence among countries, but in other cases diversity is rooted in systemic differences and should not only be accepted but perhaps even be reinforced. Globalisation does not call for regions within nation states to become more similar, and there is no reason why standardising national systems in all respects should be a part of European integration.
- When delimiting the field to be covered by the benchmarking exercise it is important to avoid a narrow focus on single variables. It is important to include in the benchmark exercise performance indicators referring to economic dynamics and indicators referring to distributional issues. Before drawing conclusions from benchmark exercises, it is important to take into account the systemic features of the national economy.
- Benchmarking innovation systems, learning effects and social cohesion presuppose the development of more reliable indicators for the quality and intensity of relationships, interactions and networks. The same is true for indicators referring to the characteristics of learning organisations and experience-based tacit knowledge. In these and other areas there is a need for new conceptual work before meaningful indicators can be constructed. This is especially true for social capital, which is a crucial element in the formation of intellectual capital.
- To use simple analytical tools such as benchmarking and operating with intuitive and very incomplete models of reality may be a more realistic strategy than aiming at more ambitious grand econometric models, but there might be alternatives in between these two extremes. Using simple statistical techniques to map systemic characteristics may be helpful in order to avoid naïve and ideologically biased benchmarking.

COMPETENCE BUILDING AND LEARNING IN THE NEW ECONOMY

Information and communication technology (ICT) has had a major impact on the workings of the economy. Some of the changes involved have been condensed into the concept the new economy (OECD, 2000b; OECD, 2000c). The new economy refers to a scenario where the rapid development and widening use of information and communication technology give rise to new opportunities for economic growth and job creation. The classical trade-off between low unemployment rates and low inflation rates is assumed to be becoming less stringent. One weakness with this scenario, and the policy recommendations based upon it, is that the transformation mechanisms from the development, diffusion and use of information and communication technology to a better macroeconomic performance remain extremely fuzzy.

One way to get a better understanding of the transformation mechanism is to focus on how information technology affects the production, diffusion and use of knowledge and to analyse how changes at the micro level—new modes of knowledge management at the level of the firm—interact with changes at the level of the whole economy. One major impact of the information technology revolution on the economy goes through its impact on knowledge and learning.

Can Information Technology Substitute for Human Competence?

There is a strong normative bias in Western civilization in favour of explicit and well-structured knowledge and there are permanent efforts to automate human skills. One historical example is the effort to transfer the knowledge of skilled workers into machinery connected with Taylorism. Current efforts to develop general business information systems and expert systems go in the same direction.

So far automating human skills has proved to be economically successful only in relation to relatively simple repetitive tasks taking place in a reasonably stable environment. Highly automated process industries may be extremely cost-efficient but at some time when the products lose their competitiveness because of the appearance of more attractive substitutes, they leave behind them rust-belt problems difficult to solve.

Let us take a closer look at how IT affects different elements of knowledge. It is claimed that the increased use of information technology enhances both the incentives and the possibilities to codify knowledge (David and Foray, 1995). We suggest that this is only one side of the coin. The other is a speeding up of change that makes it less meaningful and attractive to engage in the development of codification and information systems.

While some skills will be transformed into a codified form, demand will grow for complementary tacit knowledge. The very growth in the amount of information which is made accessible to economic agents increases the demand for skills in selecting and using information intelligently. For this reason experience-based learning might become even more important than before. The major impact of the information technology revolution on the process of learning might, however, be that it speeds up the process of change in the economy. The codification, standardization and normalization of certain parts of the knowledge stock increase the rate with which some stages in the innovation process are progressing, and the diffusion of this kind of knowledge might also be accelerated.

The most fundamental problems of using ICT as a substitute for human skills and tacit organisational competence have to do with difficulties to absorb, allow for and promote change. In a stable environment characterised by a high degree of standardisation in inputs and outputs, it would be possible and economically attractive to build information systems which substituted for at least some of the functions which had so far been pursued by skilled labour and human intelligence. But when materials, processes, product, markets and regulations all change, these efforts will often prove to be counterproductive—they will become barriers to flexible adaptation. It will also be difficult to pursue innovative activities in an organisational environment where human skills were highly automated[9].

A Different Perspective

If the main impact of ICT is a speedup of specific phases of the innovation process, the use of information technology may be regarded from a different perspective where the emphasis is upon its potential to reinforce human interaction and interactive learning. Here the focus is not upon its capability to substitute for tacit knowledge but rather on how it can support and mobilise tacit knowledge. E-mail systems connecting agents sharing common specific codes of communication and frameworks of understanding can have this effect, and broad access to data and information among employees can further the development of common perspectives and objectives for the firm. In the future, multimedia exchange and virtual reality applications may actually change the transferrability of tacit knowledge. Increasingly it may become possible to transfer elements of tacit knowledge by, for instance, using combinations of voice, pictures and other virtual reality elements in an interactive mode. In connection with product development, flying prototypes may in the future be transferred electronically rather than through physical transport.

DIFFERENT GOOD PRACTISES OF KNOWLEDGE MANAGEMENT: ALTERNATIVE MODELS FOR HARNESSING TACIT KNOWLEDGE

This section examines three institutional models with capabilities for harnessing tacit knowledge as a source of learning and innovation: the Japanese, the US and the Danish model. The presentation is highly stylised and the main objective is to illustrate that the new economy has to be realised in systemically different institutional context. At the same time it illustrates the limits of benchmarking where the assumption is that there is one and only one best-practise when it comes to cope with the challenges of the new economy.

The Japanese Organisational Community Model

Nonaka and Tageuchi (1995) give a series of examples of how Japanese firms organize the process of product innovation in ways which explicitly take into account the important role of tacit knowledge. Japanese managers do not give detailed instructions in order to tell in what direction the search of their innovative teams should go. Instead they promote the search for innovative solutions by formulating metaphors and analogies. These are based on the intuition of management and they leave ample room for creativity and the formation of new intermediate concepts. An intermediate layer of project team leaders makes these open concepts interact with the tacit knowledge of skilled workers and engineers. They formulate somewhat more concrete slogans and gradually the conceptualization of the new product takes place.

All through the process face-to-face interaction and hands-on experimenting are given high priority. There are several examples of how information technology is used to give all participants more easy access to banks of information in order to support the knowledge creation, but these efforts are always combined with direct human interaction, they are not regarded as substituting for them.

It is argued that the organizational model best suited to the creation of new knowledge is a hypertext organization where there is one regular divisional structure which is overlayered with ad hoc horizontal teams directly aiming at creating new products and new knowledge. Members of these teams should be completely taken out of their regular function and division.[10]

The focus of the analysis is limited to management strategies in connection with product development in big knowledge-based firms. It is, however, possible to extend the basic perspective in order better to understand the Japanese model and innovation systems by locating firms' knowledge management strategies within the wider national competence building system. Lam's (2000) analysis of the J-form organisation within the organisational community model depicts how the learning strategies of Japanese firms are rooted within a firm-based internal career structure and a broad-based education system. The model favours the accumulation of tacit knowledge through collective learning. It is geared to competence preservation and incremental innovation.

A U.S.-like Model: The Experimentally Organized Economy

Eliasson's (1996) notion of the experimentally organized economy is rooted in liberal market institutions which favour entrepreneurial forms of organisation with competency destroying and radical innovation capabilities and it may be seen as being close to the U.S. system. Eliasson stresses the importance of tacit competence and the limits of the usefulness of universal information business systems. His analysis covers a broader set of topics relating to the management of the firm than the analysis of the Japanese firms just referred to. It covers, for instance, the need for financial information systems that support efficiency in the short term as well as the incentives to create new knowledge and to innovate.

It is interesting to note that the analysis at some point ends up recommending organizational solutions which are similar to the ones proposed by Nonaka and Takeuchi (1995)—this includes the idea that horizontal teams independent of the divisional structure are necessary to provoke change (p. 60). It also includes the perspective on information technology where the emphasis is on facilitating communication rather than substituting for human skills.

But it is also interesting to note that the broader institutional environment of the experimentally organized economy is very different and sometimes appears as the polar opposite of the Japanese system (pp. 109-110):

- In product markets low entry barriers and fierce competition are creating the best environment for promoting experiments and to get rid of inefficient noninnovative firms. Little is said about long-term interfirm cooperation.
- In the labor market the emphasis is on top management as an authority, selecting competent teams and designing material incentives to stimulate the top teams in the firm. If anything it is assumed that the bias in compensation goes against the most competent participants. The idea that social cohesion could promote learning and innovations is not considered.
- The most important function of the financial market is to intervene and enforce a shift in top management when it becomes incompetent or conservative. U.S. types of capital markets combining takeover threats, junk bond markets and venture capital are presented as the ideal in this context. Little is said about the problem of short-termism in Anglo-Saxon financial markets.
- Governments should not intervene in the market mechanism because governments are assumed to be systematically incompetent when it comes to recognize and correct their own mistakes—this is a key competence for the successful firms. There is no reference to historical cases where active governments have stimulated economic development by indicating broad trajectories for industrial development.

Eliasson's (1996) experimentally organized economy is characterised by competence-destroying institutions in the areas of both labour markets and corporate governance. This is in contrast with the Japanese model rooted in a competence-preserving and incremental innovation institutional environment.

Another significant difference lies in the definition of competence. Eliasson has a hierarchical understanding of competence—there is a lot of competence at the top and very little at the bottom. Operators at the bottom have a very limited role to play in the process of learning and job-creation. This perspective may be explained by the focus on tacit knowledge exclusively as business decision competence and the corresponding neglect of tacit knowledge as connected to direct and indirect physical human action. The Western separation of mind from body, which is major element of the Western Model, according to Nonaka and Takeuchi, is not at all confronted by Eliasson. Perhaps it is fair to say that the

Eliasson revolt against neoclassical rationalism remains rooted in the Cartesian tradition.

Denmark—A New Economy Characterised by a Weak Formal Knowledge Base and by an Extremely High Degree of Income Equality

There has been a tendency to see the United States as the Mecca of the new economy, and in Europe there has been a growing opinion in favour of importing elements of the U.S. institutional setup. It is interesting to note that in the ongoing OECD analysis the Danish economy is also presented as a success story in the context of the new economy in spite of the fact that Denmark in many respects deviates from the prototype of such an economy.

Denmark remains one of the most equitable societies in the world and has become even more so over the last decade. It is actually the only OECD country where the unskilled workers have not experienced a worsening of their position in the labour market. In the United States, on the other hand, high employment growth has been linked with extreme degrees of growing polarisation in incomes and wealth. While Denmark offers a high degree of collective security for workers, very little is offered in the United States.

The Danish economy has a rather weak formal knowledge base. This is true if we compare the rating of students in international tests, it is true in terms of the proportion with higher education and with Ph.D. degrees and it is also true for the private R&D effort. And the incentives to engage in formal education are weak since the private rate of return on investment in human capital is low. In spite of these weaknesses the Danish economy is extremely successful. We ascribe this apparent contradiction to certain systemic features that make it successful as a learning economy:

- the system is abundant in terms of social capital,
- there is a positive attitude to organisational and technical change,
- processes of incremental innovation involve a close collaboration between engineers and skilled and unskilled workers,
- there are dense networks including networks involving both private and public organisations.

This is why Maskell (1998) has characterised the Danish economy as a village economy. Social capital that makes it possible to engage in learning and uncertain transaction without too much fuzz may be the secret behind the success of several of the small welfare states in Europe—including Denmark.[11] But village-like institutions may also be at the core of the exclusion of immigrant workers and minorities from the labour market.[12] In the era of globalising processes it is a major challenge to try to make village economies open so that they can allow for the coexistence of cultures and ethnicities, without undermining the social capital that keeps them together.[13]

It is important to note that Denmark combines the village economy characteristics with certain strengths in more traditional knowledge and ICT-related areas:

- Together with the other Nordic countries Denmark is using ICT more widely than other countries.
- The proportion of total R&D investment allocated to services is high in Denmark.

Even more important is perhaps the public effort regarding investment in adult and vocational training. In most other countries life-long learning has been referred to without taking it too seriously in terms of common commitment. In Denmark there is need for reform of this part of the training system, but there are no good arguments for downsizing the effort.

We see a clear connection between the national system and the firm strategies that thrive in the system. Denmark has designed a national system of innovation and competence building that is oriented toward competence-intensive low-tech small and medium-sized firms. It gives a good environment for firms that want to build learning organisations and move ahead fast and adjust rapidly to new challenges in low- and medium-tech areas. This includes certain knowledge-intensive services. It is less ideal for big-scale enterprise and for firms engaged in strongly science-based and formalised activities (here the major Danish Pharmaceutical company Novo and pharmaceuticals is an important exception).

CONCLUSIONS

The chapter is based on a hypothesis that we have entered a specific phase of economic development (which we refer to as the learning economy) where knowledge and learning have become more important than in any earlier historical period. In the learning economy, individuals, firms and even national economies will create wealth and get access to wealth in proportion to their capability to learn.[14] This will be true regardless of their present level of development and competence. This implies, of course, a broad definition of knowledge and learning. Wealth-creating knowledge includes practical skills established through learning-by-doing as well as competencies acquired through formal education and training. And it includes management skills learnt in practice as well as new insights produced by R&D efforts.

As illustrated by the Danish case, learning is an activity going on in all parts of the economy, including so-called low-tech and traditional sectors. As a matter of fact, for most countries, learning taking place in traditional and low-tech sectors may be more important for economic development than learning taking place in a small number of insulated high-tech firms. The learning potential (technological opportunities) may differ between sectors and technologies, but in most broadly defined sectors there will be niches where the potential for learning is high. This is important in a period where new economy-euphoria threatens to result in the neglect of competence building in sectors outside the core of ICT.

In the learning economy there is a need to integrate and coordinate management and policy strategies. Promoting processes where citizens move between different functions and sectors and between local and global activities may be one way to counter the myopic and narrow perspectives fostered by a highly developed division of labour. To neglect the social ecological foundations of the economy is not an acceptable long-run strategy, and finding ways to combine instrumental rationality with the broader issues is a major challenge.

As management and policy makers have become aware that the working of the economy has changed and that knowledge and learning now are more important than ever, the search for methods to adapt institutions, organisations and management styles has intensified. One expression for that common concern is the explosive diffusion of what has been called benchmarking. Benchmarking in the very strict sense of the concept is not directly applicable either to knowledge management or to national innovation policies. It is only in extreme situations that a sample of organisational units can find and agree upon one definite best practise to benchmark against.

While benchmarking in the strict sense may be based on crude assumptions, a broadly defined process of *learning by comparing* is a useful activity. Confronting what you are doing and the institutional context within which you operate with what others are doing in different contexts, you are forced to reflect on your own practice. And reflection is a key element in learning. Taking a step back and thinking about the possibility of doing things differently enhances the results of learning by doing, and it may be a way of making tacit knowledge more explicit. So, while naïve ideas about benchmarking may give negative results, learning by comparing in order to enhance performance is both a legitimate and useful activity.

The strong focus on the concept new economy should be welcomed because it signals a wider insight that there have been fundamental changes in the working of the economy that challenge both management strategies and public policies. But there are risks with the diffusion of the concept. An exaggerated emphasis on information and communication technology as a *solution* and on unhampered individualist entrepreneurship as the *only social response* to the new context goes fundamentally wrong. Analysing the impact on competence building and learning and developing models that combine the production of intellectual capital with the reproduction of social capital is what is called for.

NOTES

1. There is a growing recognition among managers and policy makers that we are in a period of radical socioeconomic transformation. This is reflected in the wide use of the term "new economy". At the end of the chapter we will relate our analysis to the new economy perspective. Basically we see information and communications technologies as major factors that lie behind the development of the learning economy.

2. Recent developments where evolutionary economists have become more ambitious in building mathematical models and where neoclassical economists have focused explicitly on knowledge as a factor behind economic growth have brought the different camps closer together on the research arena. But it has not necessarily weakened the animosity between the two styles of research. Standard economists are unwilling to recognise evolutionary economists as real economists and evolutionary economists argue that the standard economists have become "prisoners within their own tool shed", to quote the Swedish Schumpeterian, Erik Dahmén.

3. Reality is, as always, more complicated than such a bipolar presentation indicates. While digging deep into the limitations of their own models, some of the leading neoclassical economists have become strongly aware of a need for a broader perspective. The fact that it was Kenneth Arrow who first introduced learning-by-doing as a central concept in economic growth models illustrates this fact.

4. Furthermore, in his Richard T. Ely lecture at the American Economic Association, Kenneth Arrow raised doubts about methodological individualism when it comes to analyse knowledge and learning (Arrow, 1994).

5. Kenneth Arrow has pointed out that "you cannot buy trust—and if you could it would have no value whatsoever" (Arrow, 1971).

6. To take a well-known example from the history of technology, the transfer from sailing techniques to steam power was not as simple as just moving from an old inferior to a new and much better practice. The two technologies coexisted for many years because the pressure from the new technology stimulated substantial progress in the old. Even today, for some specific purposes sailing might become the better technique. The same could of course be true for a new marketing technique or competing management information systems.

7. A similar point is relevant for the European integration process. In the European discourse diversity among member states should be regarded as a potential source of innovation and growth, and valorising diversity has been pointed to in contrast to strategies aiming at convergence of institutions and policies. On the other hand the single market project and the strong focus on reducing barriers to competition has in practice stimulated processes of convergence. The distinction between legitimate national differences and barriers to the free movement of goods, services, labour and capital is becoming less clear as competition policy is extended to cover services and procurement procedures related to the public sector.

8. At first sight this seems to correspond to the Hawthorn experiments that showed that any kind of trivial change—such as variation in the light—in the factory might stimulate productivity growth. In the context of a learning economy it should also be taken into account that benchmarking gives room for *reflection* on practices, and reflection is a key element in any learning process. It corresponds to the establishment of what is called double-loop learning in the literature on learning organisations (Argyris and Schön, 1978).

9. The difficulties with automating tacit knowledge do not rule out that new attempts will be made to formalise and structure tacit knowledge, and it is also reasonable to assume that the growing importance of information technology will further stimulate such attempts. Already one can see a number of new applications which change the character of knowledge creation at certain stages of the innovation process. Developing and testing drugs or aircraft with the help of computer power and the use of computer-aided design in many other areas illustrate a successful transfer of problem-solving from human skills to computers (Foray and Lundvall, 1996, pp. 14-15).

10. The analysis is much more complex than indicated by these short remarks. For instance, a model of knowledge creation that assumes this process to be a spiral moving

from tacit to explicit and than back to tacit knowledge is developed. The conversion between these forms plays a crucial role in the theory. This is a point worth critical attention. In some of the examples it is not obvious to what degree what is illustrated is an interaction between the different forms of knowledge rather than a conversion of one into the other.

11.   Classical contributions to the analysis of the success of small countries are Kuznets (1960), Svennilson (1960) and Katzenstein (1985). The strength and weaknesses of small innovation systems have been analysed in Freeman and Lundvall (1988) and in Edquist and Lundvall (1993).

12.   While the Danish labour markets show good performance in terms of high participation rates, and low rates of youth and long-term unemployment, its record in terms of integration emigrants and refugees has been quite deplorable. This has to do with a mode of production and innovation that is intensive in terms of informal communication giving a broader responsibility to workers (Lundvall, 1999).

13.   Actually, one of the most difficult dilemmas to be coped with in the future may be the growing tension between intraefficient but somewhat exclusive nation-state institutions on the one hand and cosmopolitan market-oriented institutions that are more universally inclusive but offer very little security for workers and ordinary citizens on the other. The fact that the most successful economies, besides the United States, in the new economy seem to be small homogeneous economies with a strong welfare state such as Norway, The Netherlands and Denmark (OECD, 2000b; OECD, 2000c) indicates that there may be great costs associated with a cosmopolitan (Anglo-Saxon driven) project aiming at full institutional convergence.

14.   Similar perspectives have been developed by others. Among the most well-known is Peter Drucker (1993). Earlier contributions spelling out the characteristics of the learning economy include (Lundvall and Johnson, 1994; Lundvall, 1996, 1998).

REFERENCES

Amable, B., Barré, R. and Boyer, R. (1997), *Les systémes d'innovation a l'ére de la globalization*. Paris: Economica.

Archibugi, D. and Lundvall, B.-Å. (eds.) (2000), *The globalising learning economy: Major socio-economic trends and European innovation policy*. Oxford: Oxford University Press.

Argyris, C. and Schön, D. A. (1978), *Organisational learning: A theory of action perspective*. Reading, MA: Addison-Wesley.

Arrow, K. J. (1962), "The Economic Implications of Learning by Doing", *Review of Economic Studies*, Vol. 29, No. 80.

Arrow, K. J. (1971), "Political and Economic Evaluation of Social Effects and Externalities", in Intrilligator, M. (ed.), *Frontiers of Quantitative Economics*. Amsterdam: North-Holland.

Arrow, K. J. (1974), *The Limits of Organisations*. New York: W. W. Norton & Company.

Arrow, K. J. (1994), "Methodological individualism and social knowledge", Richard T. Ely Lecture, in *AEA Papers and proceedings*, Vol. 84, No. 2 (May).

Bogan, C. and English, M. J. (1994), *Benchmarking for best practices: winning through innovative adaptation*. New York: McGraw-Hill.

Carter, A. P. (1989), "Know-how trading as economic exchange", *Research Policy*, Vol. 18, No. 3.

Carter, A. P. (1994), "Production workers, metainvestment and the pace of change", paper prepared for the meetings of the International J.A. Schumpeter Society, Munster, August.

David, P. and Foray, D. (1995), "Accessing and expanding the science and technology knowledge-base", *STI-review*, No. 16. Paris: OECD.

Drucker, P. (1993), *The Post-Capitalist Society.* Oxford: Butterworth Heinemann.

Edquist, C. and Lundvall, B.-Å. (1993), "Comparing the Danish and the Swedish systems of innovation", in Nelson, R. R. (ed.), *National Innovation Systems: A Comparative Analysis.* New York: Oxford University Press.

Eliasson, G. (1996), *Firm Objectives, Controls and Organization.* Amsterdam: Kluwer Academic Publishers.

Ernst, D. and Lundvall, B.-Å. (1997), "Information technology in the learning economy—challenges for developing countries", *DRUID Working Paper*, no. 97-11. Aalborg, Denmark: Aalborg University, Department of Business Studies.

European Commission (1999), *First report by the high level group on benchmarking. Benchmarking Papers No. 2.* EC Directorate General III—industry.

Fagerberg, J., Verspagen, B. and Marjolein, C. (1997), "Technology, Growth and Unemployment across European Regions", *Regional Studies*, Vol. 31, No. 5.

Foray, D. and Lundvall, B.-Å. (1996), "The knowledge-based economy: from the economics of knowledge to the learning economy" in Foray, D. and Lundvall, B.-Å. (eds.), *Employment and Growth in the Knowledge-based Economy.* Paris: OECD Documents.

Freeman, C. and Lundvall, B.-Å. (eds.) (1988), *Small Nations and the Technological Revolution.* London: Pinter Publishers.

Gjerding, A.N. (1996), "Organisational innovation in the Danish private business", *DRUID Working Paper*, no. 96-16. Aalborg, Denmark: Aalborg University, Department of Business Studies.

Hatchuel, A. and Weil, B. (1995), *Experts in Organisations.* Berlin: Walter de Gruyter.

Helgason, S. (1997), "International benchmarking: experiences from OECD countries". Paper presented at the International Benchmarking conference, Copenhagen, 20-21 February.

Katzenstein, P. (1985), *Small States in World Markets—Industrial Policy in Europe.* New York: Cornell University Press.

Kuznets, S. (1960), "Economic growth of small nations", in Robinson, E. A. G. (ed.), *Economic Consequences of the Size of Nations.* London: Macmillan.

Lam, A. (2000), "Tacit knowledge, organisational learning and societal institutions: an integrated framework", *Organization Studies*, Vol. 21, No. 3, pp. 487-513.

Lund, R. and Gjerding, A. N. (1996), "The flexible company, innovation, work organisation and human resource management", *DRUID Working Paper*, no. 96-17. Aalborg, Denmark: Aalborg University, Department of Business Studies.

Lundvall, B.-Å. (ed.) (1992), *National Systems of Innovation: Towards a Theory of Innovation and Interactive Learning.* London: Pinter Publishers.

Lundvall, B.-Å. (1996), "The Social Dimension of the Learning Economy", *DRUID Working Paper*, No. 1. Aalborg, Denmark: Aalborg University, Department of Business Studies.

Lundvall, B.-Å. (1998), "The Learning Economy—Challenges to Economic Theory and Policy", in Johnson, B. and Nielsen, K. (eds.), *Institutions and economic change.* Cheltenham: Edward Elgar.

Lundvall, B.-Å. (1999), *The Danish System of Innovation* (in Danish). Copenhagen: Erhvervsfremmestyrelsen.

Lundvall, B.-Å. and Borras, S. (1998), *The globalising learning economy: Implications for innovation policy*. Brussels: DG XII-TSER, the European Commission.

Lundvall, B.-Å. and Johnson, B. (1994), "The learning economy", *Journal of Industry Studies*, Vol. 1, No. 2 (December), pp. 23-42.

Lundvall, B.-Å. and Nielsen, P. (1999), "Competition and transformation in the learning economy—illustrated by the Danish case", *Revue d'Economie Industrielle*, No. 88, pp. 67-90.

Maskell, P. (1998), "Learning in the village economy of Denmark. The role of institutions and policy in sustaining competitiveness". In Braczyk, H. J., Cooke, P. and Heidenreich, M. (eds.), *Regional Innovation Systems: The Role of Governance in a Globalized World*, pp. 190-213. London: UCL Press.

Nelson, R. R. (ed.) (1993), *National Innovation Systems: A Comparative Analysis*. New York: Oxford University Press.

Nonaka, I. and Takeuchi, H. (1995), *The Knowledge Creating Company*. Oxford: Oxford University Press.

OECD (1994), *The OECD Jobs Study—Facts, Analysis, Strategies*. Paris: OECD.

OECD (1996a), *Science, Technology and Industry Outlook 1996*. Paris: OECD.

OECD (1996b), *Transitions to Learning Economies and Societies*. Paris: OECD.

OECD (1999), *OECD science, technology and industry scoreboard 1999: Benchmarking knowledge-based economies*. Paris: OECD.

OECD (2000a), *Knowledge management in the learning society*. Paris: OECD.

OECD (2000b), DSTI contribution on the New Economy, DSTI/IND/STP/ICCP(2000)1/REV1. Paris: OECD.

OECD (2000c), Document for the OECD Council meeting summer 2000 on the New Economy. Paris: OECD.

Polanyi, M. (1958/1978), *Personal Knowledge*. London: Routledge and Kegan Paul.

Polanyi, M. (1966), *The Tacit Dimension*. London: Routledge and Kegan Paul.

Porter, M. (1990), *The Competitive Advantage of Nations*. London: Macmillan.

Putnam, R.D. (1993), *Making Democracy Work—Civic Traditions in Modern Italy*. Princeton: Princeton University Press.

Samuels, M. (1998), *Towards Best Practice: An Evaluation of the First Two Years of the Public Sector Benchmarking Project 1996-8*. London: Cabinet Office (Office of Public Service). January.

Senge, P. (1990), *The Fifth Discipline: The Art and Practice of Learning*. New York: Doubleday.

Smith, A. (1976), *An Inquiry into the Nature and Causes of the Wealth of Nations*. Oxford: Oxford University Press.

Svennilson, I. (1960), "The concept of the nation and its relevance for economics", in Robinson, E. A. G. (ed.), *Economic Consequences of the Size of Nations*. London: Macmillan.

Whitley, R. (1996), "The social construction of economic actors: institutions and types of firm in Europe and other market economies", in Whitley, R. (ed.), *The Changing European Firm*. London: Routledge.

Woolcock, M. (1998), "Social capital and economic development: toward a theoretical synthesis and policy framework", *Theory and Society*, No. 2, Vol. 27, pp. 151-207.

World Bank (2000), *Entering the 21st century; World Development Report 1999/2000*. New York: Oxford University Press.

3

Innovation Policies in the Knowledge Era: A South American Perspective

Helena Maria Martins-Lastres and José Eduardo Cassiolato

INTRODUCTION

Since the early 1980s, the world environment has gone through significant transformations—encompassing political, productive, technological, organisational, informational, commercial, financial, institutional, social and cultural dimensions—that are dynamically related. The analysis of the characteristics and consequences of such transformations are among the group of questions most discussed in the academic, entrepreneurial and policy-making environments in the 1990s.

It is recognised that the transformations associated with the setting up of a new world order implys important adaptations and restructuring that deeply affect, among others:

- political and economic hierarchy of different segments within national societies, as well as of these societies (and their associations) in the world scene;
- various economic activities and productive sectors (e.g., the higher dynamism of the services sector and other information-intensive activities);
- different institutions (particularly regarding their role and forms of organisation, articulation and functioning); these include firms, centres of education and training, R&D and the state with its different organisms and levels; and
- people (as workers, consumers and citizens).

In the so-called Information and/or Knowledge Era, the emergence of the ICT (information and communication technologies) paradigm and the increasing economic competition and acceleration of the globalisation process are producing significant impacts in the way industrial and technological development is produced with important consequences to economic development. As a result, the concepts of production and innovation, competitiveness, firms' organisation, management and strategies, nationstate, public and private strategies and policies are being reviewed and new approaches are needed.

One of the aims of this chapter is to understand how these transformations are inducing new forms of insertion of developing countries in the globalised world. First, we will summarise some of the main characteristics of the Knowledge Economy, highlighting the main challenges it poses; second, we will discuss the hypothesis of globalisation of technologies, focusing particularly on the role played by less developed countries; third, we will briefly examine the role of innovation in the Knowledge Economy; and fourth, we will present the findings of a research project aimed at understanding how productive and innovation systems in Mercosur countries are facing the changes of the 1990s. The final target of the chapter is to draw policy implications from this discussion and to explore what would be the main requirements for the promotion of systems of innovations in the South.

INFORMATION AND KNOWLEDGE ECONOMY

Of course information and knowledge have always been important in human evolution. The notion of a Knowledge Economy relates to the observation that, since the postwar period, the economy has increasingly relied on knowledge-based activities than ever before. More than that are the new possibilities of generating, processing, transmitting and diffusing knowledge offered by the development of new information and communication technologies. Therefore, three interrelated main characteristics of the world economy nowadays are: (1) the proportion of labour that handles tangible goods has become smaller than the proportion engaged in the production, distribution and processing of knowledge; (2) there is increasing share of knowledge and information in the value of many products and services; (3) there has been rapid growth of knowledge-intensive activities, which have become the heart of recent economic expansion.

Since the early 1980s, the world has witnessed a very rapid increase in the technical means of codifying knowledge and of incorporating information into a variety of good and services.[1] Of course the acceleration and deepening of both codification of knowledge and spread of information are related to the development and rapid diffusion of ICTs. The new possibilities of converting information to a digital format help to explain the radically different features of the so-called Information/Knowledge Economy. Most of the compass of human experiences—voices, images and even smells—can now be captured in various degrees of verisimilitude; and all representations can be codified (reduced to the

Esperanto of 1s and 0s). Once digitised, information acquires the digital advantage (Davis and Stack, 1997): a universal rendering that is resource-conservative, cheap to store and transport and easy to copy, meter and manipulate. Among other things, digital rendering liberates information from the constraints of any particular medium; and raises the possibility of the liberation of information from the constraints of scarcity and rationing by price: easy and cheap replicability means that whatever can be digitally rendered can be made universally available.

In this discussion it is important to note the trend towards dematerialisation; that is the reduction—both in relative and absolute terms—of the importance of the material component in the production of goods and services. An obvious example is *software*, which can be developed, produced, bought, distributed, consumed and discharged without ever assuming physical formats. The conversion of different types of codified knowledge and information into an electronic format offers the possibility of a minimum dependency on matter. One consequence is the reduction of costs associated with the consumption of energy and physical resources, into its development, production and consumption. Additionally, virtual reality is increasingly being used in activities where human physical presence used to be considered as indispensable, as in the case of teaching, conferences, consultancy, medical visits and even operations.[2]

It is important to stress that these transformations are dramatically challenging traditional economic concepts. They are also urging the development of new theories and instruments to deal with them. As pointed out in other chapters, the production and distribution of knowledge have some features, which are not compatible with approaches, models and indicators prevalent within orthodox economic theory. One small example of this rather important discussion refers to the obvious fact that information and knowledge are resources which, different from energy and materials, are (more than abundant) inexhaustible. Their consumption does not destroy them; and when they are sold, transferred or given, this does not mean that they are lost. This is partially why the specificities of the new patterns of economic development expose even more the restrictions of traditional economic approaches, theories and correlated indicators and statistics systems.[3] A related point refers to the difficulty these approaches have in providing sufficient conditions to (1) measure and evaluate the impacts of such transformations on different economies and societies; (2) analyse the new forms of insertion and roles of these countries in the international division of labour associated with the new pattern of economic accumulation.

The development of this new pattern can be seen as an alternative to the restrictions imposed by the energy and material-intensive mass-production (and highly polluting) Fordist paradigm of the 1950s and 1960s. In this sense, the recent transformations offer important new opportunities. However, at the same time, they also present new challenges for firms, sectors, countries, regions and people. It is now recognised that—together with the new possibilities offered by the increasing diffusion of the ICTs—new forms of economic and social polarisation (and exclusion) can be created. These are linked to digital illiteracy, as well as to unequal access to both new products and services and the opportunities to acquire and renew knowledge basis and skills. In this sense, it should be stressed, the point made by those authors who prefer to use the term Learning Economy, claiming that in this new era, "knowledge is the main resource and learning is the central process" (Lundvall and Johnson, 1994).

As seen above, one of the aims of this chapter is to reflect on the consequences of these transformations on developing countries. The analysis will focus on their new mode of insertion and role in the international division of labour, which is being conformed. In this sense it is worth analysing some of the most visible characteristics of the new economy—the diffusion of ICTs and the acceleration of the globalisation process—focusing particularly on their repercussion on least developed countries.

GLOBALISATION OF TECHNOLOGIES AND LESS DEVELOPED COUNTRIES

In the discussion of the recent phase of the globalisation process one should first point to the conceptual inconsistency and the strong ideological content of the term.[4] Second, it is important to emphasise that most analyses about the globalisation process usually do not take into account two big southern regions of the planet—which together amount to more than 60 countries—Latin America and Africa. Contrary to what should be expected about the acceleration of the process of globalisation, these two regions are not becoming more integrated into the world economy. The share of these two regions in world trade has been rapidly reduced since the early 1980s. In 1996 their share represented less than 6% of world trade.

Additionally, it is worth recalling that the available evidence shows us that (Lastres and Albagli, 1999):

- about 80% of the world production is still consumed within countries that produce it
- domestic savings are estimated to finance 95% of capital formation;
- the participation of less-developed countries in global R&D effort remains insignificant: it is estimated to be around 2%, if China and Southeast Asian countries are excluded;
- multinational corporations are responsible for two-thirds of the international commerce, with 40% of the world commerce being realised within these corporations;
- perhaps the most blatant distortion is the increase in the barriers to the mobility of people, particularly low-skilled workers.[5]

Of course the diffusion of the new pattern of accumulation based on information technologies has provided the technical means for people and institutions geographically separated to be connected in real time. Economic contacts of all types have intensified and deepened, as exchanges between actors—individually and collectively—spread all over the world. However, it should be stressed that both the upsurge and diffusion of the new techno-economic paradigm and the acceleration of the globalisation process result from and reflect political and institutional changes which have characterised the environment of the most developed countries in the second half of the 20th century.

These changes have induced a progressive movement towards liberalisation and deregulation of world markets (particularly deregulation of financial systems and capital markets) that were supposed to be associated with the needs of greater competitiveness both at national and international levels by countries and firms. To open, stabilise, deregulate and privatise have become the central targets of macroeconomic policies. As a consequence, both a reconcentration of power and an unequal diffusion of the benefits of the changes have been observed in the last decades of the century.[6]

On the one hand, a reinforcement of the so-called tripolar economic international policentrism (United States, European Union and Japan), and particularly the hegemonic position of the United States inside the Triad has been noticed. This process presents a trend towards incorporating markets that have relevant dimension and that adopt labour, environment, tax norms deemed to be the most attractive, flexible or competitive. As a consequence, the condition of development and subdevelopment of a country and its hierarchical position in the international division of labour have become even more dependent on the form and degree of integration with economic blocs (particularly those of the most advanced countries).

On the other hand, it is pointed out that the lack of evidence showing significant changes towards either deconcentrating the appropriation of revenues or the improvement of the present situation in terms of international intelectual division of labour. When analysing the case of globalisation of information, knowledge and activities that are strategic to firms and countries—related to planning and decision control and to R&D, for example—a number of authors note that the evidence available points out not only the importance of localisation (contrary to globalisation) but to a reconcentration of such activities, information and knowledge.

In this discussion it is recognised that the transformations associated with the coming of the Knowledge Era (particularly those related to the diffusion of the new ICT paradigm and the acceleration of the process of economic globalisation) are (1) allowing for both the setting up of integrated research on a world scale and the efficient diffusion of technologies and knowledge; and therefore (2) leading up to a supposed technoglobalism (implying that the generation, transmission and diffusion of technologies are becoming increasingly global in scope).

As argued in previous works (Cassiolato, 1996; Lastres, 1997), implicitly technoglobalists assume that technologies are commodities and propose that, in a borderless world, international technologies are accessible by firms and could be transferred internationally under market mediation via price mechanism. However the hypothesis of technoglobalism has been contradicted by the findings of evolutionary studies on innovation carried out since the late 1960s, which emphasise:

- the tacitness and cumulativeness of knowledge, and
- that knowledge (differently from information) is hardly transferred.

These findings point out the importance of localised learning, of externalities associated with proximity and, from the mid-1980s onwards, of national and local systems of innovation. It is worth mentioning that it is not by chance that the concept of a national (sub- and supranational) system of innovation was developed in the 1980s as a response to the spread of the idea that the world was going through a process of economic and technological globalisation. One main objective there was to deny the hypothesis that in the new economy, local and national specificities would disappear and the role of policies (in general, and government policies, in particular) would have no relevance. Therefore, two main arguments of the group of authors working with this framework relates to the importance that the agenda for active innovation policy is extended rather than made obsolete by the new developments.

Research developed in the 1990s has also helped to understand the scope and depth of globalisation of technologies, covering three different categories: international exploitation of technology, technological collaboration and generation of technology (Archibugi and Michie, 1995). The result is that the only important and rapidly growing form of globalisation is exploitation abroad of technologies generated in the home country. A significant growth in technological collaboration (aimed at getting access to relative strongholds abroad) was noticed. However, more important was the conclusion that the frequency and growth of global creation of new technology—where multinational firms locate their development of new products and processes abroad—remain relatively low compared to the dynamics of the other two forms. The conclusion is that national systems of innovation take on more (rather than less) importance in this context.

One important point in this discussion refers to the concept of globalisation used in the research and analysis developed in the last two decades on this issue. Most of the authors who argue in favour of the technoglobalism argument (but not only them) are referring, in fact, to a phenomenon which—whenever observed—is concentrated within the group of the most advanced countries. One of the arguments here is that the trends observed both in the case of the increase in collaboration and joint generation of technology should be better understood as the result of the acceleration of the process of triadisation, *but never globalisation*. It is worth noting that this process is largely occurring between countries of the same economic bloc, as mainly in the case of the European Union. Also it is important not to forget that the countries within the two biggest continents in the South have now a more marginal insertion in the global economy than they used to have in the early 1980s.

Data from MERIT's Cooperative Agreements and Technology Indicators (CATI) information system provide evidence of the sharp contrast between advanced countries and less developed countries' (LDCs) involvement in strategic cooperative technology arrangements. As pointed out in an analysis of the involvement of LDCs in the 1980s, when an explosion in the number of such alliances was observed (Lastres, 1993): (1) more than 95% of the strategic technology partnerships involved only advanced economies, while only 4.3% involved firms from LDCs; (2) while 72% of the arrangements involving advanced countries concentrated on new generic technologies, arrangements involving LDCs concentrated on mature technologies; (3) over 50% of the alliances between companies from the developed economies were oriented towards R&D, however, only less than 13% of the agreements involving LDCs were R&D-oriented; (4) considering the agreements in which technology transfer is a major objective, nearly 90% were made between companies from advanced economies. While this share increased throughout the decade, the share of LDCs fell from 5.3% in the first half of the 1980s to 4.8% in the second half. This situation offers important reasons for reflection. Data for the 1990s show that developing countries are still participating only marginally in international technology alliances, transfers and innovation networks.

Therefore, the evidence available shows in fact that the economic space where information and knowledge are produced and used is becoming even more confined within very specific boundaries. Then, opposed to the view that science and technology efforts and results are becoming globalised, the observed pattern is of a marked concentration of such activities within firms and national boundaries, and particularly those in the most advanced countries. Recognising this trend, the report of the European Commission (1996) on the Information Society pointed out the responsibility of Europe in the definition of both the pattern which separated the industrialised and subindustrialized countries, after the Industrial Revolution, and the new pattern associated with the Information Revolution. Acknowledging that, after more than 200 years of Industrial Revolution, the so-called developing countries seemed to have acquired a chronic state of subdevelopment, the main recommendation was to avoid conforming to a new pattern that would reinforce even more the distance between countries.[7]

Of course the allusion here is to the possibility of reinforcing the so-called digital divide. More serious than that is the risk of also reinforcing the learning divide (Arocena and Sutz, 2000). Concerned with the social side of globalisation, authors like Freeman (1995) point to the need of designing policy strategies (1) targeting not only the access to information and new technologies and equipment, but also the access to knowledge; (2) to push forward the process of knowledge generation to allow all societies (and the different segments within them) to make use of the information technologies and products available; and (3) to promote the social changes required by the new era. In a similar line, Foray and Lundvall (1996) provide an alert for the risk of threatening the social cohesion of economies if policies promoting information infrastructures neglect the social and distributional dimension, as well to the importance of promoting capabilities and competences (particularly learning capabilities) as central elements in any strategy aiming at limiting the degree of social exclusion. They also pinpoint the risk of IT becoming an abbreviation for "intellectual tribalism" instead of "information technology".

A number of other authors have forcefully pointed out that when we talk about the recent changes and about globalisation we cannot ignore the important exclusion produced by them; and most importantly their victims.[8]

INNOVATION IN THE KNOWLEDGE ECONOMY

There is an increasing convergence of visions among different schools of thought when trying to understand the factors underlying a better competitive and innovative performance of firms, sectors and countries. The interaction among firms and other agents has been recognised as a fundamental source of competitive advantage in the Knowledge Era. As a result, the investigation about different clusters, relations between firms and between firms and other institutions and their forms of networking and their local environments has significantly grown since the 1980s.

Terms such as synergy, collective efficiency, economies of agglomeration (clustering), associational economy and learning-by-interaction express the main preoccupations of this debate. Different concepts—such as industrial districts and poles, clusters, milieu, networks and others—have been used to account for the need to focus on a specific set of economic activities and interactions. Also new and old approaches have been developed to encompass these activities and agents as well as to allow the examination of the relationship between them, namely: *milieu innovateur* and national (local) systems of innovation. It is noteworthy that the main motive behind all the research effort on these issues relates to their policy-making implications and the investigation of what new policies and policy tools would be more adequate in stimulating, nowadays, industrial and innovative development.

One main argument here is that, far from losing relevance, this research effort focusing on national and local capacities and characteristics will probably increase with the intensification of the globalisation process. The clustering of firms and the collective synergies generated by their interactions, and between them and their environment, make it feasible to innovate, strengthening their chances of survival and growth, which constitute an important source of advantages in the increasingly globalised economy.

To summarise: among the main questions contributing to a better understanding of the innovation process of the last few years we can single out:

- the recognition that innovation and knowledge are, more than before, central elements of the dynamics and growth of nations, regions, sectors and institutions and that there are marked differences among agents and their capacities to learn;
- the growing international competition led firms to centre, even more than in the past, their strategies on the development of innovative capacity, which is essential for them even to participate in the flows of information and knowledge (as in the cooperative arrangements and all sort of innovation networks) that characterise the present stage of world capitalism;
- the understanding that innovation—as a process of search and learning and dependent on interactions—is socially determined and strongly influenced by specific institutional and organisational formats; therefore, far from a trend towards

homogenisation, important differences among innovation systems of countries, regions, sectors, organizations and so on persist to reflect the social, political and institutional context where they are embedded;

- the vision that, if information and codified knowledge present growing conditions of "transferability"— given the efficient diffusion of IT — tacit knowledge (which has a localised character) continues to have a prime role for the innovative success and is still very difficult to be transferred.

PRODUCTIVE AND INNOVATION SYSTEMS IN SOUTH AMERICA IN THE 1990s

This section starts by summarising some of the main findings of a research project on how local and national productive and innovation systems in Mercosur countries have coped with the changes of the 1990s,[9] aiming at comparing them with other Latin American data.

The main conclusions of the first phase of the project have provided evidence that, with few exceptions, instead of being reinforced:

- productive and innovative efforts are decreasing and this of course affects both their core capabilities as well as their learning processes;
- productive and innovative networks are being disarticulated (and there is no new significant articulation between the new investments and the local R&D infrastructure);
- the level of employment of specialised personnel has decreased (and this has been followed by the downgrading of the occupations of some of the specialists who remained employed).

In Brazil these trends were particularly observed in the case of the high-tech clusters (and particularly ICTs), despite the important role they have played, and are expected to play, as promotors and diffusors of technical progress to other sectors and activities. The research showed that the most successful arrangements in the 1990s were those dealing with traditional technologies and in particular those that are natural resources intensive.[10] Of course it was very important to learn how these arrangements were positively responding to the new challenges of the 1990s. However, the disarticulation of important high-tech arrangements that had accumulated both productive and innovative capabilities and the fact that Brazil's share of world exports of high-tech products decreased from 0.6% in 1985, to 0.26% in 1991 and 0.19% in 1995[11] assume particular importance in this new coming era entitled the Knowledge Era.

This and other difficulties uncovered by the research were related to the policies adopted by Mercosur countries (as well as by most countries in Latin America) which give a predominant role to short-term financial objectives and follow the imperatives of structural reforms. One of the main problems in terms of industrial and innovative development of the competitive insertion model, adopted in Brazil, is that it assumes that:

1.   the opening of the economies and the attraction of multinational corporations would be the best way to improve the form and degree of integration of these economies in the world market;
2.   technology, innovation and knowledge can be globalised; and like a commodity, could be acquired internationally under market conditions.

It is also important to stress the influence of the resultant macroeconomic scenario in these countries on the innovative propensity of local firms. Many of them, seeking primarily to survive, have abandoned most long-run concerns. Above all, the predominance of short-term financial objectives (in both public and private strategies in most of these countries), associated with high levels of instability, high exchange and interest rates, impose in real terms important restrictions to any policy aiming at promoting industrial and technological development.

On the whole the results of the research project show, first, that a simple exposure to international competition is not at all sufficient to foster firms to increase their innovative activities and their competitiveness. They also show that these policies and efforts should be based on strengthening local resources, organisations and institutions, as well as on the quality and intensity of their interactions. Both these cases and those where local innovation systems were negatively affected by the changes of the 1990s suggest that the setting up of local systems of innovation in development conditions in a globalised world requires a type of intervention that should be much more sophisticated than a simple attraction of foreign direct investment (FDI).

In a broader point of view, it is important to stress that the attempts to accelerate integration of Latin America into the global economy by attracting foreign capital resulted in an effective increase of FDI. The flows of foreign investment increased from US$8 billion in 1990 to US$67.3 billion in 1998, and the stock of FDI grew 60% in the period (Mortimore, 1999). Although inflows in the 1990s were approximately 13 times of what was observed during the 1970s, economic growth was 50% lower. One of the main reasons for that performance is that FDI in the 1990s was mostly directed to mergers and acquisitions of existing firms (and their accumulated capabilities) rather than greenfield investment. Also, and contrary to what was planned, these new investments are basically import-intensive and are not geared to exports, targeting mostly local markets. An additional great concern relates to the sustainability of this situation.[12]

Therefore, the appropriateness and effectiveness of the policies adopted in the region can be questioned on many points, including their very basic target: the increase of international competitiveness and an effective insertion in the globalisation process. In the case of the Mercosur economies, although the export profile has evolved in such a way that industrialised goods are increasingly important, their insertion in the international market is still characterised by the exports of *commodities* that are intensive in natural resources and/or energy and in low wages. These *commodities* have shown a tendency to low dynamism, excess supply and price stagnation.

More important than that is the loss of world market shares as indicated by the data shown in Table 3.1. Export growth of these countries has been much slower than the increase in world trade. Brazil, for example, accounted for 1.5% of world exports in 1984; in 1993 and 1996 the same figures were 1% and 0.93% respectively. However, the situation is even worse if intra-Mercosur trade is excluded. In this case, Brazil's share declined from 1.42% in 1984 to 0.79% in 1995; Argentina's from 0.31% in 1986 to 0.28% in 1995; and Uruguay from 0.037% to 0.022% over the same period. In the globalisation period these countries are all losing importance in export markets.

Table 3.1
Mercosur Countries' Share in World Trade

Selected	Brazil (%)		Argentina (%)		Paraguay (%)		Uruguay (%)	
Years	including Mercosur (1)	excluding (2)	including Mercosur (1)	excluding (2)	including Mercosur (1)	excluding (2)	including Mercosur (1)	excluding (2)
1984	1.500	1.426	NA	NA	NA	NA	NA	NA
1986	1.116	1.026	0.342	0.315	0.012	0.005	0.056	0.037
1990	0.943	0.903	0.371	0.316	0.029	0.017	0.051	0.033
1995	0.916	0.794	0.413	0.279	0.016	0.007	0.042	0.022

1 - Total Exports of Country over Total World Exports
2 - Same as 1 less Exports to Other Mercosur Countries
Source: Cassiolato and Lastres (2000).

Even worse than these poor and negative economic results is the deterioration of social conditions in Latin America. A recent CEPAL report (Stallings and Perez, 2000) conclude that the reforms had negative impacts not only on the level of employment and of the posts offered, but also on the income distribution of the region. Emphasing this points Humbert point out that "pas d'effets miraculeux à constater, bien au contraire nous dit l'économiste en chef de la Banque Mondiale: 'the gap between the developed and the less developed countries is widening' (Stiglitz, 1999). Quel horrible résultat à afficher face à l'insolente bonne santé des Etats-Unis: la théorie aurait-elle des préférences où récompenserait-elle les seuls bons élèves ou encore serait-elle fausse?" (Humbert, 2000).[13]

THE NEED FOR PRIVATE AND PUBLIC POLICIES TO PROMOTE PRODUCTIVE AND INNOVATIVE SYSTEMS IN SOUTH

It is acknowledged that there are many important destructions occurring all over the world, as well as in Latin America. However, the serious economic and social situation confronting the former calls for an urgent start of a concerted and continuous phase of creation. As the international experience shows us, the improvement of the processes of learning and innovating (instead of allowing only for their deterioration) are at the centre of the new public and private strategies and policies targeting the promotion of the capacity to acquire and use knowledge.

This chapter argues that one of the main problems facing Brazil and other Latin American countries, at the turn of the millennium, results in part from a very unclear understanding about the nature and consequences of the present transformations the world economy is going through. Policies adopted by most Latin American countries reflect these misunderstandings. One of the main points here refers to the conceptual inconsistency and the strong ideological content of some of the most important concepts and arguments used in the characterisation and analysis of these transformations. This is particularly true and worrisome because we are living in a phase of radical transformations; some of them being still very difficult to measure. This fact can be seen as contributing to the creation of an ideal intellectual environment for the adoption of neoliberal guidelines, sometimes, in very uncritical ways.

Perhaps the most dangerous issue here relates to the assumption that we are living in a globalised world characterised by irreversible movements and the domination of uncontrollable international forces. In this case, globalisation is seen as a myth that annuls the search for alternatives and tends to paralyse national initiatives, particularly in those less advanced countries. In fact, one main neoliberal argument regarding the acceleration of the globalisation process is of course that the role of local and national projects and policies would be reduced and even bound to disappear. In a totally opposed way, this chapter agrees with the idea that one main reason for the continuation of the crisis of adaptation to the new paradigm refers precisely to the delay in better understanding their specificities and in designing appropriate policies to cope with them.

The chapter also argues that it is worth advancing towards an adequate understanding of the Knowledge Economy: the real characteristics and impacts of the new socio-techno-economic pattern of accumulation, role of information technologies, innovation and learning processes, and so on. There are two main related points here. The first is that, as history shows us, even if some important specificities and trends of this new pattern are still invisible and seem uncontrollable, they should not be taken as permanent obstacles. The second refers to the acknowledgment that, progressively, new indicators, regulations and procedures are being developed, aiming at attending to the new requirements.

Therefore, in an effort towards developing a conceptual framework to understand the recent transformations, their consequences on productive and innovative systems in the south, as well as to base the formulation of policy suggestions to respond to the new challenges, it is necessary to understand that:

- conventional indicators, models and approaches (mainly in policy making, economics and administration terms) do not capture the complexities brought about by the growing importance of knowledge in economic activities;
- far from being an integrated and borderless world, the so-called globalisation requires the attaining of local, national and regional specificities;
- traditional visions that define competitiveness in terms of prices, costs (particularly labour costs) and exchange rates have become obsolete;[14]
- sustainable forms of industrial and technological partnership require levels of qualification and capabilities that are much higher than in the past;
- local learning processes are fundamentally based on dynamically improving human resources and stimulating different forms of articulation among agents;
- the stronger the human resources base, the greater is the possibility of accelerating learning processes, the stronger the potential for innovation and the broader the chances of the system to face competitive pressure;
- all of this requires the design and implementation of more sophisticated forms of promoting industrial and technological development, taking into account the present transformations and also the changes associated with new forms of governance at world level (which of course include the conditions established by the World Trade Organization (WTO)[15]).

A last point refers to the recognition that national and local conditions may lead to completely different development paths and to a growing diversity instead of the standardisation and convergence suggested by the more radical thesis about the influence of globalisation on national and subnational systems. As emphasised by Celso Furtado, "globalisation is very far from conducting to the adoption of uniform policies. The mirage of a world behaving under the same rules dictated by a super International Monetary Fund (IMF) exists only in the imagination of some people. The disparities among economies are due not only to economic factors but, most importantly to diversity in cultural matrices and historical particularities" (1998, p. 74).

NOTES

1. See, for instance, the very interesting reflection made by Marques (1999)—a well-known Brazilian researcher who participated of the development of the first computers in the country and who was one of the presidents of Cobra Computers—on the intensification of these processes in the production of tomatoes.

2. The main point here is to stress the new opportunities offered by these changes. Without intending to discuss in this chapter these rather important cases, I would only note my preference for the arguments that emphasise the complementarity of virtual and human direct contacts, rather than the displacement of one by the other.

3. See, for instance, Lastres and Ferraz (1999). This work discusses the difficulties that neoclassical theory has always had when dealing with technology (and the possibilities of transferring it) particularly because it assumes that information and knowledge are equivalents commodities.

4. In a previous work, this issue is discussed in a more detailed way (Cassiolato and Lastres, 1999).

5. At the same time qualified workers, from developing countries, in high-tech areas are increasingly being attracted to work in developed countries, characterising an important brain drain.

6. See Fiori (1993 and 1995); Ianni (1996 and 1997); Girvan (1996); Tavares and Fiori (1997); Amin (1997); Furtado (1998); Cano (1999).

7. Its interesting to note that this position appears requalified in the final report of Information Society Technologies (IST) Conference in Helsinki (EC, 1999): "Particular attention is needed for the developing countries. The potential of IST should be used to reduce rather then to widen the already existing gap with the North. *Moreover, the South can only become a market for IST from the moment its population is (IST) literate and has access to required infrastructure*" (p. 53).

8. See among many others Celso Furtado and Samir Amin. It is also worth pointing out the impressive exposition launched in the year 2000 by the Brazilian photographer Sebastião Salgado disclosing the social side of globalisation and illustrating its exclusion and victims. This includes also people who have their transit blocked by the so-called new wall separating two countries within the biggest economic bloc of the world, NAFTA: Mexico and the United States. The exposition was shown simultaneously in different capitals of the world.

9. See Cassiolato and Lastres (1999 and 2000). The project, Globalization and Localized Innovation: Experiences of Local Systems in the Mercosur and S&T Policy Proposals, involves a network of researchers of more than 50 people in Brazil, Argentina and Uruguay. The reports produced for the first phase of this project were published in Cassiolato and Lastres (1999). They are also available in the home page of the project: http://www.race.nuca.ie.ufrj.br/gei/gil/shtml. The results of the second phase of the project can also be found on this home page.

10. This is in agreement with the observed trend towards the specialization of most Latin American economies in natural resources (Katz, 1999).

11. Argentina's share decreased from 0.08% in 1985 to 0.04% in 1995. In both cases, the relative decline was accompanied by an absolute one, while total exports of these goods more than doubled during the period. For more details see Cassiolato and Lastres (2000).

12. For more details see Cassiolato (2000).

13. Aiming at explaining his argument, Humbert points out that: "toutefois on doit remarquer que les pays économiquement dominants ont tendance à prôner le libre échange, comme le font surtout les Etats-Unis aujourd'hui, comme le fit aussi le Royaume-Uni pendant une bonne partie du 19ème siècle. En revanche il est facile de comprendre que *les pays* qui ne sont pas leaders, en un mot qui sont *dominés*, et pour qui un déficit de la balance commerciale est difficile à gérer, pour qui le rééquilibrage automatique ne survient pas facilement malgré les dévaluations et les réductions de salaires exécutées selon les recommandations du FMI, *ne peuvent afficher le même optimisme que les Etats-Unis* qui traînent un déficit commercial croissant et permanent. En outre ils peuvent douter qu'une spécialisation selon leur actuelle structure de prix relatifs et les productions dans lesquelles ils devraient s'investir durablement soient le garant d'une croissance future. Si, suivant la bonne idée de Krugman, on quitte à nouveau, pour un instant le monde fictif de la théorie et la relique vénérée de Ricardo, et si l'on en croît les chiffres, les trente dernières années ont été très bonnes pour le libre-échange et les Etats-Unis en ont profité sans que l'égalisation des rémunérations avec les pays non-indutrialisés les aient menacés: c'est que réciproquement les pays non-industrialisés n'ont guère obtenu de bénéfices.

14. See the concept of spurious competitiveness introduced by Fanjzylber (1980).

15. See, for instance, the penalties imposed by the WTO on the Brazilian producer of small and medium-sized aircrafts—*Embraer*—in the case brought by its Canadian competitor.

BIBLIOGRAPHY

Amin, S. (1997), *Los Desafíos de la Mundialización*. Madrid: Siglo XXI.

Amin, S. (2000), *L'hégémonisme des Etats-Unis et l'effacement du projet européen.* Paris: L'Harmattan.

Archibugi, D. and Michie, J. (1995), "The globalisation of technology: a new taxonomy", *Cambridge Journal of Economics*, no. 19, pp. 121-140.

Arocena, R. and Sutz, J. (2000), "Knowledge, innovation and learning: systems and policies in the north and in the south, globalisation and localised innovation in the MERCOSUR", Institute of Economics of the Federal University of Rio de Janeiro.

Cano, W. (1999), "América Latina: do desenvolvimento ao neoliberalismo". In Fiori, J. L. (org.), *Estados e Moedas no Desenvolvimento das Nações*. Rio de Janeiro: Vozes.

Cassiolato, J. E. (1996), "Innovation and the dynamic competitiveness of Brazilian industry: the role of technology imports and local capabilities", *Texto para Discussão*, no. 366, IE/UFRJ, Rio de Janeiro.

Cassiolato, J. E. (2000), "National systems of innovation in the south", South-South-South Conference, organised by AEGIS/University of Western Sydney, May.

Cassiolato, J. E. and Lastres, H. M. M. (1999), "Inovação, globalização e as novas políticas de desenvolvimento industrial e tecnológico". In Cassiolato and Lastres (eds.), *Globalização e Inovação Localizada: Experiências de Sistemas Locais do Mercosul.* Brasília: IBICT/MCT.

Cassiolato, J. E. and Lastres, H. M. M. (2000), "A inserção da América Latina na economia do conhecimento", paper presented at the International Seminar *Hétérodoxie et Orthodoxie dans les problématiques actuelles de l'économie internationale et de l'économie du dévelopement: de la division internationale du travail au système industriel mondial.* Lisbon, May.

Chesnais, F. (1995), *La Mondialisation du Capital.* Paris: Syros.

Davis, J. and Stack, M. (1997), "The digital advantage". In Davis, J. Hirschl, T. and Stack, M. (eds.), *Cutting Edge: Technology, Information, Capitalism and Social Revolution.* New York: Verso.

European Commission (1996), *Building the European Information Society for Us All.* Brussels.

European Commission (1999), *Exploring the Information Society: Business, People, Technology*, Final Report of the Information Society Technologies Conference, Helsinki.

Fanjzylber, F. (1980), *Industrialización e Internacionalización en la America Latina.* Mexico: Fondo de Cultura Económico.

Fiori, J. L. (1993), "Globalização, estados nacionais e políticas públicas", *Ciência Hoje*, vol. 16, no. 96, pp. 24-31.

Fiori, J. L. (1995), "A globalização e a novíssima dependência, *Texto para Discussão*, no. 343, Instituto de Economia da UFRJ, Rio de Janeiro.

Foray, D. and Lundvall, B.-Å. (1996), "The knowledge-based economy: from the economics of knowledge to the learning economy", In Foray, D. and Lundvall, B.-Å. (eds.), *Employment and Growth in the Knowledge-Based Economy*, OECD Documents. Paris: OECD.

Freeman, C. (1995), *Information Highways and Social Change*. Ottawa: International Development Research Centre.

Furtado, C. (1998), *O Capitalismo Global*. São Paulo: Paz e Terra.

Girvan, N. (1996), "Exclusion, learning, and information technology: some lessons from the Caribbean", mimeo, Intech/UNU, Maastricht.

Humbert, M. (2000), "Le système industriel mondial", paper presented at the International Seminar *Hétérodoxie et Orthodoxie dans les problématiques actuelles de l'économie internationale et de l'économie du dévelopement*. Lisbon, May.

Ianni, O. (1996), *A Era do Globalismo*. Rio de Janeiro: Civilização Brasileira.

Ianni, O. (1997), *Teorias da Globalização*. Rio de Janeiro: Civilização Brasileira.

Katz, J. (1999), "Structural reforms and technological behaviour: the sources and nature of technological change in Latin America in the 1990s", Intech International Conference *The political economy of technology in developing countries*, Brighton.

Lastres, H. M. M. (1993), "New trends of cooperative R&D agreements: opportunities and challenges for Third World countries", ECIB, IE/Unicamp & IEI/UFRJ, Rio de Janeiro.

Lastres, H. M. M. (1997), "Globalização e o papel das políticas de desenvolvimento industrial e tecnológico", *Texto para Discussão*, no. 519. IPEA, Brasília.

Lastres, H. M. M. and Albagli, S. (eds.) (1999), *Informação e Globalização na Era do Conhecimento*. Rio de Janeiro: Campus.

Lastres, H. M. M. and Ferraz, J. C. (1999), "Economia da informação, do conhecimento e do aprendizado". In Lastres, H. M. M. and Albagli, S. (eds.), *Informação e Globalização na Era do Conhecimento*. Rio de Janeiro: Campus.

Lundvall, B.-Å. and Borrás, S. (1998), "The globalising learning economy: implications for innovation policy", Targeted Socio-Economic Research—TSER Programme, DG XII European Commission. Luxembourg: European Communities.

Lundvall, B.-Å. and Johnson, B. (1994), "The learning economy", *Journal of Industry Studies*, vol. 1, no. 2, pp. 23-42.

Mortimore, M. (1999), "Contribuye la inversión estranjera directa al crescimento económico?", *Notas de la CEPAL*, no. 5 (July), p. 2.

Ortiz, R. (1996), "Anotações sobre a mundialização e a questão nacional", *Sociedade e Estado*, vol. 11, no. 1. UNB, Brasília.

Perez, C. (1983), "Structural change and the assimilation of new technologies in the economic and social systems", *Futures*, vol. 15, no. 5, pp. 357-375.

Stallings, B. and Perez, W. (2000), *Crecimiento, empleo y equidad. El impacto de las reformas económicas en América Latina y el Caribe*. Santiago de Chile: CEPAL.

Stiglitz, J. (1999), "Trade and the developing world: a new agenda", *Current History*, November, pp. 387-393.

Sutz, J. (ed.) (1997), *Innovacion y Desarrollo en America Latina*. Caracas: Editorial Nueva Sociedad.

Tavares, M. C. and Fiori, J. L. (eds.) (1997), *Poder e Dinheiro: Uma Economia Política para a Globalização*. Rio de Janeiro: Editora Vozes.

4

Development Policy and the Economy: Whither the State?

Robert H. Wilson

INTRODUCTION

The end of the twentieth century brought great change in the nature and roles of the state in many nations[1]. Rapid technological and economic change and expanded international commerce, prominent features of today's world economy, provided part of the context in which this redefinition occurred. Macroeconomic and fiscal policies promoting trade and adhering to monetary stability are now found in countries throughout the world. Some argue that these policies reflect the ascendancy of neoliberal philosophies. Others have cast economic policy changes as a reflection of a battle between government and the market, a battle in which the market is prevailing.[2] Regardless of the accuracy of these views, one should expect the role of the state and public policies to adapt to periods of great economic and social change.

The neoliberal framework, evident in the United Kingdom and the United States of America during the 1980s, but also imbedded in the policies of the International Monetary Fund (IMF), suggests that government should not influence capital investment decisions and that other forms of economic intervention should be minimized. These decisions are best left to entrepreneurs prepared to accept the risks of uncertain investments for their potential returns. Free markets will eventually determine the most efficient utilization of capital and the generation of wealth will be maximized. This view finds that the developmental state model, observed in many Latin America countries, and the planned economy of communist bloc countries produced inefficient economic outcomes and, as a result, achieved suboptimal levels of wealth. Furthermore, excessive taxation and regulation, an unstable monetary regime, and impediments to the international flows of capital and goods can prevent national economies from reaching their true potential. The implementation of the appropriate set of neoliberal policies is expected to unleash the entrepreneurial spirits of citizens, leading to increased investment, economic growth, and the creation of wealth.

Underpinning the set of neoliberal prescriptions is a minimal state. By divesting itself of publicly owned enterprise, maintaining low rates of taxation, especially on businesses, and avoiding deficits in public spending in order to maintain a stable currency, the state minimizes its role in the economy. This chapter will examine, in an exploratory fashion, the extent to which countries have followed these policy prescriptions.

One arena in which adherence to the neoliberal view has not been found historically is technology policy. Technological change and systems of innovation have been increasingly viewed as essential building blocks for the emerging economy. Economic growth has become increasingly dependent on the creation and application of knowledge, leading to what some have called the knowledge-based or learning economy.[3] The free market approach and minimal government action have been found to be inadequate policy frameworks since knowledge, especially its creation, has significant public good dimensions. Technological innovation has often benefited, historically, from government support through the mobilization of scientific and technological resources to address questions of national interest and through creation and operation of educational systems. Private investment and free markets alone would not have produced the breakthroughs in solid-state physics necessary for the development of integrated circuits or optic-based microprocessors. Public-sector support for science and technology, however, often takes the form of the development of underlying infrastructure systems rather than directing capital investment or producing goods in advanced technology sectors.

The evolving nature of the state, especially its role in the economy, is the concern of this chapter. Although examples of dramatic change in the nature of the state can be observed among formerly communist bloc countries and in the abandonment of state-led development in Latin America, neoliberal policies have been pursued in many countries. In addition, the consolidation of democratic political systems has led to state reform and new economic roles. The chapter will first assess empirical trends on the size and spending patterns of national governments. Further detail will then be provided on government support, by level and form of support, for science and technology. The impact of technological change on the provision of development infrastructure, especially the role of government, will be discussed. Following a discussion of state reform and decentralization, challenges to existing systems of governance will be explored. Worldwide trends will be examined when possible, but developed countries, especially the Organization for Economic Cooperation and

Development (OECD) countries, will receive closer examination due to data availability.

THE SIZE OF GOVERNMENT

One premise of the neoliberal view is that the state should interfere with the market and private sector as little as possible; it argues for a minimal state. When measured as expenditures by the public sector, governments have tended not to diminish in size in recent years (see Table 4.1). World Bank data on public expenditures as a share of gross domestic product (GDP) actually show an increase from 15% to 16% (weighted averages for over 120 countries) between 1980 and 1998. The increase in government expenditures was greatest among middle-income countries, but the level of government consumption was highest among high-income countries (16% in 1980 to 17% in 1998).

Table 4.1
General Government Consumption (share of Gross Domestic Product) and Public Expenditures on Education (share of Gross National Product)

	General Government Consumption (in %)		Public Expenditures on Education (in %)	
	1980	1998	1980	1996
Low Income	12	12	3.2	3.9
Middle Income	12	14	4.0	5.1
Lower Middle	NA	14	4.2	5.3
Upper Middle	11	11	4.0	5.0
High Income	16	17	5.6	5.4
World	15w	16w	4.0m	4.8m

w - Weighted average.
m - Median.

Source: The World Bank, *Entering the 21ˢᵗ Century, World Development Report 1999/2000* (New York: Oxford Press), pp. 241 and 255.

The composition of government expenditures in these two time periods varied in important ways. Military expenditures as a share of gross national product (GNP) decreased from 5.2% to 2.8% (weighted average for over 120 countries) between 1980 and 1995.[4] A principal contributor to increasing total public expenditures is found in education. Among the nations of the world, the median public expenditures on education, as a share of GNP, increased between 1980 and 1996 from 4.0% to 4.8% (see Table 4.1). Among the high-income countries, expenditure levels were substantially higher, 5.4% of GNP in 1996, than in other countries.

The expenditure pattern reveals a well-established relationship between level of development and size of government. As development progresses, processes such as urbanization generate new societal needs that produce governmental responses leading to increased expenditures. For example, economic and social infrastructures become increasingly important and more costly as countries develop. Furthermore, higher income countries have greater capacity to finance such expenditures. Basic demographic change, especially the aging of population, will likely lead to yet higher public spending. Development has invariably led to an increase in governmental expenditures and the size of government. Whether this pattern will continue into the future for the advanced economies remains to be seen, but certainly for less-developed countries, increased government expenditures are likely.

Another prominent element of neoliberal policies argues that deficit spending by governments should be avoided. To test for compliance to this policy recommendation, IMF data on revenues and expenditures by country and the resulting deficit (or surplus) for 83 countries in 1988 and 1998 are utilized. For purposes of this study, the annual deficit or surplus is measured as a percentage of total expenditures and lending minus repayments (negative for deficits and positive for surpluses).

The trends in average national deficits across regions of the world demonstrate quite distinct patterns (see Table 4.2). Countries in Africa have the highest average deficits and those in the Western Hemisphere the lowest. The average deficits in Europe were quite high in early years of the period but substantial improvements began in 1993. By 1998 Europe had the lowest average deficit of all regions. The deficit across all regions tended to decline over time but the year-to-year fluctuations and significant regional variations suggest that other factors need to be examined before drawing conclusions concerning trends in deficit spending.

A multiple regression model, with fixed effects, was utilized to further isolate the time trends (see Appendix A for variable definitions). The model estimates the deficit/surplus as a function of the annual growth in GDP in a country, a dummy variable for each country (the fixed effects variable) and a time variable. The estimated model was highly significant (see Table 4.3) and the annual growth of GDP had a positive effect on deficit spending, as expected. Of the 83 countries, the coefficients of the dummy variable for 31 countries were significant (see Appendix A). Of greatest interest in terms of trends in deficit spending, however, is the statistical significance of the coefficient of time. Holding constant other factors, the estimated model indicates that deficit spending improved by .3 percentage points per year. For the entire 11 years, the estimated deficit spending declined by 3.2 percentage points. Although a more complete model of deficit/surplus spending would require a structural equations approach, in which revenues and governmental budgets are modeled separately, these results provide strong evidence that national governments are becoming more fiscally disciplined.

Table 4.2

Regions/Number of Countries

Year	Middle East 9	West. Hem. 20	Europe 23	Asia/Pacific 17	Africa 14	All Countries 83
1988	-13.6	-8.5	-6.7	-8.5	-15.6	-10.2
1989	-14.1	-11.2	-5.7	-10.3	-14.9	-10.4
1990	-5.6	-8.0	-7.8	-8.9	-12.2	-8.6
1991	-8.4	-0.2	-11.7	-9.7	-13.5	-10.9
1992	-5.3	-3.0	-12.2	-9.5	-13.1	-10.6
1993	-7.0	-5.9	-14.5	-7.9	-14.5	-10.9
1994	-9.5	-7.4	-12.1	-5.3	-12.3	-9.3
1995	-8.6	-5.8	-10.8	-5.1	-11.3	-8.3
1996	-6.9	-4.0	-10.3	-5.8	-9.8	-7.4
1997	-0.8	-5.9	-7.3	-5.4	-13.0	-6.3
1998	-7.4	-9.2	-3.0	-14.3	-5.5	-8.8

Average National Deficits/Surplus by Region, 1988-1998 (Deficit or Surplus as Share of National Budget, in %)
Source: The International Monetary Fund, *Government Finance Statistics Yearbook*, Vol. XXIII (Washington, D.C., 1999), pp. 10-11.

Table 4.3
Estimation of National Deficit/Surplus in 83 Countries

Model Definition

DEFSUR = f (GDP annual growth rate, TIME, C1, ...C82)

Variables

DEFSUR = Overall Deficit/ Surplus as a Percentage of Total Expenditure and Lending minus Repayments

GDP = Gross Domestic Product annual growth rate (%)

TIME = Values of 1, 2, .., 11 (for 1988, TIME = 1; for 1988, TIME = 2; for 1998, TIME = 11)

C = Country Dummy Variables: (significant coefficients are reported in Appendix A)

Results

Variable	Slope Coefficient	P-value
Intercept	-9.27	0.0016
GDP	0.32	0.0001
Time	0.29	0.0081
C		

$$F = 16.98, \text{P-value} < 0.0001$$
$$R^2 = 0.65$$

Under neoliberal philosophy, government should not own productive enterprise since the political imperative will be to protect them from competition, thus distorting capital allocation and diminishing economic performance. The World Bank provides data from two time periods for 28 countries on the relative size of state-owned enterprises (see Table 4.4). State-owned enterprises tend to comprise a larger share of national economies in countries found in the less-developed regions, such as in Africa and Asia and the Pacific. In addition, the share of domestic investment made by state-owned enterprises tends to be greater than these enterprises' share of GDP in virtually all countries. Given that the creation of public enterprise is often justified in terms of mobilizing investment in key economic sectors, the larger role in investment is not surprising. In terms of changes over time, state-owned enterprises represent a declining share of GDP and of domestic investment in most countries. This trend is strongest in Africa (as measured by the average for countries within the region) and in Latin America. Although the decline in the role of state enterprise could result from higher rates of growth in the private sector, substantial privatization of public enterprise has occurred in many countries of the world, as advocated by the neoliberal position.[5]

Table 4.4
State-Owned Enterprises in Selected Countries (Share of GDP and of Domestic Investment)

Country or Region by Income	Value of Product as share of GDP		Investment as share of gross domestic credit (in %)	
	1985-90	1990-97	1985-90	1990-97
LOW AND MIDDLE INCOME				
Africa	**10.30**	**7.18**	**31.10**	**17.65**
Botswana	5.60	5.50	16.20	12.40
Cameroon	18.00	5.40	NA	NA
Tanzania	9.00	8.60	46.00	22.90
Zimbabwe	8.60	9.20	NA	NA
Asia and Pacific	**6.00**	**6.03**	**20.97**	**19.39**
Bangladesh	2.30	2.50	16.80	11.90
China, Mainland	NA	NA	37.00	27.60
India	13.40	13.40	35.40	32.40
Indonesia	NA	NA	8.90	15.70
Pakistan	NA	NA	28.80	28.20
Philippines	2.30	2.20	8.40	9.90
Thailand	NA	NA	11.50	10.00
Europe & Central Asia	**6.50**	**5.00**	**27.10**	**13.80**
Turkey	6.50	5.00	27.10	13.80
Western Hemisphere	**7.29**	**5.81**	**12.09**	**8.28**
Argentina	2.70	1.30	9.40	3.10
Bolivia	13.40	11.40	21.10	18.00
Brazil	7.70	7.40	13.10	8.20
Chile	14.40	8.30	15.50	6.70
Costa Rica	NA	NA	8.50	11.50
Ecuador	NA	NA	12.70	13.90
Guatemala	1.90	2.00	6.70	4.80
Mexico	6.70	4.90	14.40	10.30
Panama	7.60	7.30	9.70	4.60
Paraguay	4.80	4.60	11.20	5.50
Peru	6.40	5.10	10.70	4.50
HIGH INCOME	**3.60**	**2.80**	**8.28**	**7.16**
Australia	NA	NA	15.00	12.00
Japan	NA	NA	5.80	6.50
Sweden	NA	NA	10.30	8.70
United Kingdom	3.60	2.80	6.40	4.60
United States	NA	NA	3.90	4.00
TOTAL	**7.49**	**5.94**	**15.79**	**11.99**

Note: Change in regions is defined as the unweighed average for those countries with available data.
Source: World Bank, 2000, *World Development Indicators 2000*, Table: 5.8 State-Owned Enterprises, http://www.worldbank.org/data/wdi/statesmkts.htm

The international trends concerning the size of the public sector reveal only partial compliance with the neoliberal policy prescription. The role of public enterprise appears to be on the decline, and national governments are trending toward lower deficits in national budgets. On these results the neoliberal view on the role of the state seems to be broadly adopted. However, in terms of the relative size of government, the neoliberal expectation of smaller government has not been met. Furthermore, a positive correlation between the relative size of expenditures and level of development in a country was suggested in Table 4.1. Although this observation is not necessarily inconsistent with the minimal state concept, it does suggest that the role of the public sector expands as countries develop. This discussion indicates that national governments will continue to exercise an enormously important economic role in all countries. The ability to make wise choices in fiscal policy and in expenditure patterns is in part a question of governance systems, a topic to be addressed below.

SCIENCE AND TECHNOLOGY SYSTEMS

As the role of science and technology in today's economy has become better understood, governments have reassessed public funding for and perform-ance of this infrastructure.[6] This section will first examine government support for R&D itself and then government support of education, a critical element in a country's science and technology system.

Government funding for R&D has declined in recent years in most of the industrialized countries, although the level of funding varies substantially across countries (see Table 4.5). This somewhat surprising result is due to lower levels of defense-related spending, following the end of the Cold War, and to the grow-ing private-sector R&D investment. Private enterprise recognizes the importance of R&D investment to innovation competitiveness in global markets. In some countries, such as the United States, a decline in government-funded R&D, as a share of GDP, represents tight fiscal policy as well as declines in military research. The one exception to this broad pattern of government R&D funding is found in basic research because of its public good characteristics. The uncer-tainty concerning rates of return for this type of investment would likely lead to inadequate funding if left to private-sector investment decisions.[7]

Government as performer of R&D has also been on the decline in the OECD countries during the 1990s (see Table 4.6). This trend was particularly sig-nificant in the United States and Japan. The more moderate decline in govern-ment performance of R&D in countries of the European Union (EU) has been compensated by an increase in university performance of R&D. In the United States, performance by higher education has remained at the same level, while in Japan a decline has occurred.

The OECD countries have quite varied R&D priorities (see Table 4.7). The United States places a high priority on defense R&D, although defense's share of total R&D fell substantially between 1991 and 1997, as did its share in the EU. In terms of civilian R&D, the United States targets its funding to specific objectives, especially to health, environment, and space, while Japan and the EU provide high levels of funds to general university research, presumably allowing university communities to set priorities.

Table 4.5
Governmental R&D Expenditures in the OECD

A. As percentage of total R&D in Country or Region

	1981	1989	1991	1993	1995	1997
United States	49.2	45.6	38.7*	37.7	35.6	31.9
Japan	24.9**	16.8**	16.4**	19.7**	20.9**	18.7*
European Union	46.7	40.5*	41.2*	40.0*	39.0	38.3
Total OECD	45.0	38.9	35.5*	35.1	33.8*	31.4

B. As percentage of GDP in Country or Region

	1981	1989	1991	1993	1995	1997
United States	1.20	1.25	1.09	.99*	.93	.86
Japan	.53	.46	.46	.53	.58	.53*
European Union	.80	.80*	.80*	.77*	.72	.70
Total OECD	.90	.90	.83*	.79	.73*	.69

* Break in series from previous year for which data are available.
** Underestimated.
Source: OECD, *OECD Science, Technology and Industry Scoreboard, 1999* (OECD, 1999), pp. 126 & 127.

Table 4.6
Performance of R&D by Government and by Higher Education (as percentage of total R&D performed)

	Government					
	1981	1989	1991	1993	1995	1997
United States	12.1	10.7	09.8*	10.2	09.6	08.2
Japan	12.0**	08.6**	08.1**	10.0**	10.4**	08.8*
European Union	18.9	16.6*	17.0*	16.5*	16.2	15.3
Total OECD	15.0	12.6	12.4*	12.8	12.6*	11.3

	Higher Education					
	1981	1989	1991	1993	1995	1997
United States	14.5	15.5	14.1*	15.5	15.3	14.4
Japan	17.6**	12.5**	12.1**	14.0**	14.5**	14.3**
European Union	17.4	17.4*	18.8*	20.4*	20.8*	21.0
Total OECD	16.5	16.1	16.0*	17.4	17.3*	16.9

* Break in series from previous year for which data are available.
** Underestimated.
Source: OECD, *OECD Science, Technology and Industry Scoreboard, 1999* (OECD, 1999), p. 128.

Table 4.7

Government Budget Appropriations or Outlays for R&D by Objective, 1991 and 1997

	Defense as a percentage of total R&D Budget		Percentages of civilian R&D budget									
			Economic Development		Health and Environment		Space		General Nonoriented Funds		University	
	1991	1997	1991	1997	1991	1997	1991	1997	1991	1997	1991	1997
United States	59.7	55.3	22.1	19.7	43.5	46.6	24.5	24.5	9.9	9.2	NA	NA
Japan (adjusted)	5.7	5.8	33.5	34.8	5.7	7.3	7.2	6.7	8.5	11.5	45.1	39.7
European Union	21.0	15.8	30.3	23.3	14.3	15.3	7.2	7.2	15.7	16.3	30.8	35.2
Total OECD	37.3	31.4	28.6	24.3	22.3	23.1	12.2	11.6	13.4	13.3	NA	NA

There seems to be little convergence across countries on these priorities, suggesting that policy and political processes in countries produce quite different outcomes.

Governments in the OECD countries promote R&D by the private sector through direct support (e.g., grants) or indirect support (e.g., tax subsidies) (see Table 4.8). Tax subsidies give firms greater flexibility in R&D investment decisions than does direct funding. Although the United States had one of the highest levels of government-funded industrial R&D (as share of industrial domestic product) for most of the 1990s, the trend was toward lower funding, reflecting the country's tight fiscal policy. If Mexico is excluded, convergence can be observed among the other nine countries reported. In 1991 the high-spending countries—the United States, Germany, and the United Kingdom—lowered spending through the decade, while the initially lower-spending countries increased spending. In terms of tax treatment of R&D expenditures, four countries provided R&D subsidies in 1990, but this number had increased to seven in 1998. These patterns suggest an increasing predisposition for publicly supported industrial R&D, especially through tax systems.

A country's capacity to engage in R&D and to meet the demands of the emerging economy depends to a large extent on the education and skill levels of its workforce. Although the specific role of education on economic performance is disputed, there is little doubt that a country's scientific and technological capacity is directly related to its education and training system. The recent pattern of public spending on education systems in the nations of the world is striking (see Table 4.9). The level of spending per pupil, as measured by per capita GNP in 109 countries, increased by just less than 20% between 1990 and 1997. The increase in spending was substantially higher for higher education than for primary education. Although there are some troubling trends in these expenditure patterns (e.g., the great disparity between per capita expenditures in primary versus higher education in developing countries), the overall increase in effort clearly reflects that an important priority is placed on education in many countries.

The focus on higher education can also be observed in terms of earned doctoral degrees in science and engineering. Although the United States holds a considerable lead over other countries in the total number of doctorates granted in a year, Europe as a whole (including Russia) produces almost twice as many (see Table 4.10). The relationship between training of research scientists and intellectual productivity in countries is a question of great interest to research communities and policy makers.[8] In some countries, such as Russia and India, the expected impact of very sizable research and scientific communities on national economic performance has not been realized. Some developing countries, such as Brazil and China, have encouraged nationals to seek advanced training abroad as they develop their in-country research and training facilities. Furthermore, research scientists are highly mobile, thus complicating the analysis of the relationship between advanced scientific training and national economic performance.

Table 4.8
Government Support to Industrial Technology

	Expenditures (percentage of Industrial Domestic Product)									Tax Subsidies*	
	1989	1990	1991	1992	1993	1994	1995	1996	1997	1990	1998
Canada	.35	.36	.38	.39	.41	.36	.32	NA	NA	.170	.173
Mexico	.03	.04	.04	.04	.04	.04	.04	.04	.04	-.018	.031
United States	.76	.76	.68	.72	.66	.63	.60	.56	.54	.090	.066
Australia	.29	.29	.33	.38	.38	.39	.36	.31	.31	.276	.110
Japan	.20	.21	.20	.21	.23	.24	.24	.25	.27	-.021	.104
Finland	.45	.50	.66	.73	.65	.53	.61	.65	.63	-.015	-.009
France	.63	.66	.74	.67	.62	.57	.51	NA	NA	.090	.086
Germany	.55	.51	.44	.46	.43	.39	.40	.40	.37	-.054	-.051
Netherlands	.38	.42	.34	.32	.29	.32	.35	.34	.38	-.020	.096
United Kingdom	.57	.56	.57	.55	.56	.45	.44	.42	.40	.000	.000

*Tax subsidies for 1 US dollar of R&D, large firms.
Source: OECD, *OECD Science, Technology and Industry Scoreboard, 1999* (OECD, 1999), p. 134.

Table 4.9
Public Educational Expenditure per pupil, 1990 and 1997 (as share of GNP per capita)

	Number of Countries	Total Expenditure (in %)		Primary Secondary (in %)		Higher Education (in %)	
		1990	1997	1990	1997	1990	1997
More Developed Regions	23	20.5	21.0	18.4	19.5	23.5	25.2
Northern America	2	20.0	21.5	19.1	20.3	21.9	26.2
Asia/Oceania	4	18.5	17.9	17.0	17.0	18.9	15.8
Europe	17	23.0	23.0	19.0	20.9	30.4	31.5
Countries in Transition	17	20.5	26.0	15.8	19.0	35.5	32.7
Less Developed Countries	69	16.6	15.5	12.6	12.0	82.8	68.0
Least Developed Countries	19	14.8	14.4	10.5	10.4	125.5	88.2
Total	109	22.2	27.0	17.7	17.9	64.2	65.7

Source: UNESCO, *World Education Report 2000*, p. 119. Available at http://www.unesco.org/education/information/wer/index.htm

Table 4.10
Earned Doctoral Degrees in Science and Engineering by Region/Country, Most Recent Year

	All Doctoral Degrees	All S&E Doctoral Degrees
Asia	32,087	15,678
China	4,364	3,230
India	9,369	4,425
Japan	13,044	5,453
South Korea	4,462	1,920
Europe	78,791	45,647
France	9,801	8,575
Germany	22,404	10,128
Italy	3,603	1,432
Spain	5,193	1,794
United Kingdom	9,761	6,512
Switzerland	3,804	1,840
Russia	14,005	10,042
North America	44,855	28,493
Canada	3,356	2,027
Mexico	488	259
United States	41,011	26,207
Total	155,733	89,818

Source: National Science Board, *Science and Engineering Indicators-1998* (Arlington, VA: National Science Foundation, 1998, NSB 98-1), p. A-83.

In addition to issues of funding of R&D, many countries have focused on the performance of national science and technology systems. In the United States, for example, a variety of programs have been adopted to encourage commercialization of federally funded R&D activities, such as the Advanced Technology Program, Cooperative Research and Development Agreements, and Small Business Innovation Research.[9] Many countries have sought mechanisms to encourage universities to undertake research with commercial potential, sometimes in collaboration with the private sector.

Government spending patterns on science and technology present a complex set of trends. Defense-related R&D spending is on the decline and private-sector spending has substantially expanded. Nevertheless, a public-sector presence in science and technology remains strong in the OECD countries. The rates of R&D investment appear to be positively correlated with levels of development, suggesting that higher income countries have greater capacity to invest in R&D. Among developed countries, R&D spending priorities are quite varied. Several countries appear to be making efforts to reach the spending levels of the leaders—the United States, Germany, the United Kingdom, and France. Education expenditures have increased throughout the world, especially for higher education. Developing countries will likely increase expenditures for R&D and education as their economies develop. In addition, governments are introducing a commercialization objective in R&D investments by encouraging universities and national research laboratories to cooperate more extensively with the private sector. European governments have shifted to a new policy framework—that of systems of innovation—from the traditional science and technology and education categories. Substantial international experience on public support for science and technology infrastructure has accumulated and individual countries can be expected to rely on these experiences in reshaping policy systems.

DEVELOPMENT INFRASTRUCTURES, TECHNOLOGICAL CHANGE AND THE PUBLIC SECTOR

The provision of infrastructure has been a crucial public-sector role in development policy in most nations. Transportation, communication, education and training, and other infrastructure systems have public good characteristics that have led to a prominent government role in their provision. Whether through planning and coordination, financing, regulation or direct provision of services, governments are principal actors in infrastructure development. The externalities inherent in these infrastructures and the very high levels of investment required have justified a leadership role for government.

The technological revolution of the late 20th century, however, led to a repositioning of the state in the provision of several infrastructures, especially telecommunications, transportation, and energy. The case of telecommunications is illustrative of the broader trend. Given the nature of the product and system of distribution, a variety of positive externalities were realized by having a single provider. The local network, the so-called local exchange loop, consisted of a physical plant distributed through a geographically defined community or city, similar to water and electricity distribution systems. To achieve service provision at minimum cost and the realization of network externalities, including interconnectivity within and among exchanges, a single company provided services in the local exchange.

Some countries chose public ownership of the monopoly provider and others chose a privately owned, but regulated, industry structure. Special provisions had to be created to insure adequate service in high-service-cost areas, such as rural areas. Some countries adopted universal service policies that obliged the regulated telephone companies to serve as large a share of the population as possible.

The nature of the new telecommunications technology led to a fundamental change in the nature of the industry, eliminating the natural monopoly characteristics of the service and leading to a change in the role for government. The interexchange markets were the first to be opened to competition. In this segment of the industry, multiple alternatives for long-distance service provision became available, including microwave or, more recently, satellite-based transmissions. If interconnection standards could be adopted and enforced, there was no longer a compelling justification for the single provider. In fact, better service, it was believed, could be obtained through competition. More recently, a similar technological process can be observed in local exchange markets and these too are being deregulated. The wireless telephone industry was never characterized as a natural monopoly (no physical distribution system was found in the local exchange loop) and competition in this market can be found in large cities in most countries, although regulatory regimes often permit only a few competitors in a single market. Competition among telephone companies, cable companies, Internet providers, and, in the near future, high-capacity wireless networks, has become increasingly viable.

The process of deregulation and privatization of telecommunications industries can be observed in countries throughout the world as the industry steadily loses its natural monopoly characteristics. In spite of technological commonalties in terms of standards and equipment, countries have established significantly different regulatory regimes.[10] Although the range of reform is clearly affected by the unique legacy of previous telecommunications policies and industry structures, the new policy regimes also reflect the importance of politics and power relationships within individual countries. In other words, no ideal, technologically defined regulatory regime has emerged across countries.

The promise of telecommunications deregulation has been lower prices and better services. The speed at which prices decline depends on the degree of competition in markets. Privatization itself does not guarantee competition. In markets of limited competition, regulatory bodies must remain vigilant to ensure the efficient and fair pricing of services. Antitrust regimes may emerge to prevent the excessive dominance of individual firms in nonregulated markets, but a regulatory or monitoring role for government will likely be found essential for this critically important industry.

INSURING PROPER OPERATION OF MARKETS

Apart from the appropriate role of the public sector as a provider of national infrastructure, the neoliberal agenda is concerned with the role of government as regulator. Regulation, it is argued, creates barriers to efficient capital allocation and increases the cost of economic transactions. The appropriate policy option is, therefore, deregulation.

National governments are encumbered with a number of public responsibilities beyond that of promoting economic growth. For example, in most countries, governments are responsible for guaranteeing human rights and protecting health and welfare of its citizens. Protection of health and welfare produces regulatory structures for many activities, including environmental and water quality, food and drugs, building codes, and safety standards. Elimination of regulation in these areas appears entirely unfeasible, given current practice in most countries, but the quality and effectiveness of regulation will undoubtedly be the subject of close scrutiny. Even in fields of economic regulation, such as product liability, intellectual property rights, and labor laws, deregulation generally means the revision of regulatory regimes rather than abandonment of a governmental role.

It is curious that in the developed world there has been a call for deregulation and regulatory reform, while in Africa a renewed appreciation for the importance of the public sector in securing well-functioning markets has emerged. A clear articulation of this significant shift can be found in the World Bank's analysis of its programs in Africa.[11] Various weaknesses in governmental institutions, including corruption, lack of accountability and denial of human rights, created conditions that prevented or impeded development. Improvement in governance, therefore, was viewed as an enabling condition for successful development. The World Bank defined governance as the exercise of political power in the management of a country's affairs, including the management of a country's economic and social resources for purposes of development.[12] A democratic society, with the rule of law, checks and balances in the public sector, and a public accounting of actions by the state, is required for markets to function properly. This position reverses the long-held consensus that democracy can be achieved only after a certain level of development has been reached.

The move from state-led to market-centered development has created the need for institutions or forums where disputes can be resolved within countries. In Brazil, for example, questionable utilization of tax incentives by state governments became an issue requiring action by Congress and several federal agencies.[13] As noted in the case of telecommunications, a government monitoring or regulatory role is required to insure that competition exists in markets. In addition, the role of governments in establishing product standards, accounting standards, intellectual property rights and disclosure requirements for banks and firms are among many governmental functions believed to be needed for markets to operate efficiently.[14]

The fundamental role of government in creating markets can be seen in the formation of common markets and international trade agreements. Only a government can represent a nation in the articulation of international agreements. Furthermore, a primary responsibility for monitoring trade patterns and managing and resolving international trade disputes falls to the government. New international trade dispute resolution mechanisms are emerging and, again, national governments are essential to their creation and effectiveness.

DECENTRALIZATION OF THE PUBLIC SECTOR

Decentralization of the public sector has been a common element of the neoliberal discourse; not only should the size of government be reduced, it should also be decentralized. The decentralization theme also appears in discussions of democracy and public-sector performance. A government close to citizens, it is argued, facilitates democratic practice and improves government performance.[15]

In many developing countries, the adoption of structural adjustment policies in the 1980s often coincided with the reform and decentralization of the state. A managerial orientation in the public sector, to replace the earlier, relatively inflexible, administrative orientation, was an element of the reform.[16] Local problem solving and locally determined deployment of resources were viewed as promising strategies. Although the management orientation represented an innovation in many countries, the notion that government should lead, not just manage, the development process was widely held. The concept of government as a neutral forum for resolving conflict, as suggested by neoliberal philosophy, was found by many to be inappropriate in most developing countries. The national state often provided the leadership believed necessary for development in countries with relatively low per capita income and limited productive capacity. Policy making and government action often became consolidated in national governments to achieve efficiencies through centralized management and strategic planning in the allocation of scarce resources. The success of the Asian tigers and the stagnation of the import substitution models in some developing countries lent credence to alternative, market-oriented strategies that were subsequently adopted in many countries, especially in Latin America. The corresponding reform for the public sector was decentralization from national to subnational governments of public decision-making and policy implementation.

The promise of decentralization, in which relatively rigid centralized systems would devolve powers and cooperate with lower levels of government, has proven difficult to fulfil in practice. Successful decentralization depends on the capacity—fiscal, organizational and human resource—in subnational government as well as a supportive set of intergovernmental relations. Local governments are linked to higher levels of government in several different ways, including through constitutional and statutory frameworks, fiscal relations, joint responsibilities of program implementation, and through politics. The specific set of linkages varies substantially across countries, and trends in decentralization can even vary within a country by policy area. In the United States, even though significant decentralization of power from the federal government to state government occurred in the 1970s and 1980s, federal control over states increased significantly in the realm of regulation.[17] Decentralization holds the potential for improved local government action, but local governments operate within a set of complex intergovernmental relations that can constrain, if not impede, their actions.

Despite broad support for decentralization, some national governments have not encouraged fiscal autonomy of lower-level governments. Given the greater efficiency in national systems for revenue collection, as compared to local government capacity, decentralization of revenue generation may not be effective. In Brazil, the enormous debt of state government banks has prompted the central bank to reassert control.[18] In countries with strong clientelistic traditions, the timely transfer of funds from national government to lower levels, as required in a truly decentralized system, is problematic because it requires fundamental change in well-established political systems. Some policy issues, due to externalities and implementation requirements, are inherently national questions. Subnational governments, for example, cannot effectively address macroeconomic and monetary policy. Some elements of science and technology policy, such as support of basic research, will be more efficiently addressed at a national level.

Decentralization generally implies the enhancement of the fiscal situation of local governments, especially in terms of creating opportunities for generating revenues locally. Although local decision-making and local control of own-source revenues are sound principles, the capacity to administer local tax systems may be absent; thus tax collection and new taxes may represent a political problem for local officials. Furthermore, the tax base in poor regions may be severely limited. From a national perspective, fiscal decentralization, especially in terms of greater local fiscal independence, will likely place poor regions at a substantial disadvantage in comparison to more wealthy regions.[19]

Decentralization of the public sector has created, however, new opportunities for regional development policy. In the United States, state governments in the 1980s became more active in development policy when the federal government chose not to intervene in the spatial adjustments to structural economic change.[20] In Brazil, following the decentralized governmental structure offered by its 1988 Constitution, state governments have become very active in promoting development and actually compete with other states for investment.[21]

The EU, through its structural funds program, has also engaged in an innovative strategy in less-developed nations. The formation of common markets and trade agreements has generated dynamic regional economies that cross national boundaries in Europe and North America.[22] A renewed interest in regional economies has also emerged as a result of research on several highly innovative regions.[23] Innovation was found to have an important regional dimension, thus a focus on systems of regional innovation has emerged in research and policy development efforts.

SYSTEMS OF GOVERNANCE

An examination of the changing role of the state should not avoid consideration of governance systems. Even if the role of the state has evolved, it nevertheless remains critical to national economies, and the quality of public policy decisions will affect economic performance to some degree. Since the early 1990s we have witnessed a significant increase in the number of nations with democratic political systems. Decentralization also created new contexts for democracy and subnational policy making. In spite of well-known imperfections, democratic politics is on the rise. Public decision-making, including the articulation of goals of the major development institutions and the formation of development policy, is the outcome of governing processes that are increasingly democratic, implying that public discourse will incorporate more actors, and consensus formation around development issues will be more difficult to achieve.

In a period of great economic and social change, traditional institutions are often forced to change and new institutions appear. The emergence of the research university in Germany in the 19th century and the adoption of social welfare policy during the great depression in many countries illustrate the potential for social creativity in times of change. The setting of appropriate standards for new technologies can also spur innovation and growth.[24] However, history holds many examples of poor economic performance resulting from inadequate governmental response or societal incapacity to change. The ability to transform social institutions depends, to a significant extent, on the nature of governance systems. Governance systems that are open, participatory and representative and which can integrate new voices and interests are likely to be the systems that produce the most effective public policy.

Universities, in particular, are viewed as critical institutions for future development, and governance systems of higher education in many countries have been found unresponsive to new social needs and economic change.[25] Higher education governance systems—the bodies that make decisions regarding size and access, programs and curricula, and standards—vary substantially across countries. Adapting universities, both public and private, to new economic and social circumstances will depend on the effectiveness and responsiveness of these governance systems.

National governments must decide the type and level of support for science and technology and education systems. Fairly narrow interests, such as scientific communities or national defense interests, have often dominated the formulation of R&D and science policy. In the case of the United States and several European countries, defense-related R&D was shielded from close public scrutiny for strategic reasons. Rarely has a broad social consensus been rallied for these types of policies, as contrasted with those of education, infrastructure or health. As policy systems become more democratic, science and technology policies will increasingly be formulated in the context of competing proposals for governmental funding and tight budget constraints. Closer scrutiny and greater public debate on the effectiveness of existing policies and institutions can be expected.

The recent attention concerning environmental and labor standards and international trade presents a challenge to governance systems in individual countries as well as to international trade bodies. As systems become more democratic, policies of national governments and multilateral organizations must be forced to respond to the demands of organized groups. The expanding use of trade agreements and common markets will create a context where international visibility and pressure will provide yet further challenges to national governance systems.

CONCLUSIONS

The role of the state in national economies has evolved in recent decades in most countries. National governments have followed several neoliberal policy prescriptions, as seen in the worldwide trend toward declining levels of deficit spending. In addition, privatization of public enterprise has reduced this public-sector role in many countries. In some sectors technological change has eliminated the public-good characteristics of products, and governments have moved from monopoly provision of services toward markets with competitive firms. However, in less-developed countries, public enterprise, as a share of GDP, has not contracted to the same extent as in other countries, perhaps reflecting the critical role played by government in mobilizing capital in these countries. In contrast, a minimal state is not emerging in that there is little evidence that national governments are reducing expenditures as a share of GDP. Substantial increases in educational expenditures were found in all but the high-income countries. The high levels of total spending, as share of GDP, in more advanced economies may well be associated with their higher levels of wealth and, consequently, greater capacity to raise revenues.

Changes in the government role in science and technology were observed but few well-defined international trends could be identified. Direct government spending on R&D is down among OECD countries, driven in part by declines in military-related expenditures following the end of the Cold War. A very substantial increase in private-sector R&D investments more than compensates for the decline in public spending. Some convergence in spending levels among OECD countries was found in that countries with lower levels of R&D spending have increased expenditures more rapidly than higher spending countries. Quite distinct priorities for and means of support for government R&D were observed across countries. In this field the role of the state country can best be described as evolving with new policy approaches being tested. Spending levels have trended downward, but an active role for the state has not disappeared, particularly if government support for higher education is considered as a element of science and technology policy. Public funding for R&D must compete with other national priorities at a time of tightening fiscal conditions.

State reform, a process defined by political processes within a country as well as external influences, has several potential impacts on the relationship between government and the economy. Although the state-led development model has disappeared, government plays critical roles in ensuring markets operate efficiently and fairly, both within individual countries as well as internationally. Decentralization of the public sector creates new opportunities for locally defined development strategies. To the extent that subnational governments have access to tax bases affected by local economic conditions, local development initiatives should be expected in the future.

Government roles in national economies have evolved, but governmental action will remain critical to economic prosperity in many ways. The vision of a minimal state is not consistent with the trends identified in this chapter. Furthermore, economic policy formulation, both at national and international levels, will occur in contexts that are increasingly democratic and open. Governance systems will be challenged to find broad consensus on policies in a period of tight fiscal policy.

Appendix A
Model of Deficit/Surplus Spending

Model Definition

DEFSUR = f (GDP annual growth rate, TIME, C1.....C82)

Variables

DEFSUR
Overall Deficit/ Surplus as a Percentage of Total Expenditure and Lending minus
   Repayments
Data Source: The International Monetary Fund, *Government Finance Statistics Yearbook*,
   Vol. XXIII (Washington, D.C., 1999), pp. 10-11.

GDP
Gross Domestic Product annual growth rate (%)
Data Source: The International Monetary Fund, "Real Gross Domestic Product (annual
   percent change)," The WEO Database September 2000, website:
   http://www.imf.org/external/pubs/ft/weo/2000/02/data/index.htm

TIME
Values of 1, 2, ..., 11 (for 1988, TIME = 1; for 1988, TIME = 2; ... for 1998, TIME = 11)

Country Dummy Variables: C1.....C82
Individual country dummy variable (the base dummy country: Venezuela)
Those countries in the IMF data with less than 3 missing values between 1988 and 1997
   are included in the data.

Results

Variable	Slope Coefficient	P- value
Intercept	-9.27	0.0016
GDP	0.32	0.0001
Time	0.29	0.0081
Botswana	27.84	0.0001
Democratic Republic of Congo	-11.74	0.0150
Ethiopia	-21.60	0.0001
Madagascar	-11.44	0.0072
Sierra Leone	-24.83	0.0001
South Africa	-8.73	0.0304
Zambia	-15.65	0.0002
Zimbabwe	-13.63	0.0010
China, Mainland	-17.88	0.0001
India	-26.22	0.0001
Malaysia	7.54	0.0706
Maldives	-9.90	0.0148
Myanmar	-21.64	0.0001
Nepal	-27.24	0.0001
New Zealand	11.85	0.0042
Pakistan	-22.58	0.0001
Singapore	59.18	0.0001
Sri Lanka	-20.90	0.0001
Thailand	18.32	0.0001
Finland	-6.70	0.1049
Greece	-25.08	0.0001
Italy	-10.13	0.0121
Luxembourg	9.14	0.0381
Romania	7.39	0.0749
Turkey	-15.91	0.0001
Oman	-13.11	0.0012
Canada	-11.64	0.0082
Chile	12.98	0.0014
Dominican Republic	8.62	0.0371
Panama	14.07	0.0007
Venezuela	-9.27	0.0016

Only significant country variables are included here.

NOTES

1. The research for this chapter was supported by the Mike Hogg Professorship of Urban Policy, University of Texas at Austin. The author wishes to acknowledge the very able research assistance from Mr. Suho Bae.

2. Daniel Yergin and Joseph Stanislaw, *The Commanding Heights: The Battle Between Government and the Marketplace That Is Remaking the Modern World* (New York: Simon & Schuster, 1998).

3. Pedro Conceição, Manuel V. Heitor, David V. Gibson, and Syed S. Shariq, "The Emerging Importance of Knowledge for Development: Implications for Technology Policy and Innovation," *Technological Forecasting and Social Change: An International Journal*, vol. 58, No. 3, July 1998, pp. 181-202.

4. World Bank, *World Development Report, 1999-2000: Entering the 21st Century* (Washington, DC: The World Bank, 2000), Table 17, pp. 262-263.

5. World Bank, *World Development Indicators 2000: States and Markets*, Table 5.8, State-Owned Enterprises. http://www.worldbank.org/data/wdi/statesmkts.htm.

6. European Union, *EU Socio-Economic Research, Project Cluster Systems of Innovation* (Brussels: European Commission, 2000).

7. OECD, *OECD Science, Technology and Industry Scoreboard, 1999* (Paris: OECD, 2000).

8. National Science Board, *Science and Engineering Indicators-1998* (Arlington, VA: National Science Foundation, 1998), pp. 6.18-6.30.

9. Ibid., pp. 4.19-4.35.

10. Judith Mariscal, *The Mexican Telecommunications Reform: A Political Economy Approach*, Ph.D. Dissertation, The University of Texas at Austin, December 1998.

11. World Bank, *Sub-Saharan Africa: From Crisis to Sustainable Growth: A Long Term Perspective* (Washington, DC: The World Bank, 1989); *Governance: The World Bank's Experience* (Washington, DC: The World Bank, 1994).

12. World Bank, *Governance and Development* (Washington, DC: The World Bank, 1992), p. 3.

13. Robert H. Wilson, "Redefining Regional Development: Decentralized Policymaking and International Markets in Brazil", in Pedro Conceição et al., eds., *Knowledge for Inclusive Development* (Westport, CT: Quorum Books, 2000).

14. World Bank, *World Development Report, 1998-99: Knowledge for Development* (Washington, DC: The World Bank, 1999).

15. World Bank, *World Development Report, 1999-2000*, op. cit., pp. 44-45.

16. Mohamed Halfani, "The Governance of Urban Development in East Africa: An Examination of the Institutional Landscape and the Poverty Challenge," paper presented at the Meeting on Urban Governance of the Global Urban Research Initiative (GURI), Mexico City, October 1995, pp. 4-5.

17. Timothy Conlan, *New Federalism: Intergovernmental Reform from Nixon to Reagan* (Washington, DC: Brookings Institution, 1988).

18. Robert H. Wilson and Lawrence Graham, eds., *Policymaking in a Redemocratized Brazil, Vol. 2: Social Policy and Exclusions*, no. 119 (Austin, TX: LBJ School of Public Affairs, 1997).

19. Ibid.

20. Robert H. Wilson, *States and the Economy: Policymaking and Decentralization* (Westport, CT: Praeger, 1993).

21. Robert H. Wilson, Antonio Rocha Magalhães, and John Cuttino, "Redefining Regional Development Policy in Brazil: State Development Policy and International

Markets in Parana and Ceara" (Austin, TX: LBJ School of Public Affairs, mimeo). For the case of the United States, see Wilson, *States and the Economy*, op. cit.

22. For the case of Europe, see European Union, *EU Socio-Economic Research*, op. cit., pp. 125-172.

23. Michael Stroper, *The Regional World: Territorial Development in a Global Economy* (New York: Guilford Press, 1997).

24. European Union, *EU Socio-Economic Research*, op. cit., p. 16.

25. Pedro Conceição, Manuel V. Heitor, and Pedro M. Oliveira, "Expectations for the University in the Knowledge-Based Economy, *Technological Forecasting and Social Change: An International Journal*, vol. 58, No. 3, July 1998, pp. 203-214.

5

Gateway Airports, Speed and the Rise of Aerotropolis

John D. Kasarda

THE SPEED IMPERATIVE

A decade ago, futurist Alvin Toffler (1990) argued that by the beginning of the 21st century one indisputable law would determine competitive success: *survival of the fastest*. In Toffler's view, producing high-quality goods at competitive prices would still be necessary but no longer sufficient for commercial success. Speed and agility would take center stage, as industry increasingly emphasized accelerated development cycles; international sourcing and sales; flexible, customized production, and rapid delivery.

How right he was. During the 1990s, the most successful companies used advanced information technology and high-speed transportation to source parts globally, minimize their inventories, and provide fast and flexible responses to unique customers' needs, nationally and worldwide. They sought international partners, just-in-time suppliers and sophisticated distributors and logistics providers. By combining flexible production systems with information systems that connected companies simultaneously to their suppliers and customers, firms reduced cycle times and customized their products to create additional value. They also offered the same speed and flexibility in the delivery process from the time the finished goods left the factory until they arrived at the customer's doorstep.

Mandating such changes are rapid and relentless worldwide political and economic transformations. Former socialist countries are entering the capitalist marketplace with vigor. What were previously known as Third World countries in Asia and Latin America have achieved much higher levels of output and are producing more sophisticated goods and services.

International customers (including those in developing countries, which many believe pose the best long-term markets) are also becoming more sophisticated and demanding. They have available an unparalleled variety of products from all over the world. They are able to assess and identify value, and are therefore highly selective in purchasing. They expect quality, reliability and competitive pricing. They also want customization of the products they buy, and they want these customized products right away, not in two to six months. For some purchases, not even two to six weeks is fast enough.

E-COMMERCE AND AIR EXPRESS FULFILLMENT

The rise of e-commerce further heightened time-based competition. As late as 1995, sales through the Internet were essentially zero. By 1999, U.S. Internet-based business-to-consumer (B2C) sales had grown to nearly $7 billion. According to Forrester Research (2000), 166 million packages were shipped in 1999 by Internet retailers (e-tailers), with approximately 70 percent going by express delivery (Gose, 2000). By 2003, e-tailers are expected to ship 1.1 billion packages annually, with overall global e-commerce approaching $7 *trillion* in transactions in 2004 (Forrester Research, 2000).

Most of this explosive growth is expected to be business-to-business (B2B), supply-chain transactions where materials and components will be ordered through the Internet and shipped to next-stage producers. Manufacturers will electronically access an international network of suppliers in order to acquire the best-quality materials and parts at the lowest possible price. The introduction of e-marketplaces (auctions, aggregators, bid systems, and exchanges) will greatly expand B2B e-commerce. Forrester Research (2000) predicts that e-marketplaces will account for up to three-quarters of B2B supply chain transactions by 2004.

However, as many U.S. e-tailers discovered during the 1999 Christmas season, as valuable as the Internet is in generating sales, the Web cannot move a box. Order fulfillment frequently broke down, and the WWW—*world wide wait*—cost e-tailers plenty.

To meet the imperative of speed in order fulfillment, e-commerce distribution centers are being built near gateway airports that have extensive flight networks, a location trend that is sure to accelerate in the decades ahead. This is especially the case at major air express hubs such as Memphis International (FedEx) and Louisville (UPS) in the United States. Air express hubs actually extend the business day for e-commerce fulfillment by allowing shippers to take orders for next-day delivery as late as midnight. Dozens of e-tailers have thus already located their fulfillment centers near Memphis International Airport, including barnesandnoble.com, PlanetRx.com, Toysrus.com and williams-sonoma.com. The same story holds for Louisville International Airport where such companies as Nike.com, Drug Emporium.com, and Guess? have sited e-commerce fulfillment centers.

Complementing airport-linked e-commerce fulfillment centers are flow-through facilities for perishables (either in the physical or economic sense), just-in-time supply chain and emergency parts provision centers, and reverse logistics facilities for the repair and upgrade of high-tech products such as computers and cell phones. The clustering of such time-sensitive goods facilities around airports is stimulating further expansion of air cargo, air express, less-than-load (LTL) trucking, freight forwarders and third party logistics providers along major arteries leading into and out of gateway airports. All of these functions and facilities are leveraging off each other.

Speed and agility have become so critical to the new economy that air commerce is quickly becoming its logistical backbone. Forty percent of the value of world trade now goes by air, and the percentage is steadily rising (Kasarda, 1998/1999). Air cargo, which represented a US$200 billion industry in 1999, is expected to triple in the next 15 years while international air-express shipments are expecting to increase at least five-fold during this period (Boeing Company, 2000). Already, air cargo and air express are the preferred modes of shipping of higher value to weight B2B transactions in microelectronics, automobile electronic components, aircraft parts, mobile telephones, fashion clothing, pharmaceuticals, optics and small precision manufacturing equipment, as well as many perishables such as seafood and fresh-cut flowers. Even lower value to weight B2B product distribution such as toys are becoming time-sensitive and increasingly shipped by air.

Further evidence that we have entered "the fast century" is offered by data showing that nearly two-thirds of all U.S. air cargo is transported via 24- to 48-hour door-to-door express shipments, with Memphis International Airport becoming the world's leading air cargo airport (World Airport Week, 2000). Billions of dollars of time-sensitive goods-processing and distribution facilities have been attracted to the vicinity of the FedEx hub, transforming the once-sleepy Memphis into a global air commerce gateway.

AIRPORTS AS OFFICE, COMMERCIAL AND KNOWLEDGE NETWORK MAGNETS

Not only time-sensitive goods-processing and distribution facilities are being drawn to gateway airports. As our service economy also shifts into fast-forward, these airports are becoming magnets as well for corporate headquarters, regional offices and professional associations that require officers and staff to undertake frequent long-distance travel. Airport access is likewise a powerful attraction to information-intensive industries such as consulting, advertising, legal services, data processing, accounting and auditing, which often send out professionals to distant customers' sites or bring in their clients by air. Business travelers, overall, benefit considerably from access to hub airports, which offer greater choice of flights and destinations, more frequent service, more flexibility in rescheduling, and generally lower travel-related costs (for example, hub airports make it easier to avoid the time and expense of overnight stays).

Such accessibility and travel flexibility hub airports offer have become essential to attracting major conventions, trade shows, and merchandise marts. Two U.S. megafacilities—Infomart and Market Center, both of which are located on the I-35 corridor between the Dallas Love Field Airport and the Dallas-Ft. Worth International Airport—offer examples of this attraction. Infomart is a huge, ultracontemporary merchandise display building for tele-communications and information technology companies. Market Center—a cluster of six large buildings that contain nearly 7 million square feet of display space for fashion clothing and home merchandise—is the world's largest wholesale merchandise mart. Hundreds of thousands of buyers and vendors fly into Dallas annually to conduct business at Infomart and Market Center. In 1999, Market Center alone attracted buyers and vendors from all 50 U.S. states and 84 countries, who purchased 300,000 airline seats and filled 720,000 nearby hotel rooms while conducting an estimated $7.5 billion in wholesale transactions.

Knowledge networks and travel networks are also increasingly overlapping and reinforcing each other. With intellectual capital supplanting physical capital as the primary factor in wealth creation, time has taken on heightened importance for today's knowledge workers. So has the mobility of these workers over long distances. Research has shown that high-tech workers, for example, travel by air between 60 percent and 400 percent more frequently than those in the general workforce (Erie, Kasarda, McKenzie, and Molloy, 1999).

Some observers have suggested that advances in Internet access, video-conferencing, and other distributed communications technologies will diminish the need for air travel. The evidence indicates that telecommunications advances often promote additional air travel by substantially expanding long-distance business and personal networking. Indeed, innovations in telecommunications technology have generated spatial mobility at least since the days of Alexander Graham Bell—whose first words over his newly invented telephone were, "Watson, come here, I need you."

HIGH-TECH AND URBAN ECONOMIC IMPACTS

In a networked economy increasingly geared to speed, mobility and global access, frequent and extensive air service has become essential to the location of many advanced information-processing and other high-tech facilities. In the United States, clusters of high-tech facilities and information technology companies are increasingly locating along major airport corridors, such as Dulles International Airport in Northern Virginia and the Chicago's O'Hare International Airport. Dulles's and O'Hare's experiences are being replicated across the United States, with centrality in aviation networks becoming a primary predictor of an area's high-tech job growth.

Apropos this logistic centrality, Kenneth Button and Roger Stough (1998) conducted a comprehensive study of the impact of hub airports on employment growth in high-tech fields. Their multiple-regression analysis (which controlled for other factors that may affect high-tech job growth) covered all 321 U.S. metropolitan statistical areas (MSAs) and generated convincing results. Button and Stough showed that the presence of a hub airport in an MSA increases the number of high-technology jobs in the area by over 12,000, and their multiple-regression model explained over 64 percent of the variation among metropolitan areas in high-technology employment growth. Additional analysis revealed that the causal link between job growth and air network centrality flowed from extensiveness of connections to the creation of high-tech employment, and not vice versa. This finding has been corroborated by a study that Michael D. Irwin and I conducted which documented that airports have pervasive effects on overall metropolitan employment growth and that the causal relationship flows from centrality in air networks to employment growth (1991).

Numerous other studies from across the United States and around the world are documenting the remarkable impact of gateway airports on urban economies and land use. Let me note just a sample:

- Los Angeles International Airport (LAX) is responsible for over 400,000 jobs in the five-county Los Angeles region; 80 percent of which were in L.A. County, where 1 in 20 jobs was found to be tied to LAX. The airport currently generates $61 billion in regional economic activity, which translates to $7 million per hour.
- Dallas-Ft. Worth International Airport has become the primary driver of Metroplex's fast-growing economy. The number of companies located within the dynamic Las Colinas area, just to the east of the airport, has expanded to more than 2,000 and includes Abbott Laboratories, AT&T, Exxon, GTE, Hewlett-Packard and Microsoft.
- In the 26-mile commercial corridor linking Washington, D.C.'s two major airports—Reagan National and Dulles International—employment grew from 50,000 in 1970 to over 600,000 by 1996. This represents a 1,100 percent increase; in contrast, overall U.S. employment growth during this period was 59 percent. Among the companies located along the airport corridor are America Online, Computer Associates, Nextel Communications, Cisco Systems and EDS.
- In the Philippines, Subic Bay Freeport is rapidly expanding around a former U.S. naval air base that was converted to commercial use in 1993. Since FedEx located its Asia/Pacific hub at Subic Bay in 1994, over 150 firms—employing 40,000 workers—have located there, generating almost $2.5 billion in investment. Between 1994 and 1999, the annual value of exports from Subic Bay jumped from $24 million in to $559 million. In late 1998, Acer opened its largest personal computer assembly facility in the world at Subic Bay; the facility relies heavily on air freight for its supply-chain management.
- By late 1997, nearly 50,000 people were employed on the airport grounds at Amsterdam's Schiphol Airport, a 7.2 percent increase over the previous year. In 1998, nearly half of the 547 companies linked to Schiphol grew—compared with 31 percent in 1995. Schiphol alone accounts for 10 percent of the European air cargo market and 1.9 percent of Netherlands' GNP. The airport forecasts that by 2015, it will generate 2.8 percent—approximately $14 billion.

The impact of airport-induced job growth on land use in the vicinity of airports is likewise substantial. An analysis of employment growth in the suburban rings of U.S. metropolitan areas showed that areas within four miles of airports added jobs two to five times faster than the overall job-growth rate of the suburban ring within which the airport was located (Neuwirth, Reed and Weisbrod, 1993). Most of the employment was concentrated around the airport or along a major access corridor within 15 minutes of the airport.

THE RISE OF THE AEROTROPOLIS

Emerging corridors, clusters, and spines of airport-induced businesses are giving rise to new urban forms as much as 15 miles from gateway airports. These represent the beginnings of the aerotropolis. In response to the new economy's demands for speed and reliability, the aerotropolis is based on low density, wide lanes, and fast movements. In other words, form is following function.

Although aerotropoli have so far evolved largely spontaneously—with previous nearby development often creating arterial bottlenecks—in the future they will be improved through strategic infrastructure planning. For example, dedicated expressway links (aerolanes) and high-speed rail (aerotrains) will efficiently connect airports to nearby and more distant business and residential centers. Special truck-only lanes will be added to airport expressways, as well. Seamlessly connected multimodal infrastructure will accelerate intermodal transfers of goods and people, improving logistic system effectiveness and further influencing business locations and resulting urban form.

The new metric for determining land value and particular business locations will be time-cost access to the airport. Firms of various types will bid against each other for accessibility predicated on the utility each gives to the related combination of time and cost of moving people and products to and from the airport. Urban structure will no longer be measured by traditional bid-rent functions that decline linearly with spatial distance from the primary mode (here, the airport) but by speed to the airport from alternative sites via connecting highway and rail arteries.

To many, this new structure will appear simply as additional sprawl along main airport corridors. Yet the aerotropolis will actually be a highly reticulated system based on time-cost access gradients radiating outward from the airport; in short, the three "A's" (accessibility, accessibility, accessibility) will replace the three "L's" (location, location, location) as the most important business location and commercial real estate organizing principles.

Air-commerce clusters and spines are already taking on distinct spatial form around major gateway airports such as Miami International, New York's Kennedy, Los Angeles International Airport (LAX), London's Gatwick and Heathrow, Paris's Charles de Gaulle, and Amsterdam's Schiphol. In the United States, even smaller, specialized air-cargo airports—such as Alliance Airport, near Ft. Worth, Texas, and Rickenbacker Airport, in Columbus, Ohio—are generating mini-aerotropoli in the form of low density cluster and spine development.

Commercial growth surrounding Southern California's Ontario Airport—which cornerstones a major logistics complex 40 miles east of Los Angeles—offers an excellent contemporary illustration of an aerotropolis in evolution. Over 12 million square feet of warehouse and distribution space was added in 1999 adjacent to the airport and on Interstates 10 and 15 radiating out from it. As of June 2000, another 10 million square feet were on the way, led by e-commerce fulfillment and distribution facilities ranging up to 1 million square feet in floor space for companies such as eToys, Toysrus.com, Staples, and Home Shopping Network.

Enhancing Ontario's position as a leading logistics and e-commerce fulfillment center is the growth of express transportation services at and around Ontario Airport. During 1999, UPS, whose West Coast hub is at Ontario Airport, handled over 700 million pounds of freight while FedEx carried over 100 million pounds. This express service was boosted by another 100 million combined pounds carried by BAX Global, Emory Worldwide and Airborne Express.

In Brazil, one can observe an emerging aerotropolis centered around Viracopos International Airport in Campinas, located 60 miles east of São Paulo, where high-tech manufacturing, distribution and logistics industries are clustering. Viracopos will likely become the air cargo and e-commerce fulfillment center of South America in the 2000s with aviation-driven urban form resulting from these logistic and high-tech clusters radiating outward from the airport.

Aerotropoli are also emerging in distinct patterns around new international airports in Asia. One example is Lantau Island, where the newly opened Hong Kong International Airport is spawning highly visible business and residential clusters directly linked to the airport. In late 1999, the Walt Disney Company announced that it would be locate its third international theme park (Hong Kong Disneyland) on Lantau Island to take advantage of the international airport and its high-speed rail and expressway links to Hong Kong. This siting decision is not unlike those Disney made earlier for Tokyo Disneyland, near Narita International Airport, and EuroDisney, near Paris's Charles de Gaulle Airport.

A major planned aerotropolis is under development at Inchon, Korea, where the government is creating a 24-hour Aviation City on Yongjong Island, about 40 miles west of downtown Seoul. The new international airport (scheduled to open in 2001) will anchor an expansive urban agglomeration composed of commercial, industrial, residential and tourism sectors. Its centerpiece will be Media Valley, Korea's version of Silicon Valley. Designed as a center for global high-tech industries, Media Valley is being constructed adjacent to the airport on a 3.6-million-square-meter site that will include a large technopark and a university research center.

As of mid-1999, 625 companies—including 49 companies from Canada, Israel, Japan, The Netherlands, Taiwan, and the United States, among others—had submitted letters of intent to move into Media Valley. Arthur D. Little predicts that by 2003, a total of 1,300 companies will be located in Media Valley's campuslike setting, and by 2005 slightly over 2,000 (Business Korea, 1999).

A new town is being developed to serve as a residential base for those employed at Media Valley and in other sectors of this emerging aerotropolis. Dedicated expressways will give both Media Valley employees and the new town residents high-speed access to Inchon Airport.

By 2004, the airport, currently 90 percent complete, will be complemented by a seaport and a teleport now under construction. The plan is to form a consolidated "triport" for 21st-century transportation, distribution, and information processing.

An even more ambitiously planned aerotropolis radiates northward from the Kuala Lumpur International Airport in Malaysia. This massive new airport will provide the aviation foundation for Malaysia's Multimedia Super Corridor (MSC), a high-tech government, commercial, education and residential zone about the size of the city of Chicago. Promoted internationally as the future information technology center of Asia, MSC will contain two new cities (Putrajaya, the relocated government capital, and Cyberjaya, or Cybercity, each of which will house about a quarter of a million residents), along with a multimedia university to train IT workers. MSC's advanced infrastructure will be complemented by laws and policies designed to create the ideal commercial environment for developing and merging 21st-century audio, video and data transmission technologies.

THE FIFTH WAVE OF DEVELOPMENT

Hong Kong's Lantau Island, Malaysia's Multimedia Super Corridor, and South Korea's Inchon-Aviation City demonstrate that gateway airports will cornerstone dynamic new forms of 21st-century urban and regional development. Put in historical perspective, they really represent the fifth in a continuum of transportation infrastructure-induced development waves that have catalyzed and spatially shaped commercial growth over the centuries.

The world's first great commercial centers grew up around seaports. The next wave of economic development occurred along networks of rivers and canals that formed the backbone of industrial revolutions in Europe, the Americas and Asia. Railroads generated a third wave of economic development as they opened up inland areas to manufacturing and trade: major goods processing and distribution centers emerged at rail hubs and terminal points. For example, in the United States, the South's largest city, Atlanta, first developed as a railway hub and was originally known as Terminus.

The fourth wave of economic development was fostered by the shift to cars and trucks to move people and goods. Freeways, beltways, expressways and interstate highways generated a massive dispersion of housing and firms. Large suburban malls and commercial centers, industrial research parks and office complexes sprouted as far out as 45 miles from major city centers. Some of these fourth-wave "edge cities" now have more retail and office space than the downtowns of their metropolitan areas.

As I noted above, we have already commenced the fifth, and perhaps most opportune economic era—The Fifth Wave—where aviation, digitization, globalization and time-based competition will predominate. The combined thrust of these interacting forces is creating and shaping new economic growth nodes, as gateway airports supplant seaports, rail and highway systems as logistical drivers of development and as primary job and wealth generators.

This is all happening because companies, in general, and e-businesses, in particular, have learned that they cannot meet the challenges of the speed-driven, globally networked economy without dramatic changes in how they organize their flows of information, materials and finished goods. Digitized infrastructures and air logistics have become central to their strategies as they leverage the power of information networks and global supply chains to their competitive advantage. Governments, too, are recognizing that in order to help their industries complete, boost exports and attract foreign investment they must provide modern logistics infrastructures that enable local and regional firms to rapidly and flexibly source and sell on the global stage.

Thus, whether it is futurists such as Alvin Toffler or leading economists and management professionals such as Lester Thurow and Peter Drucker, and, whether they are talking about enterprises or nation-states, they are all saying the same thing: logistics is the next frontier of competition. Strategic advantage will be gained by those companies and development advantage by those countries that fuse digital technology and air commerce through logistical infrastructure that optimizes their position in the global network of information and product flows.

REFERENCES

Boeing Company (2000). *World Air Cargo Forecast* (Seattle, WA: The Boeing Company).

Business Korea (1999), "Miracle on Han is Moving Down River: Inchon Nurtures its Strategic Central Location," July, pp. 18-22.

Button, Kenneth, and Stough, Roger (1998). "The Benefits of Being a Hub Airport City: Convenient Travel and High-Tech Job Growth" (Fairfax, VA: The Institute of Public Policy, George Mason University), November.

Erie, Steven P.; Kasarda, John D.; McKenzie, Andrew; and Molloy, Michael A. (1999). "A New Orange County Airport at El Toro: Catalyst for High-Wage, High-Tech Economic Development" (Irvine, CA: Orange County Business Council), September.

Forrester Research (2000). *E-Marketplaces Will Lead US Business eCommerce to $2.7 Trillion in 2004* (Cambridge, MA: Forrester Research).

Gose, Jose (2000). "Thanks to E-Commerce, Warehouses Aren't Just for Storage Anymore," *Barron's*, March.

Irwin, Michael D., and Kasarda, John D. (1991). "Air Passenger Linkages and Employment Growth in U.S. Metropolitan Areas," *American Sociological Review*, August, 56(4): 524-37.

Kasarda, John D. (1998/1999). "Time-Based Competition & Industrial Location in the Fast Century," *Real Estate Issues*, Winter, pp. 24-29.

Neuwirth, Roanne M.; Reed, John S.; and Weisbrod, Glen E. (1993). "Airport Area Economic Development Model," paper presented at the PTRC International Transport Conference, Manchester, England.

Toffler, Alvin (1990). *Powershift: Knowledge, Wealth, and Violence at the Edge of the 21st Century* (New York: Bantam Books).

World Airport Week (2000). "Traffic Continues to Increase at World's Airport," May, p. 5.

6

Institutions and Knowledge Networks: The Chinese Experience

Leslie Young

HISTORICAL BACKGROUND

Once upon a time, an economy that was already the richest and largest in the world developed a new system of communication that facilitated new forms of commerce. The scale economies that then became possible permitted new kinds of economic specialization and drove an industrial revolution: a surge of technical innovation that lifted economic growth to levels never before seen in a large economy. Through international trade and finance, this surge of growth pulsed outward and eventually uplifted economies on the other side of the world.

That economy was China one thousand years ago during the Sung Dynasty. The new form of communication was an integrated nationwide canal system that permitted bulk transport of commodities among large population centres that had never previously been economically connected. The new technologies were in rice and silk farming, printing, iron smelting and machine-driven cotton spinning and weaving. The economic stimulus was carried out by ocean-going junks that carried trade to Southeast Asia and into the Indian Ocean, eventually stimulating the economies of the Middle East and Europe, leading to the medieval economic revolution that set the stage for the European Renaissance.

There are remarkable parallels between the position of the United States today as technology leader and that of China one thousand years ago. Both were continental-scale economies blessed with high agricultural productivity, political unity, internal peace, excellent internal communications and effective administration. Perhaps more important for their world leadership was their intellectual openness, market-friendly leadership, technological dynamism and openness to international trade. To understand why China lost that leadership, we must examine the differences between the two societies.

Whereas the United States is a relatively new, diverse nation of immigrants, held together by the democracy and the rule of law, Sung China was a homogenous nation that already had a thousand years of unified history behind it. This long-term unity was based on a unique system of governance. The autocratic emperor was restrained by the doctrine of the Mandate of Heaven: that the right to rule depends on performance. He ruled through a civil service selected through written examinations on the Confucian classics that enjoined righteous behaviour by rulers. This civil service provided the social order required for small-scale commerce, but never tolerated the development of large-scale capitalism that would have challenged its monopoly on power and status. The civil service never gave priority to security of mercantile property, nor state enforcement of contractual relations. Consequently, the only types of enterprises that could develop were those that could rely upon ties of trust based on long-term relationships of kinship, dialect and region.

The commercial skills and informal institutions that evolved in this large, orderly economy were deployed to good effect in modern times when they enabled the Overseas Chinese to dominate commerce in Southeast Asia. From that base they used their capital and skills to spearhead the recent resurgence of economic growth in China. Capital can be mobilized with extraordinary speed by quick-thinking patriarchs to exploit new opportunities acquired from private information networks. The Asian Miracle resulted from this way of doing business, within the framework provided by a resurgence of the classical Chinese system of governance. The modern autocracies of East Asia sought legitimacy—the Mandate of Heaven—through economic growth and achieved it with the assistance of a meritocratic civil service.

Their success led to a brief burst of rhetorical triumphalism on Asian values that was cut short by the Asian financial crisis. This revealed that behind the extraordinary East Asian dynamism and efficiency in product markets were grotesque inefficiencies in capital markets. The basic problem was that the East Asian shortcut past the Rule of Law had also bypassed the development of the capital market institutions. The concept of a corporation, an entity that exists only in legal and accounting space, provided a perfect vehicle for expropriating minority shareholders and creditors when the supporting institutions of law and accountancy had not been developed. By stacking corporations in pyramids, it was possible for the controlling shareholders atop the pyramid to control companies at the base with very little capital investment. Since they therefore would have received only a small proportion of any dividends paid out by a company at the base, they had an incentive to channel wealth up the pyramid by intragroup transactions at unfair prices. The controlling shareholders found that they could make more money internally by exploiting minority shareholders and creditors than by creating external value.

Such corporate pyramids are incompatible with development and exploitation of new technology. This is illustrated by the unfortunate fate the Growth Enterprise Market that was launched in Hong Kong in 1999 at the crest of the dot-com wave and almost immediately became the world's worst-performing market. By contrast, China has developed a successful high-tech industry based on unique adaptations of the institutional structures of socialism.

HONG KONG'S GROWTH ENTERPRISE MARKET

The Growth Enterprise Market (GEM) was launched on 15 November 1999 by the Stock Exchange of Hong Kong Limited (SEHK). Preparation for the launch began with the May 1998 Consultation Paper on a Proposed Market for Emerging Companies (the Consultation Paper). GEM's mission is to provide a channel for growth enterprises to raise capital for their business development and expansion. As of November 2000, there are 49 companies listed on the GEM, a majority being related to high tech.

Those who favour the establishment of the GEM regard Hong Kong as having an internationally recognized financial market with excellent infrastructure, strategically placed in a high-growth region with plenty of opportunities for growth enterprises to flourish, especially local enterprises in the Greater China region. GEM fits in with Hong Kong's policy to promote the development of high technology and high value-added industries.

The Main Board of the SEHK is poorly placed to exploit these opportunities. The stringent listing rules require the listing company to have a three-year record, with HK$20 million profit in the most recent year, and HK$30 million profit in the preceding two years. Many growth companies would not qualify as they have a short history and would not be generating sales, let alone profits, in their early years of operation.

The Main Board is weighted towards financial, property and conglomerate giants. Industrial and technology companies are often ignored by ordinary investors so such companies can be relatively under valued.

Regulatory Philosophy

The philosophy of GEM is "buyers beware" and "let the market decide" supported by a strong disclosure regime. Investors are expected to take the initiative to understand the disclosed information of GEM and make its own assessment of the merits of the GEM, before deciding on whether to invest or continue to invest.

Unlike on the Main Board, the SEHK is not responsible for the assessment of the commercial viability of GEM companies. GEM targets growth companies in Hong Kong, as well as the Greater China region. Companies of all industries and sizes are welcome. It offers a market for entities with just a two-year history of active business and with no profit track record. Profit forecasts are no longer mandatory and there is no asset-backing requirement.

In place of these traditional requirements, GEM imposes post-listing requirements such as the two-year continuous sponsorship and quarterly financial reporting. GEM imposes stringent requirements on disclosure and corporate governance.

In addition to half-yearly and annual accounts, a GEM company must publish quarterly accounts within 45 days after the relevant period end, which is more stringent than the Main Board. GEM requires frequent and timely disclosures while the electronic trading and communication facilities allow investors convenient and timely access to order processing and information.

GEM companies must disclose in more detail past business history and future business plans. After listing, a GEM company has to make a comparison every six months of its actual business progress with the business plan for the first two full financial years.

Corporate Governance

Under GEM, compliance with Listing Rules is the dominant objective of corporate governance. A company must appoint a full-time qualified accountant to supervise its finance and accounting functions. There must be a designated senior compliance officer and two independent directors with relevant experience. There must be an Audit and Compliance Committee, chaired by an independent director, which meets quarterly. Senior management compensation must be approved by this Committee and disclosed in the annual accounts.

To be eligible for listing, a new applicant must appoint a qualified GEM sponsor to submit its listing application. Therefore, the quality of the GEM companies depends to a large extent on the due diligence of sponsors in their selection process for sponsoring companies to be listed on the GEM. In order to qualify, a sponsor must satisfy the SEHK that it has the requisite experience and professional competence. As at 31 January 2000, there were 37 qualified sponsors and one qualified cosponsor.

A GEM stock issuer must retain a sponsor in an advisory capacity for two years after listing. Sponsors have to make due and careful enquiry to satisfy themselves that the applicant is suitable for listing and that all the information in listing document is accurate and complete in all material aspects. After listing, the sponsor serves as a channel of communication between the company and SEHK informing the company on compliance issues and reporting on non-compliance.

Target Investors

As the easing of listing rules would imply much greater risks for investors, the GEM is not targeting ordinary investors but professional investors such as venture capitalists and institutional investors. Such investors have the knowledge, expertise and experience necessary to understand what is going on with the companies and are also better at assessing both the growth potential as well as the risks in investing in such companies. Each investor is required to sign a statement with its broker to acknowledge understanding of the risks.

The GEM has publicized its "buyers beware" philosophy to warn investors. For example, posted prominently on its Website are the following words: "GEM: A Buyers Beware Market".

Market Performance

Table 6.1 shows the performance of GEM: the worst performing market in the world. Table 6.2 shows the monthly turnover of newly listed stocks. Table 6.3 shows the daily turnover ratio (the daily turnover value/day-end market capitalisation of a stock). For the first five trading days this is over 20 times its turnover in the fourth month of trading!

Table 6.1
Performance of the GEM Market

	GEM	Main
Price Fall 11/99-10/00	66%	1.7%
Index volatility = average monthly standard deviation of daily returns of the index	4.69%	1.95%.
Monthly turnover = monthly equity turnover value/ month-end market capitalisation	14.12%	5.39%
October 2000 turnover	4.30%	3.52%

Table 6.2
Monthly Turnover Ratio of GEM and Main Board

Month	GEM	Main Board
12/1999	30.11%	5.08%
1/2000	15.30%	7.92%
2/2000	16.83%	8.25%
3/2000	38.83%	6.71%
4/2000	9.85%	4.14%
5/2000	6.57%	5.07%
6/2000	5.93%	4.19%
7/2000	8.11%	5.70%
8/2000	12.52%	4.76%
9/2000	6.93%	3.97%
10/2000	4.30%	3.52%
Average 12/99-10/00	14.12%	5.39%

Table 6.3
Average Daily Turnover Ratio (%) of newly listed stocks on the GEM

(25 November 1999 to 31 May 2000)

Period after listing	Turnover ratio
Day 1 - 5	5.41
Month 1	2.05
Month 2	0.51
Month 3	0.39
Month 4	0.25
Month 5	0.22
Month 6	0.26

Corporate Structure of GEM Companies

At the end of September 2000, there were 47 companies listed on GEM with a total market capitalization of US$10,237 million. The top three companies account for 48.5% of the market capitalization; the top six companies account for almost 70% of market capitalization. This resembles the Main Board, which is also dominated by a few major groups of companies. In fact, the controlling owners of the top six companies of GEM also dominate the Main Board.

Thus, the corporate giants of Hong Kong have established information technology (IT) venture companies and listed them on the GEM with minimal track records in that field. The core competencies of the controlling shareholders are not in IT, but in property and finance. Their GEM companies serve as investment holding companies and venture capitalists to invest in Internet start-ups, content providers and other infrastructure companies. With little core technology development in Hong Kong, many IT-related companies are focusing on applications rather than technology development/acquisition.

Such listings defeat the stated purpose of GEM, which was meant for nascent enterprises in need of cash—not as an extra means for tycoons to raise money for their investments. The assessment of these companies was not based on assets, cash flows or skills of the managers, but on the reputation of tycoons (and their listed companies) and high-profile financiers, movie stars and popular radio commentators. The thousands of taxi drivers and amahs lining up for blocks for the tom.com initial public offerings (IPO's) were hardly "professional or well-informed" investors. They did not understand what is written in the prospectus. Instead, they were drawn by the reputations of the business leaders of the parent conglomerate.

Controversies Surrounding the Gem Market

GEM relaxes the lock-up period for controlling shareholders from two years to six months after listing, and reduces the track record requirement from two years to one year of active business pursuits prior to listing. Thus, a company may be only 18 months old when the management starts to unload its shares. Lifting the ban on new issue of shares within the first six months of listing means that there is a risk that listings will be "engineered" to establish a shell into which business or assets which have been excluded from IPO can be injected immediately after the new issue, thus escaping the scrutiny of a prospectus. Moreover, there has been selective waiver of GEM Listing Rules for companies such as tom.com.

There is no enforceable legal liability for the accuracy of the listing document. By contrast, under U.S. law, the prospectus is the only information that issues may distribute about the offering. As a result, it has to serve both as a disclosure document and a selling document. The company, its corporate officers and members of the board of directors are absolutely liable for any misstatement or omission of information in the registration statement, even if there was no intent to deceive, so the narratives and accounting sections must be clear, accurate and complete.

Some companies have not used the proceeds from listings as laid out in their prospectuses. Timeless Software was the first stock listed on the GEM, touted as an exemplary technology company. One month later, it announced that it would use 40% of its listing proceeds on property! It paid HK$178,375,500 for 13,213 square feet on the 79th floor of The Center (i.e., HK$13,500/sq. ft. or US$1,731). Most software houses rent office space with good communications for about HK$10 to 15/sq. ft per month. What is more, the company is now entering into property development, a new business area not mentioned in its prospectus. That announcement set the stock price plunging and it has not recovered since. As of October 30, 2000, it had fallen by 85% to $1.26 from its high of $8.5.

Companies such as Panda-Recruit (chaired by the chairman of GEM's listing committee, Mr. K. S. Lo himself) and 36.com have been rumoured to have undisclosed related parties holding shares in the company. This suggests that the funds initially gathered from the IPOs may not really be public money.

Some 31 out of the 49 companies that are now listed initially sold their shares entirely by private placement. The rest privately placed 83% of their shares on average. The real price at which the shares are actually placed is not transparent since sweeteners may have been offered in private.

CHINA'S TECHNOLOGY

Common Features of Successes in Chinese Technology

Table 6.4 displays the first four Chinese technology companies that achieved national prominence. It is striking that they are all collectively rather than privately owned. They originated in universities and national science/technology institutes that were oriented toward producing scientific achievement, not commercialization. Prior to being developed into commercial products, research results and prototypes developed in the state sector were intangible. It was difficult to define property rights over them and assign a value to them. They were available for free.

The company founders were entrepreneurial scientists with strong roots in the research institutes, so their companies had easy access to high-quality research, and to many young, talented technicians and researchers. The start-up companies were sponsored by a state institution (university, academy, ministry) or collective institution (township) that provided the legal framework for the company. The initial financial contributions from the sponsoring institutions were hoarded while working capital was generated from trading activities (even selling roller skates!). The companies' initial small commercial loans were repaid within months by initial successful products. Thus, most of the companies' capital was retained earnings. The companies were never beholden to the state and remained free from state interference. Sponsoring institutions respected the autonomy of enterprise management.

Personal friendships, alumni networks and family relationships were deployed to get the companies organized and legalized. Each company had a market niche protected by the difficulties of adapting the Chinese language to computers. This blocked entry by foreign companies and led to a mass-market product that took off and established each company's brand and finances. In Legend, researchers applied artificial intelligence to Chinese character inputting technology. In November 1985 Legend marketed the "Associative Chinese Word-Processing System Version I", an add-on card and supporting software for the IBM PC. Starting from world-class mastery of technology, the company integrated backward into manufacturing.

By contrast, the smaller East Asian tigers learned simple manufacturing technology, gradually built technical capability but never attained technological leadership. Because business opportunities were much greater, most talent went into business rather than research. Also, research talent went to the United States and never came back until very recently. By contrast, research attracted talent in China because there were no business opportunities, politics was dangerous and, until recently, there was no opportunity to exit to the United States.

Each company adopted a group structure, listing only the core businesses on a stock exchange. Controlling shares in the listed companies were held by the group. Thus they raised substantial amounts of capital without losing control over the core businesses.

Table 6.4. China Technology

	STONE	LEGEND	FOUNDER	GREAT WALL
Technology Source/ Sponsor	Qinghua University	Institute of Computing Technology, Chinese Academy of Sciences	Beijing University	Ministry of Electronics Industry
First Success	1985, Printer for Chinese characters	1986, Chinese word processing add-on card	1988, Chinese electronic publishing	1985, IBM-compatible computer with Chinese word processing
Start-up loan	RMB20,000 from Evergreen Township	RMB200,000 from ICT	RMB300,000 from Beijing University	RMB 300,000 from R&D budget of Computer Bureau, MEI
Working capital	Short-term loan by credit co-operative, retained earnings	ICT loan, short term loans from local Bank of China	Loans from Yuyuan Township, Industrial and Commercial Bank	Prepayments for Great Wall 0520 computers
Joint venture partner	Mitsui	Daw (HK)	Cheung Family (HK)	IBM
Stock Exchange Listing	Hong Kong 16/93 HK$300 million	Hong Kong 10/93 HK$220 million	Hong Kong 12/95 HK$277 million	Shenzhen 6/97 RMB649 million
Shareholding Structure				
Shares held by group	58%	38.8%	55.6%	64.05%
Long-term partners	15%	36.2%	16.9%	
General public	27%	16.9%	27.5%	35.95%

Ownership and Corporate Governance: The Stone Company

In early years, Stone Company earnings were allocated to three funds: 50% for enterprise development, 30% for the employee welfare fund and 20% for the employee bonus fund. Salaries and bonuses had to be kept low to minimize the jealousy of neighbouring firms. High salaries would have been defined as profits and taxed at a high rate. Retained earnings were used as interest-free working capital. On the company balance sheet, the undistributed bonus fund was accounted for as debts by the company to employees. Thus, a large part of the value of the company was collectively owned by the employees and could not be paid out due to institutional constraints.

In 1988, about 50% of the company's collectively owned assets were put into a joint stock company which was 40% owned by the Stone Group. All employees were counted as founders of the company and jointly given 10% of the total shares as capitalization of their intangible human resources. These were distributed to employees who joined Stone before 31 December 1988 according to seniority and merit. The remaining 50% was issued to the public.

At listing, the management wanted to distribute all accumulated assets as shares to individual employees. They argued that Stone originally had no equity capital, all loans had been repaid so all value was retained and earnings were attributable to employee efforts. The government disagreed, arguing that there was no precedent and that company assets were in part the accumulation of tax privileges.

In the 1993 listing on the Hong Kong Stock Exchange, 58% of the shares in the Hong Kong listed company were retained by Stone Holdings. These holdings were kept as collective assets, as were the Stone businesses not injected into the listed company. Chinese regulators insisted on keeping the company's collective assets intact as a condition of permitting the public listing. China's 1993 statute governing urban collective enterprises banned the partition of collective assets into private hands. Therefore, Stone used its accumulated bonus fund to buy stock and distribute to employees.

After listing, collective ownership was retained because the Stone Group remained in control. Collectively accumulated productive assets were held as shares by the Group. Under Chinese law[1] the company is collectively owned by its employees. The equity cannot be divided among individual employees. This means that the employees owned the company insofar as they constituted a collective; individual employees who left the company could not take equity shares with them. It was also prohibited by law for the company to sell its assets and distribute the proceeds among its members.

The question of the "ultimate owner" of the company is meaningful only upon liquidation. Pursuant to the laws governing collectively owned enterprises, the surplus assets upon liquidation shall be applied toward providing subsidies to the unemployed and retired employees and training employees for re-employment. Thus, in normal times, the company itself as a legal entity is the de facto owner of the company. The Group (the collective of the workers) owns the Group (the enterprise)! However, managers have full autonomy in running the enterprise. The managers control operations but cannot change ownership of the capital. This resembles the Chinese law on land, which can be leased by farmers for use but cannot be privately owned.

PRIVATE VERSUS COLLECTIVE PROPERTY
IN THE NEW ECONOMY

In East Asian corporations, the introduction of Western concepts of private property, the corporation and limited liability, without supporting institutions like the accounting and legal professions, opened massive opportunities for expropriation within corporate pyramids by the controlling shareholder/ manager. This phenomenon reemerged in a virulent form when existing corporate pyramids were extended to new economy companies. Since the output and activities of these companies leave so little trace and are subject to such subjective valuations, the ineffectiveness of the accounting profession and of regulators has left minority investors especially vulnerable to expropriation. For example, one Hong Kong conglomerate set up a public cyber company, then had it buy consultancy services from a controlling shareholder's private company at a price which was ten times that of comparable services available elsewhere.

By contrast, under collective ownership, intragroup transactions would yield no incremental benefit to the manager. Those expropriated would not be strangers but co-workers. This constrains expropriation, both psychologically and practically. Retaining collective ownership has thus been invaluable in constraining expropriation when the institutional framework is insufficiently developed to constrain private actions. Collective ownership works well when the collective is not an arbitrary set of individuals, but a living family that has interacted continually over many years and has evolved its own mechanisms for enhancing group solidarity, for internal surveillance and allocating rewards: deeply rooted mechanisms of social cooperation which limit expropriation by the leader. The state was too large a group for those mechanisms to be effective.

By retaining collective ownership of high-technology companies when the law and accountancy professions had not been developed, China was able to tap a deep pool of research and research talent to foster a successful high-technology sector. By contrast, the freewheeling style of capitalism in Hong Kong, where the expropriation of minority shareholders had been masked by a boom in manufacturing and international trade, was fatal to its attempt to develop a high-technology industry when the well-designed rules went largely unenforced.

NOTE

1. Article 8 of the Constitution of the People's Republic of China (PRC) 1982 and the Law Applicable to the Urban (Cities and Towns) Collectively Owned Enterprises of the PRC (1991).

7

Regulatory Shields and Firms' Conduct: Application to Telecommunications Companies in Chile and Mexico

Alejandro Ibarra-Yunez

INTRODUCTION

Regulatory reform in Latin America and the Caribbean (LAC) has a different meaning than in developed economies. Whereas it is assumed that regulations exist and are substitutes for privatization in economies of Western Europe, the United States, and even Asia, regulatory frameworks are needed complements in LAC economies in their process of privatization and modernization, where overregulation coexists with weak institutional oversight mechanisms and agencies, and also where regulatory reform can foster or hinder business development. Moreover, lack of professionalized vision among policy makers, generally hinders LAC's capacity to converge towards international practice in the overall technology promotion front.

Sectors such as telecommunications, banking, electricity and gas, which account for main LAC privatizations, are characterized by monopoly power and concentration, where agency problems are frequent. Governments' pragmatism about privatization entailed little or no development of institutional change and oversight effectiveness, which adversely affects allocative and internal efficiency. Moreover, regulatory agencies are subject to a problem of multiprincipals, as has been pointed out by Martimort (1996), and Estache and Martimort (1997), where the pace of designing autonomous and independent oversight institutions on firms' conduct often is held up by other areas of government. The same institutional challenges exist for overall technology and financial policies. It is not entirely clear how a second-best result of privatization with regulatory shields granted to privatized companies with monopoly power, and underdevelopment of regulations, can work its way to regulatory modernization. Such is the analysis pursued in this chapter, applied to the telecommunication sector in two countries: Chile, where regulatory modernization began at the beginning of the 1970s and was in place by the time of the privatization of telecommunications; and Mexico, where regulatory change was not present before privatization occurred. The analysis applied to the telecommunications sector, even if characterized by particularities, can be used to reflect challenges in other spheres of policy making and regulations in emerging economies that face the need to improve and modernize governance structures.

Regulations extend from before privatization to after privatization. It is apparent that *ex ante* regulations include breaking up state-owned enterprises (SOEs) when needed, changing tax and other fiscal treatments to particular SOEs, modifying the structure of labor contracts and finally the general preparation of SOEs to be sold (valuation, method of sale, timing and determination of bidders' profiles). Along with focused regulatory adjustments to SOEs, there are others, such as those to change the rules of foreign ownership and property rights (foreign investment laws and liberalization), empowerment and modernization of capital markets, debt restructuring and other macroeconomic policies. As for regulatory changes needed after privatizations, they entail mainly normative aspects of prices, coverage or investment, as well as normative aspects against predatory conduct by privatized SOEs with monopoly power, and procurement as well as sectoral integration. This chapter emphasizes competition policy and rules established to curb anticompetitive conduct. Even if the various facets of regulatory change are interlinked, it is possible to make separations of policies and institutions in charge of them, not only for analytical purposes but also because each policy concern has particular aspects and applies to a different realm of business dynamics and planning (Moran and Prosser, 1994).

It is assumed that *ex ante* regulatory changes are preconditions for a first-best solution to privatization. Another assumption is that independent and modern agencies exist to prevent predatory conduct and determine prices and/or rate of return for legal monopolies. In developed economies, regulatory adjustments are seldom necessary, and attributes of agencies are only marginally changed to cases of predation, mainly because they exist as part of government settings. However, for LAC cases, regulations and regulators have to be created from scratch or, when agencies exist before privatization, need ample restructuring and retooling of their attributes (Guasch and Spiller, 1994). The next section presents a recount of oversight mechanisms in LAC. A third section analyzes main problems of agency in regulatory institutions in LAC. Then the fourth section presents a model of effort and commitment variables in telecommunications in Chile and Mexico, to show the differences between regulatory shields. Conclusions and policy implications end the chapter, and lessons are derived for other economic sectors in emerging economies.

THE DEVELOPMENT OF INSTITUTIONAL OVERSIGHT MECHANISMS IN LAC

According to Buscaglia (1996), most of civil law in LAC was implemented in the 19th century, adopting the French codified law on civil, commercial and criminal codes (in Argentina, Bolivia, Brazil, Chile, Colombia, Ecuador, Mexico, Peru and Venezuela). During the 20th century and for those countries with more modern and ampler economic bases, economic structure and behavior moved faster than legal structures, for which adjustments have been increasingly pressing. Argentina, Brazil and Chile established courts of justice and implemented civil force to commercial contracts. More recently, harmonization of commercial law has been sought by conferences in which Panama, Mexico and Peru have joined the above governments. Argentina, Brazil, Chile, Mexico and Peru have the highest number of commercial amendments on codes, to align structures to modern legal dispute settlements as well as to oversight business conduct in increasingly open economies. Trade agreements were therefore important as push forces towards revision of codes, mainly NAFTA and Mercosur.

Regulatory changes have been apparent on foreign investment aspects, both for regional free trade areas (FTAs), and also as bilateral liberalizations in a network of countries. For example, there are around 15 FTAs in LAC, but more than 30 cross-country foreign investment treaties in the region. Also, main FTAs, for their depth and coverage, are characterized by large countries establishing agreements with less-developed ones, where policy objectives are important, mainly to lock in such policies as commitment and reputation games among signatories of FTAs. Foreign investment accords have been characterized by liberalization and convergence of treatment of investment flows. Some of the investment agreements have been impacted by more general FTAs in their lock-in character (DeRosa, 1998). However, for the LAC region, it is argued that agreements between a developed and a less-developed area contain clearer elements of oversight of commitments than regional agreements among developing countries. Hence, it is argued that North-South FTAs are better than South-South ones in enhancing government commitment and governance structures (World Bank, 2000).

On its part, privatization has been implemented with little recourse to regulatory reform, where there are needed elements to modernize regulations and institutions mainly on business conduct. On problems in property rights, governments' recourse is constitutional and institutions are mainly of the federal or central level, except for Brazil, where there are many SOEs at the state or nonfederal level. The main problem with this aspect has been red tape and low professionalization of government officials. For example, Holden and Rajapatirana (1997) show the massive red tape as a case of bad regulations.

Buscaglia (1996) shows that amendments of commercial codes were furthered by those countries with more economic growth and a wider economic base, mainly in manufacturing. Out of a total of 1,618 amendments in a sample of nine LAC economies between 1930 and 1995, Brazil was first with 32% of the total. Then came Argentina and Chile, with 32% and 29%, respectively, and Peru was next with only 3.4% of the total. For the case of Mexico, commercial codes were mostly modified during the liberalization and privatization period between 1985 and 1995, with around 91 amendments of regulations to sectors and general activities, registered in the Official Gazette, which if put in perspective with Buscaglia's analysis, would make Mexico the fourth most active economy in this respect.

Now, the timing of regulatory bodies and oversight mechanisms has been off in many cases (exceptions are Chile overall, Mexico and Argentina in banking). For example, competition laws have existed in Argentina since 1919 (revised in 1980); Chile (1959 and revised in 1973); Colombia (1959 and revised in 1992); and Mexico (1934 and revised in 1993). More recently, Peru (1991), Venezuela (1992) and Brazil (1994) passed new laws and established oversight agencies (Rowat, 1995). Relating regulatory change to years when privatization began, it can be shown that Argentina, Chile and Peru had adjusted competition laws before privatization, although the number of cases overseen by the Argentinean authority are very limited, for which it could be argued that its role is almost nonexistent. Colombia and Venezuela made the adjustments coincident with privatization, and the oversight authority is active. For Mexico, competition policies and institutions were created after the peak of its privatization, but the federal competition commission (CFC) has increased its activity fourfold since its creation in 1991 (CFC, 1997).

Sectorally, Chile enacted regulatory oversight bodies for telecommunications and electricity in 1978. The framer (or regulatory agency or body) had the task of price adjustments mainly in electricity, and bodies continue to operate under privatized SOEs and competition. Chile has an agency for oversight of telecommunications with explicit arbitration capabilities and its own financial resources since the 1970s (Guasch and Spiller, 1994). For Argentina, the passage of an Electricity Law to restructure the sector occurred simultaneously with announcements of privatization in 1991.

As for telecommunications, Argentina shows that regulations faced repeated changes between 1989 and 1992, at the same time that the telecommunication sector was privatized in 1991. However, codes and legislation are very detailed and specific. More recently, the Argentinian agency on telecommunications regulations has moved forward in liberalizing the sector. Mexico privatized telecommunications in 1990, when regulatory bodies depended on the respective ministry. Regulations were extensive, but the regulatory agency COFETEL was created as an independent agency in 1995, as the sector was deregulated. The deregulation of the sector has been characterized by litigious stances of players in concessions, interconnection and open access. Such an environment of disputed decisions of the regulator have provoked a market that is underinvested, where access seems insufficient and where insufficient participation and technology upgrading have affected the country's competitive position. On its part, Peru's privatization of telecommunications took place in 1994 and an independent regulatory agency was formed the same year. Venezuela's oversight mechanism for telecommunications is done by an independent agency for tariff rebalancing, but any other changes have to pass approval by Congress. The mechanism has existed since before the 1991 privatization. Bolivia is opening to competition in electricity, beginning in 1994. However, its Electricity Code dated from 1968 precedes privatization but has not been updated to generate oversight agencies or empowering ministries. Table 7.1 summarizes the state of regulations and timing in LAC.

Table 7.1
State of Regulatory Mechanisms in LAC

Country	Codified commercial law	Courts of justice	No. of amendments 1930-95	No. of trade related non-agricultural sectors	Years of experience with comp. policy to 1997	Years of experience with comp. policy to privat. year	Sectoral agencies for telecom./energy	Index of property rights protection (inverted)	Index of red tape (inverted)
Argentina	yes	yes	515	16	17	6	yes	½	½
Bolivia	yes	no	19	1	0	-2	no	1/3	1/3
Brazil	yes	yes	521	14	3	-4	yes	1/3	1/3
Chile	yes	yes	467	11	24	12	yes	1	½
Colombia	yes	no	9	0	5	-1	no	1/5	¼
Ecuador	yes	no	14	1	0	-3	no	¼	¼
Mexico	yes	no	91	20	4	-5	yes	1/3	¼
Peru	yes	no	56	0	6	0	yes	1/3	¼
Venezuela	yes	yes	5	2	5	-2	yes	½	½

Sources: Buscaglia (1996), Center for International Private Enterprise (1997), Rowat (1995), and own calculations.

AGENCY PROBLEMS AND DYNAMICS OF POWER IN COMPETITION POLICIES

Much change in regulatory frameworks has been the result of economic opening. Beginning with the need to restructure after economic crises in the 1980s, Mexico, Chile and Venezuela were subject to financial rescue packages and government restructuring from IMF/WB since 1983, for which clearer rules of the game implied revisions of commercial codes and practices around government subsidies, performance of SOEs, credit growth to modern sectors and private-sector development. In the cases of Argentina, Peru and Brazil, foreign-sector strategies began later. Even if important, Ramamurti (1996) shows that conditioning by international institutions was not the only determinant of institutional change and regulatory reform, but that many regulatory and institutional changes emanated internally as a means to increase government effectiveness and to improve rules for economic agents. Unilateral and bilateral trade, and investment liberalization moved countries—such as Mexico, Chile and Venezuela in the 1980s—toward modernizing laws that guaranteed non-discriminatory treatment of much-needed foreign investment (National Treatment regulations). It could be argued that the demand for foreign investment is positively related to modernization of laws governing property and intellectual rights, contracts and vertical restraints, but not necessarily in competition regulations.

Role of Economic Opening in the Design of Oversight Institutions

On trade agreements, both NAFTA and Mercosur take account of rules for foreign investment limits in SOEs, after redefinition of salvaged sectors was enacted. Competition policy issues are addressed in Chapter XV, Art. 1501 of NAFTA (Mexico, 1992). Emphasis is placed, however, only on cooperation and mutual advising, but leaves signatories exempt to pursue their own application of the respective laws and regulations. Of the myriad bilateral foreign direct investment treaties among LAC governments, it is assumed that through international investment, not only technology transfer but more competition will be generated, to also affect positively more modern regulatory frameworks in the region. However, given the lack of lock-in obligations on transnational treatment of regulatory mechanisms, discretionary decisions by authorities or even the scope of regulations on market conduct are prevalent. Moreover, an anticipated aspect of economic opening is a set of considerations on how to make the basis of codes convergent with international standards. This entails not only procedural modernization, or even the empowerment of dispute-resolution mechanisms such as arbitrage, but also a more paradigmatic evaluation of the normative bases of laws governing economic conduct and rules of the economic game. Such considerations seem far from settled. Oversight institutions—say telecommunication regulators—then become overwhelmed by petty cases, hence impeded from becoming modern regulators. If an incumbent with monopoly power faces new competition, it would be interested in overwhelming the regulator both directly and indirectly through never-ending disputes, such as interconnection, liquidation disputes, or other predatory conduct that would reach gray areas of the law and become stalemate in modernization objectives of the sector. The same would occur for other players, such as banks, energy, or petrochemicals. One could argue that these aspects of regulations are preconditions for other, high-end promotion strategies such as any relating to technology growth and networked industries. On the other hand, a dirigiste government has been criticized as ineffectual or even welfare reducing in the economics literature.

Independence, Autonomy and Conflict among Regulatory Agencies

Aspects of independence, decision and budgetary autonomy and enforcement need to be analyzed. Conflicts in LAC are evident, if privatization preceded regulatory change and a sort of duality exists for the regulator between a shielded privatized SOE with monopoly power, and other residual private firms. The same result would occur if the privatizing entity (central government) establishes delegated power to the sectoral oversight agency instead of the general antitrust agency. This is the case of so-called agency problems with multiple principals. Some literature has recently received theoretical attention for its policy implications. Agency and capture exist not only between a regulated private entity and the framer, but also between government agencies. Capture could exist at various levels that divert efficiency in favor of agency incentives for interest groups' objectives. A question is raised whether a centralized or imported set of regulations would reduce moral hazard problems brought about by multiple agents (Estache and Martimort, 1997; Laffont and Martimort, 1994; Martimort, 1996).

For example, Mexico's competition law did not make any provision for the review of privatization transactions and concessions, generating ambiguity in the treatment of sectors such as banking, airlines, telecommunications and other infrastructure. Chile exempts from its competition law (D.L. 211/1973) all SOEs or public institutions that have been granted legal monopoly, hence conflicting with sectoral regulators. In the case of Venezuela, its new law to promote free competition, which was enacted in 1992, is broad and does not make provisions for privatized SOEs. Peru has been very active in updating its competition law enacted in 1991, which covers both private and public firms. Brazil recently created an antitrust framework in 1994, by a new law on repression against economic order (8.884), but enactment on privatized SOEs is minimal. Argentina's amendments and update of competition law are also recent, and according to Rowat (1995) the new draft of the law does not create enough operational enactment, even if criminal prosecution is included in its law, as opposed to other LAC legislations.

Referring now to independence, not all administrative agencies in charge of investigation, prosecution and adjudication, are independent from the executive or the ministry doing policies. Rowat (1995) points out nonindependence in Mexico, Venezuela, Peru and Brazil in terms of budget determination and appointment by the executive, instead of the agency being financed by a percentage of fines collected. Most agencies implement fines but are somewhat separated from the judiciary. For example, penal codes in Mexico apply to economic wrongdoing basically on fiscal fraud, but not on predatory market conduct, where fines are the instruments of the agencies. The relationship between regulated firms and government agencies is also characterized by massive amounts of appeals or *amparos*. As presented above, criminal elements are present in the Argentinean law, as well as in Chile. However, no criminal prosecution has been enacted in Chile (Rowat, 1995). As for Venezuela, its antitrust or competition law is implemented by an autonomous agency not dependent from the executive, and typically sends decisions to the courts to prosecute and adjudicate. Peru's agency is independent to the point of integrating its own court of appeals, separated from the judiciary.

Enforcement problems stem mainly from the privatization-then-regulatory reform structure. However, Chile has the longest record of enforcement on conduct (mainly discrimination problems); Mexico's case is recent, but enforcement on monopolization in telecommunications, tied sales in banking, and the airline merger conditioning are salient; in Argentina, the respective commission has been quite passive in enforcement, since only 49 out of 285 investigations led to prosecution between 1980 and 1992, according to Rowat (1995). Only the Mexican competition commission has acted as advisor in privatization issues affecting competition.

As a conclusion to this section, it is apparent that regulatory frameworks are lacking in many cases of privatization and networked industries with market power. Moreover, competition policy and institutions are only recently moving in the direction of oversight and dispute resolution. Countries with wide economic bases and where trade and capital liberalization have occurred are more pressed to modernize regulatory frameworks. Determinants of regulatory advance would include, apart from economic pressing bases, the depth of privatizations, sectors covered, degree of international opening to serve as a precondition or push towards modernizing regulations and, finally, political settings.

THE CASE OF TELECOMMUNICATIONS SECTORS IN CHILE AND MEXICO

In analyzing incentive regulations, the power of incentives of regulatory contracts affects the rent-seeking behavior of the regulated economic entities. High-powered incentives promote price reduction, investment and cost effort, quantity produced and quality in the market, but also leave informational rents to the regulated firm. In contrast, low-powered incentives do not favor a firm's effort but possibly will reduce informational rents accrued to the firm. A trade-off then exists between framer and regulated company as to what type of rent should be allocated.

However, transaction costs exist not only because markets in LAC and other emerging economies are small to exert market discipline, but also because regulators participate in contract renegotiations (with high administrative and informational costs), are subject to capture by dominant firms under regulations and are not traditionally accountable before consumers or even the legislative powers, as is now the case in developed countries. Moreover, institutional development to face market conduct around the wave of privatizations is uneven or lacks know-how, independence of decisions and budgetary autonomy. Such an environment is even more complicated if one accounts for transactions costs of multiple principals, as emphasized by Baron and Besanko (1992) within the New Regulatory Economics literature.

This section analyzes how regulatory processes have developed in Chile and Mexico, two of the main open LAC economies. They differ in that in the first case, the central government during the 1970s implemented a broad institutional and regulatory reform, preceding its second privatization wave; whereas in the latter, no such strategy was planned. Then an evaluation of how transactions costs of regulating privatized SOEs in telecommunications is done, and a model of performance variables of this sector compares the differences in regulatory approach.

For the telecommunications sector, the process of regulation is a complex one. The main reason is that the sector has emerged as one with multi-applications and multitechnology, and where the traditional viewpoint of natural monopoly has been eroding. Indeed, regulation of telecommunications SOEs had focused on the wire-based telephony and the network providing point-to-point communications. Other segments were emerging in the 1980s, and a system of multicarrier services with no limitations on market access either for infrastructure or services has replaced the concept of natural monopoly. Regulators then had to design regulatory policies that promoted competition across technologies, some integration of the various segments and piecing out of services and accounting methods to reduce cross-subsidization by leading firms (OECD, 1997).

The lessons from introducing competition in telecommunications have shown that market restructuring, redefinition and the elimination of entry barriers is a process where regulators require a transition period to attain competitive markets. For this reason, time inconsistencies, contract renegotiation and credible commitments of framers versus firms give rise to transaction costs and deviations from internal or allocative efficiency. Hence, the institutional and procedural setting of regulations is important, and extends beyond the existence of a regulatory agency with powers, to amendments to constitutional, commercial or civil laws, and for agencies to solve disputes and adjudicate rights. For example, courts exist in Chile for dispute resolutions, but the total budget assigned to them by the central government is less than 1%. For Mexico, it is higher but concentrated in police matters, not in economic cases (Manriquez-Reyes, 1995). Regarding the so-called autonomy of the respective oversight commissions, Chile distinguishes between an oversight agency or *comisión preventiva* and a resolution commission, and uses a well-developed system of regional courts for adjudication. In Mexico, no such separation exists at the three levels. However, cases of illegal conduct on entry conditions in Chile's recent history have involved simple price discrimination or exclusive dealing, whereas in Mexico, recent cases have focused on more complex matters of predation. This testifies to the complicated dynamics of regulatory change in Chile and Mexico, realities that only can be accounted for indirectly by models of economic conduct of firms subject to regulations.

Nature of the Data

Chile privatized its telecommunication sector in 1986-1987, once the state monopoly was divested into two regional service companies (CTC and Entel), and privatization was implemented after its regulatory reform in 1978. Mexico's Telmex was privatized in 1991, maintaining a regulatory shield on the privatized enterprise. Using the International Telecommunication Union (ITU) database for these countries in annual data before and after privatization as a structural change point (date of announcement and date of implementation), a time-series model of lag responses was used comparatively, for two sets of variable vectors: an "effort" vector, and a "commitment" vector. The two sets of variables respond to agency problems in the imperfection of contracts brought about by the regulatory change. Two variables account for effort: a direct measure of teledensity, which implies market access; and an indirect measure of cost effort, accounted for by the ITU measure of waiting lines. The hypotheses are that competition pressure and risk of takeover will increase investment and market coverage, which will also become prevalent as a company's strategy. As for variables reflecting commitment by firms subject to deregulation—or lack of it—the percentage investment on sales, a variable often used by both ITU and OECD as a measure of strength of commitment, is used. The hypothesis is that under deregulation, commitment works as a competitive signal by the incumbent firm, and will show few jumps in such a variable. The second variable relating to commitment is internal efficiency pursuit, represented by net worth. It is hypothesized that shareholders value the conduct-performance of the company mainly referring to this variable.

Analysis of Results

With these tentative hypotheses, a set of models with leads and lags was run, under unrestricted distributed lag econometric methods, where response functions are directly tested for structural change after privatization choice was implemented. A Vector Autorregression model (VAR) directly tests for the response in each dependent variable, to changes in its structural determinants and a time variable that takes the leads and lags. Other structural variables can also present a lead or lag structure. Then the VAR model tests for the time response of the dependent variable in each run, called impulse VAR decomposition, and also the cointegration of variable responses under various assumptions (Greene, 1993). The model pursued here only tests for the impulse response of each dependent variable in each country, under the heading of effort, and commitment.

The sample between 1985 and 1996 was used with data for Chile and Mexico from the International Telecommunication Union data base (ITU, 1997). F-tests and Akaike's information criterion were used to check for the time responses determined (Greene, 1993). Table 7.2 shows the t values and the statistical significance of the responses for both cases, since the quantitative impact is of less importance.

Table 7.2
Comparative Analysis of Response to Privatization: Telecommunications in Chile and Mexico (lag structure of responses by variable)

PERIOD/ VARIABLE	after			year of privatization		before	
	t-3	t-2	t-1	t-0	t+1	t+2	t+3
EFFORT VARIABLES							
TELEDENSITY							
Chile	2.48**	1.92	1.64	1.63	1.24	-	-
Mexico	2.96**	4.82**	6.07**	5.58**	4.97**	3.72**	3.10*
WAITING LINES							
Chile	-.69	-.24	-.16	0.24	0.46	-	-
Mexico	-2.82*	-6.78**	-5.11**	2.78*	-1.86	-.94	-.56
COMMITMENT VARIABLES							
INVESTMENT							
Chile	1.97	4.28**	2.87**	2.18*	1.45	-	-
Mexico	0.18	1.08	3.13**	4.41**	4.39**	3.59**	2.12
REVENUES							
Chile	2.17	2.33*	2.08	1.86	1.17	-	-
Mexico	1.96	3.82**	5.74**	7.42**	6.02**	3.53**	2.32*

Note: t values. * significant at 5%; ** significant at 1%.

For the first two rows in the table, teledensity in Chile as a response to privatization, once regulatory reform was in place in that case, shows a positive significant increase with a lag of three years, whereas for Mexico, teledensity increased significantly around the year of privatization, but extends in the lead variable before privatization was implemented. The significant increase could be taken as the impact of the regulatory shield for TELMEX, implying that a credible shield might work as an effective trigger for teledensity, or that the shield effect was stronger than possible rent-seeking behavior of the legal privatized SOE with monopoly power. After the sample, Chile has continued to increase teledensity, but Mexico has lagged behind Chile and other LAC economies.

As for the results of effort, represented by the waiting list of mainline reduction, Chile shows no significant correlation of the variable with respect to privatization choice, whereas Mexico's behavior is markedly different, as significant values show after privatization, with one- and two-period lags. One rationalization is that Chile's telecommunications companies showed a low percentage level of waiting lines before, during and after privatization, given that competition and regulatory reform were implemented before, whereas

Mexico again implied the use of the regulatory shield along with compulsory concession clauses that significantly reduced waiting lines. It is, however, unexpected that the response is highly significant only in two years, implying tentatively that improving behavior is short-lived. Indeed, after the regulatory shield was granted, the incumbent firm could have been aware that new competitors would not be strong rivals in the servicing of new or pent-demand, and that waiting-line contests would not occur. Again, the regulator has been passive in pushing firms for contests of servicing the market, but has concentrated on dispute adjudication.

Now, investment response proves to be highly significant in Chile immediately after privatization (two period lags), whereas in Mexico the company significantly invested even before privatization (two period leads), at the same time that the variable was only significant in the year of privatization and the following year, again showing a short-lived effort with respect to the Chilean case of market competition and regulatory oversight. Finally, the proxy for internal efficiency and performance (revenue response) shows significantly in Chile with a five-period lag (not shown), but significant at the 5% level in only the two-year lag. In contrast, the Mexican sector/company shows significant revenue increases even two years before privatization, and after privatization, with a two-year lag.

CONCLUSION AND IMPLICATIONS

In situations already experienced of privatization-deregulation reversals, credible commitment to policy changes after a temporary regulatory shield might converge towards economic efficiency, as established by traditional theory, but where imperfection of contracts raised by less-than-optimal regulations towards market behavior are an important element of the privatization dynamics. Regulatory oversight agencies in Chile and Mexico differ. In the former case, cross-controls between preventive and resolution commissions exist, but the type of cases around simple price discrimination, consumer protection and others have overwhelmed the agencies, hence making them inactive around dominant firm problems, entry conditions, disputes on imperfect contracts that need a framer, and oversight of a disciplined market. In Mexico, the respective commission is recent and operates after privatization. CFC comprises prevention, resolution and even adjudication. These facts shed new light on the dynamics and powers of a regulatory framer of telecommunications, as one of the key sectors subject to privatization in LAC.

The last analysis takes account of the different leads and lags of economic conduct and performance of the two sectors in Chile and Mexico. In the recent process of multilateral negotiation, the quest is for regulatory mechanisms across the globe, to be moving towards convergence. However, multilateral institutions such as the World Trade Organization's General Agreement on Trade in Services (GATS-WTO), World Bank, or even the Free Trade Area of the Americas (FTAA) initiative, can only be tangential in overcoming regulatory lag and extensive shields in recent privatization efforts in LAC. Problems of moral hazard and multiprincipals are extensive, making it costly for new firms entering networked industries to play under standard rules of the game, and where dispute-resolution mechanisms are effective to allocate contract imperfection problems. A policy strategy would then make it necessary to apply both at the domestic level, enhancing market allocation and disputes as well as promoting competition through new entry.

Internationally, multilateral efforts for convergence of government regulations towards networked industries or industries with market power seem to be key but insufficient. A system of courts has been considered a first important step for LAC regulatory modernizations. The argument is, however, whether incentives to modernization would come through domestic modernization policies, regional treaties, or through multilateral agencies. That the decision will depend on the depth of impact by each of these options (something that is out of the scope of this research) is key in designing the regulator's rules. In sum, however, time inconsistencies are prevalent in the game between a government principal offering temporary regulatory shields after privatization and incumbent firms. Moreover, courts and the very process of investigation, adjudication and settlement of predatory conduct is the first needed step in regulatory modernization in LAC economies. The use of telecommunications as a case, showing the privatization-regulatory changes as key ingredients affecting both effort and commitment variables of networked sectors of economic activity, can be extended to other important industries characterized by externalities and dominant firms. Such is the case for transport, other physical infrastructure, energy, and some technology-intensive sectors. These are fundamental in the dynamics of developing economies that seek to position themselves among leading countries in the beginning of the 21st century.

REFERENCES

Baron, D. and D. Besanko (1992), "Information, Control, and Organizational Structure," *Journal of Economics and Management Strategy*, 1: 237-276.

Buscaglia, E. (1996), "An Economic Analysis of Legal Integration in Latin America" (mimeo). Palo Alto: Stanford University, Hoover Institution.

Center for International Private Enterprise (1997), "Measuring Economic Freedom", Washington, DC.

CFC (1997), *Informe Anual de la Comision Federal de Competencia 1996 (Annual Report of the Federal Commission for Competition)*. Mexico, online at www.cfc.gob.mx

DeRosa, D. A. (1998), "Regional Integration Agreements: Static Economic Theory, Quantitative Findings, and Policy Guidelines," background paper for World Bank (Policy Research Report), *Regionalism and Development*. Washington, DC: International Trade Division.

Estache, A. and D. Martimort (1997), "Transaction Cost Politics and Regulatory Institutions," Preliminary Version (mimeo). Washington, DC: The World Bank.

Greene W.H. (1993), *Econometric Analysis*. New York: Macmillan.

Guasch J.L. and P.T. Spiller (1994), "Regulation and Private Sector Development in Latin America." Washington, DC: The World Bank, internal document, November.

Holden, P. and S. Rajapatirana (1997), "Emerging Economic Issues in Latin America: A Second Generation Agenda." Washington, DC: The World Bank Special Report, June.

ITU (1997), ITU World Telecommunication Indicators Database (disk). Geneva: ITU.

Laffont, J. J. and D. Martimort (1994), "Separation of Regulators against Collusive Behavior" (mimeo). Washington, DC: DEI.

Manriquez-Perez, L. (1995), "Modernization of Judicial Systems in Developing Countries: The Case of Chile," in M. Rowat, W.H. Malik, and M. Dakolias, eds., *Judicial Reform in Latin America and the Caribbean: Proceedings of a World Bank Conference*. World Bank Technical Paper Number 280. Washington, DC: World Bank.

Martimort, D. (1996), "The Multiprincipal Nature of Government," *European Economic Review*, 40: 673-685.

Mexico (1992), *Tratado de Libre Comercio de America del Norte (North America Free Trade Agreement)*, Mexico, Secretaria de Comercio y Fomento Industrial.

Moran, M. and T. Prosser (eds.) (1994), *Privatization and Regulatory Change in Europe*. Buckingham: Open University Press.

OECD (1997), *Communications Outlook*, Vols. I and II. Paris: OECD.

Ramamurti, R (1996), *Privatizing Monopolies: Lessons from the Transportation and Telecommunication Sectors in Latin America*. Baltimore: Johns Hopkins University Press.

Rowat, M.D. (1995), "Competiton Policy in Latin America: Legal and Institutional Issues" (mimeo). Conference on Good Government and Law, London, March 27-28.

World Bank (2000), *Regionalism and Development*. Washington, DC: World Bank.

PART II:
NETWORKING FOR REGIONAL
ECONOMIC DEVELOPMENT:
LOOKING FOR INCLUSIVENESS

Introductory Note

Manuel V. Heitor

The previous chapters of this book show that it is thus legitimate to question the traditional way of viewing the role that contemporary institutions play in the process of economic development and to argue for the need to promote systems of innovation and competence building based on learning and knowledge networks. This broad concept has motivated the material included in Part II of this book, which integrates a set of new contributions addressing complementary aspects of relevance towards improved understanding of local conditions favouring learning societies.

Chapter 8 by Bessant addresses the specific context of small- and medium-sized enterprises (SMEs), in that they represent an important element in any economy for at least two reasons: (1) they are responsible for growth in employment and activity; and (2) they are support to large firms. But the author focuses on the specificities of SMEs in terms of the conditions affecting their growth and the gap still existing for their full enrolment on national systems of innovation. In this context, six different mechanisms are reviewed, as experiments around the question of integrating SMEs in innovation systems, namely: networks and clusters, learning networks, supply chain learning, intermediaries/innovation agents, structural connections, and transfer of skills.

In chapter 9 Baptista examines the significance of localized agglomeration externalities and increasing returns, and how they relate to spatial differences in labour productivity. He uses data from 64 counties in the United Kingdom in the late 1980s and early 1990s and estimates that doubling employment density in a county would increase average labour productivity by almost 7%. Locally increasing returns explain more than 25% of the variation in labour productivity across counties, after education level of the work force is controlled for.

Different institutional approaches to regional economic growth are further presented in the following chapters. Chapters 10 and 11 address issues related with European regional policy over the 1990s, with emphasis on related instruments and incentives. While Guth in chapter 10 focuses on transnational issues related with the increasing gaps across Europe and the relative failure of European regional policies, Corvers in chapter 11 describes and discusses one of the most important incentives (the RITTS scheme). Both authors claim for the need to understand innovation beyond a linear perspective, but also for a better consideration of local specific conditions. The continuous involvement of local actors is presented as an important challenge facing the making of regional innovation systems.

In Chapter 12 van Geenhuizen discusses the specific case of the Emmen region in the northern part of the Netherlands. The analysis is focused on the use and mobilization of information and communication technologies (ICTs), which has led the author to argue about the need to improve the learning capability of local actors through a bottom-up interactive way.

In chapter 13 Kim discusses the dynamics of research networks in Korea, namely with SMEs, and argues for the stringent need of valuing informal networking through trust and frequency of interaction. The specific context of Latin America is discussed by Bortagaray and Tiffin in Chapter 14. The authors bring to our attention the question of the necessary time process of building innovative clusters, namely in terms of the "mechanisms that promote the invisible parts of community and integration". Existing industrial clusters may play a critical role in this process, although social barriers limiting cooperative arrangements deserve special attention for further studies.

Broadening our analysis to the processes within specific firms establishing leading industrial clusters, Casanueva explores in Chapter 15 the effects of globalization and industrial restructuring in two of the most dynamic Mexican industries, namely the automotive and electronic sectors. The analysis relates to the restructuring of the organization of production, which demands a complex arrangement of suppliers, together with quality control strategies.

The three final chapters of this part of the book focus on the question of establishing regional systems of innovation in regions in initial stages of development. Quandt and Spinosa consider the case of software companies in the state of Paraná in Brazil, while Nascimento uses a survey in the Brazilian food sector to argue about the nature of existing networks and the role institutions play in the development process of industrial clusters. Heitor and Moutinho consider the process of mobilizing the information society in Portuguese regions and raise questions associated with the concept of social capital, as a relational infrastructure for collective action. Their analysis is presented in terms of a "knowledge-based view of the territory to foster institutionally organized metropolitan systems of innovation and competence building". Competence building is considered in terms of a dynamic and broad social and economic context associated with digital networks and the analysis suggests "the need for continuous public support and monitoring, as well as for the promotion of knowledge integrated communities as drivers of larger communities of users".

The various aspects summarized above include heterogeneous approaches to regional development and technological innovation, but consider *change* at the centre of the analysis. This is considered throughout the various chapters included in Part II of the book, but taking into account that firms' competencies are characterized by *stability and inertia* and, therefore, *lock-ins and competence traps* are expected to occur, in that successful firms may be driven by their success in existing technologies to disregard new alternatives. Another important aspect to take into consideration is that the phenomena of *increasing returns* and *path-dependence* affect the nature of the innovation processes and the dynamics of industries in any regional context.

Among the various aspects raised throughout this part of the book, it should also be noted that the *sectoral specificity* in the organization of innovative activities, on one hand, and the *specific characteristics of local systems of innovation*, on the other hand, are expected to play a significant role in shaping the organization of innovative activity in any region. The prevalence of one effect over another depends on history and competitiveness of firms and their degree of internationalization.

Based on the dynamics of localized technological change, as defined by Antonelli and Calderini (1999), the internal bottom-up learning process based upon the improvement of design and technological processes plays a major role in feeding the continual introduction of technological and organizational innovations. In this respect, these authors concluded *that technological knowledge is embedded in the specific circumstances in which firms operate*, and *its generation is the result of a joint process of production, learning and communication*, of which R&D activities are only a part. In these terms, current evolutionary economics has shown the importance of path dependence of economic processes, in that it is at the core of selection mechanisms between competitive firms and technologies (Metcalfe, 1994). Competition is therefore the result of *the rate of change of market share*, apart from being dependent on differences in the rates of growth of individual firms. The result is a fully endogenous process, which, in the presence of increasing returns, gives rise to a strong interdependence between *specialization and diversification*. The direct implication for regional innovation policies is the important, but limited role of demand at the firm level in assessing the amount of incentives for firms to introduce technological innovations. In more general terms, analysis call for the need to feeding all the processes of learning (formal and informal, as defined by Conceição and Heitor, 1999), implementing technological cooperation among firms and between firms and research institutions, and on the process of on-job-training of the workforce. Technological centres specifically designed to sustain localized processes of technological change might play an important role in this context. However, it is important to clearly emphasize the important role of the science and technology system, in fostering innovation, as well as the related implications for public policy.

REFERENCES

Antonelli, C. and Calderini M. (1999). "The dynamics of localized technological change", in A. Gambardella and F. Malerba (eds.), *The Organization of Economic Innovation in Europe*. New York: Cambridge University Press, pp. 158-176.

Metcalfe, J. S. (1994). "Competition, Fisher's Principle and Increasing Returns to Selection", *Journal of Evolutionary Economics*, 4, pp. 327-349.

Conceição, P. and Heitor, M. V. (1999). "On the Role of the University in the Knowledge Economy," *Science and Public Policy*, 26(1), pp. 37-51.

8

Getting the Tail to Wag: Enabling Innovation in Small/Medium-sized Enterprises

John Robert Bessant

INTRODUCTION

Small firms matter. In any economy small and medium-sized enterprises (SMEs) make up the majority of firms and account for most employment.[1] And while it is often larger firms which make the headlines their activities depend upon a wide and extensive range of SMEs supplying goods and support services.

Growth in small firms is also important. Although the evidence suggests that some types of SMEs are more significant in this (for example, new technology-based firms NTBFs) than others, the fact remains that a significant element of economic growth comes through SMEs.[2] All of which suggests that we should pay them particular attention in research and policy making on innovation—especially given the high failure rates of SMEs which try to innovate. (For example, in a study of world-class manufacturing performance of SMEs in Europe, only 1% of the firms surveyed met the world class benchmark performance criteria.)[3]

Yet the fact is that we know surprisingly little about SMEs and innovation—and although policy making has been evolving away from simplistic and universal measures addressing all SMEs towards more targeted approaches, there is still a lack of knowledge about the nature and extent of SME support needs and the mechanisms for delivering it effectively.[4] The result is that the policy environment is characterised by a wide range of experimentation, in both design and delivery mechanisms.[5]

Part of the problem is that the term is used as a very broad umbrella under which many differences exist. For example, there is clearly some distance between the kind of SME involved in simple services like hairdressing or single-shop retailing and the new technology-based startup of a biotechnology firm. Innovation in one is (almost) an incidental item, whereas in the other it is the essence of the business. Even a simple binary distinction like this excludes many SMEs and allows a clearer focus on the remainder. This problem has been recognised by many commentators, who argue for a much more differentiated approach to technology policy, building on improved understanding of specific character-istics of different sectors and firm types.[6]

A second problem in looking at SMEs and innovation is the assumption that innovation equates to formal research and development (R&D) activities. While this can be tracked in ways suited to external measurement—levels of invest-ment, number of qualified scientists and engineers (QSEs), patents registered, and so on—the problem is that this kind of innovative activity is not very common among SMEs. Most are not creators but *users* of technology and so a prime concern is that of effective technology transfer. (There are exceptions, especially in the high-technology field, but even here SMEs are likely to carry out limited and highly focused R&D; there will still be a need for technology transfer for complementary knowledge and equipment.) Too often arrangements for technology transfer assume some version of the linear model of innovation, whereby demand is clearly communicated or where new technology has enough momentum to find its way to SME users.[7] The reality is that many SMEs lack the capability to understand and articulate their needs, and rarely scan for sources of new technological opportunity.[8] Even those that have an awareness of their needs may lack the information or capability to find and access sources of technology.

Potentially SMEs have a set of advantages: fast decision-making enabled through simple and often informal structures, creative climate, and the like.[9] In practice these are neutralised or otherwise outweighed by a suite of problems including those highlighted in Table 8.1.

For this reason it can be argued that SMEs, although potentially innovative, are often weak and ineffective in their management of the process—and hence inhibited in their growth potential. But SMEs are important not only for their direct impact on growth but also because of their impact on large firms through the operation of supply chains. (In the case of inward investment it has also become clear that strong local SMEs can play an important role in helping to tie down transnational firms which might otherwise move location after the initial benefits of tax holidays and other incentives have finished.) So improving innovative capability becomes an issue of policy concern because of the relatively poor performance of many SMEs.

Table 8.1
Some Problems in SME Innovation

Problem area	Symptoms
Nonstrategic orientation	SMEs too busy fire-fighting to address long-term development issues. Often narrowly and locally focused.
Lack of management skills	Weaknesses in areas like strategic planning, operations strategy, marketing, finance, etc.
Lack of awareness	Insular and local focus means they may not see the need, direction or scale of change required. Little benchmarking or external comparison or learning.
Isolation	SMEs are unaware of external resources or how to access them, particularly in the case of the technology infrastructure
Articulation	Firms are unaware of, or unable to identify, key strategic problems, or to separate out symptoms from root cause problems
Lack of resources	Insufficient organisational slack to enable innovation

THE CHALLENGE FOR THE NATIONAL SYSTEM OF INNOVATION

It has become common to talk about the national system of innovation (NSI) and under this heading we might consider the various interlocking elements which provide support and context for innovative activities in a particular country.[10] These might include national and regional policy, the education and training infrastructure, the financial system, the network of technology resources (such as universities and research and technology organizations [RTOs]), and such. In the case of SMEs it can be argued that there is a measure of market failure, a gap between their needs and what the NSI is able to provide which needs closing. Table 8.2 gives some examples.

The existence of this kind of mismatch is a common problem across all countries, and wrestling with it from a policy perspective within the NSI is not a new concern. This chapter is concerned less with a critique of existing measures (except to flag that many of them suffer from being blunt instruments because of the lack of underpinning research perspective) but rather with an examination of some experiments towards closing the gap.

Table 8.2
Some Examples of Gaps Between SME Needs and NSI Provision

NSI element	SME gap
Financial system	Lack of availability of investment capital—much SME innovation is not on the scale of venture capital but too risky for banks
Labour market	Skills availability is limited, and scarce skills are too expensive for SMEs. If they train in-house there is a risk of staff being poached by larger (and richer) firms.
Education system	Lack of key skills in SMEs—e.g., in management and in key technological areas—but provision is often inflexible so SMEs cannot access it
Technology infrastructure (universities, RTOs, etc.)	Tradition of technology push—technology will find its own way to SMEs. Weak marketing and what there is is not SME focused. Where need pull operates, it comes from articulate players—but this often excludes SMEs who may not be aware of their needs, may not be able to articulate these and may not know to where they should address their queries.
Policy system	Although SMEs are a prime target there is often little differentiation into different types of SME—consequently policies are often blunt instruments

Arguably there is a need for innovation in the NSI and its mechanisms to help close the SME gap. Evidence suggests that most economies are characterised by a small number of high-performing world class firms and a long tail of less effective organisations (the bulk of which are SMEs) so the challenge becomes one of "getting the tail to wag".

GETTING THE TAIL TO WAG—MECHANISMS FOR IMPROVING INNOVATION WITHIN SMES

The market failure which this implies has been recognised by many policy makers and various attempts have been made to bridge the gap through the use of different mechanisms. Some of these involve intermediary structures or roles.[11] Examples include the use of innovation consultants of various kinds, whose role is to act as a counsellor, helping firms identify and articulate their needs, and also as a broker, enabling them to access the most relevant sources of support and assistance (often ranging from financial through to technological.[12]

Other options include the use of structural mechanisms to close the gap— for example, by providing specialist information centres, or specific financial support to encourage R&D or technology transfer activity within SMEs.[13] In other cases policy efforts are directed towards a particular technology and make use of several approaches to close the SME gap—often linking financial support with inputs to enhance the likelihood of successful implementation rather than simply adoption of new technology.[14]

It is also worth noting that in some cases SMEs, recognising these limitations, have organised themselves into networks and clusters which demonstrate elements of 'collective efficiency'; examples include the *consorzia* of Italy and the industrial districts in Spain, Germany, Pakistan and Brazil.[15] Several public-sector policies have been designed to promote networking of this kind.[16] Carney discusses the role of structural service agencies working within industrial clusters to help facilitate development and innovation among SMEs.[17]

In different ways these mechanisms all recognise the underlying difficulty confronting SMEs. There appears to be growing recognition of the limitations of simple linear models of innovation and there is considerable interest and sharing of learning around experiments with alternatives. Here we review six such examples.

Networks and Clusters

Based on the experience of small firms in regions like middle Italy, northern Spain and Jutland in Denmark, this model has become widely explored as a potential route for SME development. Studies of collective efficiency have looked at the phenomenon of clustering in a number of different contexts.[18] From this work it is clear that the model has diffused around the world and, under certain conditions, has proved to be extremely effective. For example, one town (Salkot) in Pakistan plays a dominant role in the world market for specialist surgical instruments made of stainless steel. From a core group of 300 small firms, supported by 1,500 even smaller suppliers, 90% of production (1996) was exported and took a 20% share of the world market, second only to Germany. In another case the Sinos Valley in Brazil contains around 500 small-firm manufacturers of specialist high-quality leather shoes. Between 1970 and 1990 their share of the world market rose from 0.3% to 12.5% and they now export some 70% of total production. In each case the gains are seen as resulting from close interdependence in a cooperative network.

The underlying principle of collective efficiency represents a viable response to the challenge that small firms are necessarily weak. As one commentator put it, the problem is not that small firms are small but that they are isolated.

The cluster model has become the cornerstone of many industrial policy approaches and it lends itself to policy making at a national strategic level.[19] Its relevance to the innovation question is clear—while firms may gather together to share resources, access to markets, and so on, there is clearly a considerable element of shared learning and experimentation which can take place. The case of CITER in Italy is a good example here. The predominantly small scale of firm operations in the fashion textile sector meant that none could afford the design technology or undertake research into process development in areas like dyeing. A cooperative research centre—CITER—was established and funded by members of the *consorzia* and chartered with work on technological problems related to the direct needs of the members. Over time this has evolved into a world-class research institute, but its roots are still in the local network of textile firms. It has become a powerful mechanism for innovation and technology transfer and has helped to upgrade the overall knowledge base of this sector.[20]

Learning Networks

A key feature of successful firms is an ability to adapt and learn to deal with their rapidly changing and uncertain environments. Research suggests that there are two important components involved in such learning; the first involves the accumulation and development of a core knowledge base—the core competence —which differentiates the firm from others and offers the potential for competitive advantage. Acquiring this is not simply a matter of purchasing or trading knowledge assets but the systematic and purposive learning and construction of a knowledge base.[21]

The second is the long-term development of a capability for learning and continuous improvement across the whole organisation. Recognition of this need has led to growing emphasis on the concept of learning organisations and on the mechanisms through which this capability can be developed.[22] One aspect is the possibility of gaining support for the learning process through working with others in what we term learning networks.[23]

There is much discussion of learning in organisations, but we can draw out a number of common themes, including the following:

- Learning can be viewed as a cyclical process (see Figure 8.1), involving a combination of experience, reflection, concept formation and experimentation.[24]
- Learning takes place only when the cycle is completed—thus much effort an activity in one or more quadrants may not necessarily lead to learning.
- Learning is not automatic—there must be motivation to enter the cycle.
- Learning needs to be purposive and can be supported by structures, procedures, and the like, to facilitate the operation of the learning cycle.
- Learning to learn—metalearning—is an important aspect of this—learning to design and operate learning systems. [25]
- Learning involves the accumulation and connection of data into information and knowledge.
- Learning involves both tacit and formal components, with the task being to capture and codify, to make explicit.[26]

- Learning may take place in adaptive mode—learning to do what we do a little better—or it may involve reframing and radical change (what some writers term a paradigm shift) in which the perception of the problems to be solved and the potential set of solutions change.

Figure 8.1
Kolb's Cycle of Experiential Learning

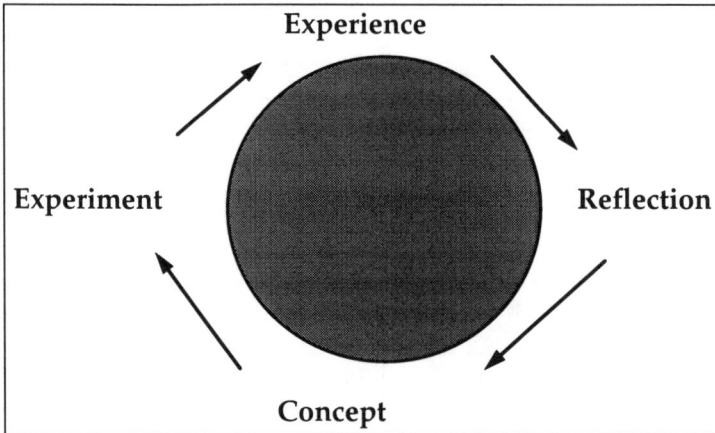

The basis of most learning literature is in individual learning, but recent years have seen a strong focus on the concept of learning organisations. There is much debate about whether organisations themselves actually learn or whether it is simply the individuals within them which do.[27] However it appears that learning organisations can exist and several mechanisms appear to help with this process of sharing and making knowledge explicit; these include exchange of perspectives, shared experimentation, display, measurements, and so on.[28] At their heart they represent ways of supporting and developing a shared learning cycle.

Learning is not automatic and there are a number of sites at which learning fails to happen unless a blockage is dealt with. For example, many firms stumble at the first hurdle by failing to recognise the need to learn, or by recognising the stimulus but choosing to ignore or discount it. (This phenomenon often gives rise to the "not invented here" problem which is commonly seen in the field of technological change.) Others may recognise the need for learning but become locked in an incomplete cycle of experiment and experience, with little or no time or space given to reflection or to the entry of new concepts. For others the difficulty lies in organising and mobilising learning skills, while in other cases the difficulty lies in making use of the rich resource of tacit knowledge—things people know about but are unable to describe or articulate.[29]

Table 8.3 summarises key blocks to learning which are particularly relevant to SMEs.

Table 8.3
Barriers to Effective Learning

Learning blocker	Underlying problem
Lack of entry to the learning cycle	The motivation problem
Incomplete learning cycle	The completion problem—understanding and support for all phases
People don't know how to learn	The skills problem
Learning is tacit, hidden, informal	The elicitation problem
Search for new solutions is too localised	The parochial/ not invented here problem
Reflection is undemanding	The challenge problem
Learning is infrequent, sporadic	The reinforcement /reward problem
Learning is not shared but localised	The sharing problem
Learning is not sustained	The motivation problem

Most of the learning literature relates to intraorganisational processes but there is a strand concerned with interorganisational learning—learning with or from others. The advantages associated with this approach are similar to those which relate to group/interpersonal learning, and can address the problems identified above. One of the major contributions here has been the development and implementation of the concept of action learning, pioneered by Reg Revans. This concept stresses the value of experiential learning and the benefits which can come from gaining different forms of support from others in moving around the learning cycle. Part of his vision involved the idea of comrades in adversity, working together to tackle complex and open-ended problems.[30]

The potential benefits of shared learning include the following:

- in shared learning there is the potential for challenge and structured critical reflection from different perspectives
- different perspectives can bring in new concepts (or old concepts which are new to the learner)
- shared experimentation can reduce risks and maximise opportunities for trying new things out
- shared experiences can be supportive, confirmational

- shared learning helps explicate the systems principles, seeing the patterns—separating the wood from the trees
- shared learning provides an environment for surfacing assumptions and exploring mental models outside of the normal experience of individual organisations—helps prevent "not invented here" and other effects

Thus it is possible to argue that there may be value in designing and building networks which offer some form of traction on the learning process which organisations need to operate. This concept of a learning network can be expressed as: "a network formally set up for the primary purpose of increasing knowledge, expressed as increased capacity to do something". This definition implies a number of features:

- formal setting up, rather than informal evolution
- a primary learning target (what learning/knowledge is the network going to enable?)
- a structure for operation, with boundaries about who is in and who is outside
- processes which can be mapped on to the learning cycle
- measurement of learning outcomes which feeds back to operation of the network and which eventually decides whether or not to continue with the formal arrangement

Learning networks of this kind are beginning to be used widely and can be organised in a variety of ways; Table 8.4 gives some examples.

Table 8.4
Types of Learning Networks

Basis of networking	Learning focus
Regional	SMEs in same area work together on shared problem issues requiring innovative response.
Sector	SMEs in same sector work together on common problems—for example, productivity improvement. This approach lends itself to benchmarking as an initiator of change.
Topic	SMEs from different sectors and possibly geographically separated work together on developing improved understanding and deployment of a particular technology or approach.
Supply chain	Variant on the above using the structure of the supply chain and the underlying sense of shared purpose as a motivator for change

As one of many examples of learning networks the Irish World Class Manufacturing (WCM) experience provides an interesting case of governments cooperating with intermediary institutions in both the academic and private sectors to promote learning.[31] The impetus behind this programme arose out of the desire to increase the adoption of WCM among SMEs in Ireland during the early 1990s. The key weaknesses diagnosed were how to transfer knowledge to SMEs, how to present complex concepts to management in easily digestible ways and how to enhance the capacity of SMEs to adopt these changes effectively. What emerged was a four-way partnership between four sets of actors:

1.   the government agency responsible for SME upgrading (IDA—Irish Development Agency),
2.   a consulting firm with extensive experience of WCM (Lucas Engineering and Systems, now part of CSC—Computer Sciences Corporation),
3.   a university department with a history of successful action-research, and
4.   a willing group of SMEs who were anxious to adopt new techniques to foster their capacity to both learn and change.

The programme which developed worked in the following way. After an initial presentation to a group of 65 SMEs, a *demonstrator firm* was identified. The *consulting company* then provided inputs into the restructuring process, which was closely observed by at least two people from five *participant teams* who then applied similar changes in their own firms. This involved visiting and observing the activities of the demonstrator plant and attendance at programme workshops. The whole process was coordinated by a *learning partnership* which consisted of a university (University of Dublin), the government department (Forbairt, now Enterprise Ireland), a private consulting firm (Lucas Engineering and Systems) and the Irish Congress of Trade Unions). The learning partnership then ensured that this experience was passed on to a wider group of SMEs.

Supply Chain Learning

A variant on the above model of learning networks makes use of the supply chain to provide an enabling structure and shared motivation for learning and specifically upgrading of capability. Despite the potential which supply chains offer for enabling learning, there is, as yet, little research-based information on the topic. Although some well-publicised examples can be cited, it appears from the available literature and other sources that the implementation of supply chain learning (hereafter, SCL) is still in its infancy. That said, the few cases where SCL is operating suggest that it does confer significant competitive advantage on the firms involved.

There are a number of reasons for suggesting that supply chains could provide additional support for learning:

- They involve an identifiable group of firms with a common concern—the competitive performance of the entire chain will depend on the extent to which all members can learn best practice.
- There is the potential to deal with the motivation problem listed in Table 8.3. Here firms can be encouraged to enter the cycle by both reward (the potential of shared gains in the event of successful upgrading) and sanctions (with increasing emphasis on preferred suppliers, those unable to reach the mark may be dropped).
- There is potential scale economy, where learning can take place across many firms sharing the same concerns and needing the same type of learning inputs.
- Major players in the chain—for example, supply chain owners as end customers— are often near the best-practice frontier in terms of their exposure to global standards and competition. They can act as teachers and examples.

In practice, however, there are important limitations to applying this concept:

- We need to be clear about the existence of different types of supply chains.
- We need to recognise that learning is not a natural feature of supply chains. It is part of the emergent new models for such interfirm arrangements which stress trust, cooperation and mutual dependence, and without such underpinning values it is unlikely to happen.
- We need to recognise that supply-chain learning depends on active governance of the supply chain—managing it as an entity.

With the development of new forms of obligational relations,[32] there has been an increasing evolution of *various forms of purposive behaviour designed to enhance systemic efficiency along the chain* to achieve what Schmitz calls "collective efficiency".[33] In the early stages this was focussed on quality assurance and the integration of logistical scheduling,[34] but as this supply-chain cooperation evolved, so the focus has changed to include cooperation in design.[35] Until recently this has been the cutting-edge of supply-chain co-operation, but in very recent years supply chains have also come to be seen as a mechanism to promote learning, including indirect suppliers—one or two positions removed from the protagonist in the supply chain. (In hierarchical jargon, the second and third tiers of suppliers. In fact tiers, with lateral as well as hierarchical positional links, do not appear to exist widely except in Japan and Korea.)

This active cooperation is usually led by a dominant party, a function which is termed supply-chain governance.[36] There are, of course (as will be shown in the following section), various styles of governance, ranging from the dictatorial imposition of standards by the governor to softer forms of exhortation. To some extent these differences are a function of firm style (see below), but in other cases they reflect the type of value chain which is involved.

It is possible to distinguish three major types of value chains:[37]

1. Buyer-pulled chains, in which the coordinating function is performed by a large buyer directly serving a final market. In the U.K. context, leading buyer-driven

chains include those led by supermarkets such as Tesco, and retailers such as Marks and Spencer and B&Q.
2. Supplier-pulled chains, where the coordination function is performed by a firm holding core designs or technology. The most well-known cases here are some of the automobile companies who coordinate logistics, quality, design and learning along their own supply chains, being themselves suppliers (i.e., of vehicles or personal transportation) to the eventual customer.
3. Supplier-pushed chains, where holders of core technologies and designs may also push change upward to customers, both intermediate and final consumers of their products. An example of this is provided in the personal computing field with hardware producers such as Intel and software producers such as Microsoft seeing their customers (e.g., IBM) as parts of their downstream supply chain—a distribution channel.

The case of leading U.K. do-it-yourself retailer B&Q is often quoted as a success story in improving the firm's business through highlighting environmental soundness in its products. In fact it represents a good example of a *buyer-pulled chain*. B&Q invested in an environmental monitoring competence, recruiting a charismatic and energetic champion, who set about tracing supply chains back to source, to discover such things as the provenance of timber (all B&Q wood products are now proven to come from sustainably managed forests) and ethical issues, such as the working conditions of people involved in manufacturing products, especially in developing countries. For example, B&Q traced a varnishing problem on brass doorknobs to poor air conditions in the manufacturing plant in India. By helping the supplier to install a clean air system, B&Q solved the varnishing problem (by removing dust in the atmosphere) and simultaneously improved working conditions, reducing the respiratory problems previously encountered. B&Q's assessment scheme for suppliers (QUEST) began with intensive analysis of practices and outputs (emissions, effluents, etc.), but now also includes a responsibility for the supplier to articulate ways in which B&Q itself might improve its ways of working (not just environmentally). This is an example of the customer—a focal firm in its network—actively preparing itself to learn from its suppliers.

An example of *supplier-pulled learning* has been the CRINE initiative in the U.K. oil and gas industry. Conceived in 1993 as a response to the 1992 oil crisis, CRINE (Cost Reduction Initiative for the New Era) was a joint effort involving government and key industry players representing contractors, suppliers, consultants, trade associations and others. The original goal was to enable, by 1996, an across-the-board cost reduction of 30% for offshore developments, and this was to be achieved by a sectorwide effort rather than individual actions.

The project was successful on a number of dimensions: for example, by 1997 the cost of field developments had fallen by 40% on a barrel/barrel basis, and attracted significant international attention and emulation. As a consequence, CRINE-based programmes are now under development or in operation in Mexico, Venezuela, India and Australia. Significantly the participants felt that the model was worth maintaining and as a result the CRINE Network was established in 1997 with the new goal of international competitiveness replacing that of cost reduction. The CRINE vision is set out as "People working together to make the UK oil and gas industry competitive anywhere in the world by the year 2000".[38]

The stretching target for the industry is to increase its share of the non-U.K. market to 5%; in 1996 this stood at only 1%, indicating a relative weakness in international competitiveness. (Significantly this position had already improved by 1998 to 2.4%, reflecting the industry's growing capabilities, partly supported by CRINE activities.)

The current mode of operation is one of supported networking, where players from regional and national government (e.g., DTI, Scottish Enterprise), major operators, trade and research bodies, and academic and other groups provide various forms of support (financial, technical, etc.) to a network made up of the main actors in the supply chain. A small coordinating group manages the network activities and the whole is steered by a representative body drawn from the above players. Activities cover a broad front, including awareness and communications via newsletters, Web sites, workshops and conferences, technical projects and other initiatives.

From an early stage the original CRINE programme sought to establish a learning and continuous improvement culture, encouraging dialogue and collaborative working between suppliers and customers, rather than confrontational modes of working. This provides a base for some more substantial initiatives; of particular relevance is the First Point Assessment programme (FPA). This is a company, owned by 11 major players in the industry, whose role is to carry out capability assessments and assist in upgrading and development of capability along the supply chain: *to provide opportunities for improvement throughout the supply chain through enhanced knowledge of strengths and weaknesses.*[39]

The Society of Motor Manufacturers and Traders (SMMT) Industry Forum is another example of a *supplier-led* learning initiative. Driven by the increasing pace of change and the global nature of the automotive assembly and component supply industry, this programme seeks to improve the performance of suppliers (especially the 2nd and 3rd tier) to major U.K. assemblers. It aims to improve competitiveness at shop-floor level by the use of training and support programmes involving engineers from the major industry firms and external sources of best practice. An important component of this work is the extensive use of measurement frameworks throughout the supply chain as a mechanism for driving improvement and learning.

An example of the way in which the Forum operates is that of Isringhausen GB, a firm in Wrexham making car seats. Although it had been involved in some improvement activities, it was able, through the input of advice from Forum engineers, to identify 60 projects which represented potential improvement projects. As a result of reconfiguration and other work, productivity in two areas was increased by 45%.

Supplier-pushed programmes are those where the technology holder works with its customers rather than its suppliers, and they represent (at least conceptually) a further form of SCL. An example here might be the U.K. surface engineering industry, where the bulk of firms involved in painting, plating and other surface treatments are SMEs and where levels of technological competence are low. Most technology transfer comes through a small group of international firms supplying key equipment and materials—and these represent a powerful potential conduit for a supply push model.

Intermediaries/ Innovation Agents

Innovation research consistently highlights the importance of key roles which enable, energise or catalyse the process of technological change. For example, Rothwell lists various kinds of "innovation champions", Witte speaks of "knowledge and power promoters" and Allen discusses the "technological gatekeeper").[40] These roles, which can be played by individuals or groups, are relevant both to innovation processes within the firm and to those which operate between firms in the wider environment. They operate in both formal and informal fashion and represent a critical *intermediary* element in the innovation process.

For example, aspects of such roles include agents acting as information brokers, as sources of particular resources or capabilities, as integrating agents, as network facilitators and as carriers of new learning about both technological competence and innovative capabilities. Carlsson and Jacobsson provide a detailed account of the ways in which such intermediaries can assist in developing technological strength within an industrial sector.[41] Realisation of the potential role such *innovation agents* (IAs) can play in facilitating the development of technological competence has increasingly made them a target for policy makers.[42]

Intermediaries may make a direct input, providing resources or capabilities to fill the gaps, but they are often involved in less direct fashion, providing a range of formal and informal services which help bridge the gap between needs and means within the innovation process. The type of bridging required may vary widely; for some firms it is about knowing where to access particular resources or capabilities, while for others it may involve articulating problems more clearly and developing a suitable strategic plan for dealing with them. Equally, the size of the gap may vary, from small gaps which can be easily bridged with a short telephone conversation, through those which require months of work with suitable intermediaries. Figure 8.2 indicates the nature of these need gaps.

To some extent there have always been innovation agents at work in the innovation process. For example, the key internal roles of project champion, business innovator, technological gatekeeper and others have been regularly reported in studies going back 50 years. Equally, the external technology marketplace has not only been filled with suppliers but also with organisations like trade associations and technical advisory groups which play an intermediating role. But recent years have seen a proliferation in the range and number of organisations which are involved as innovation agents, and at least some of this growth can be related to specific policy mechanisms designed to mobilise and deploy this resource. Several government innovation support schemes contain an element of consultancy, effectively making use of the skills and experience of consulting organizations to develop an intermediary infrastructure to facilitate innovation.

Figure 8.2
Need Gaps

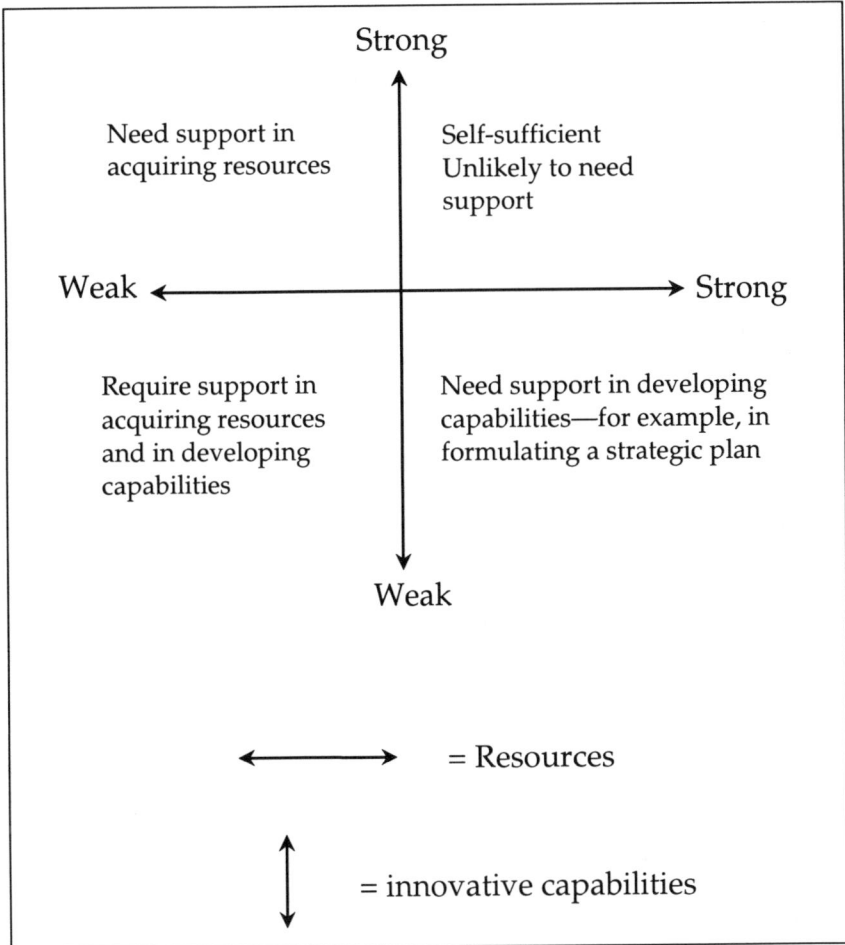

Intermediary support through innovation agents is not simply another term for consultancy activity. Traditional models of the consulting process involve a linear activity, with a transfer of expert knowledge from supplier to users. But an alternative view, identified by Schein among others, can be termed process consulting, which involves a catalytic, nondirective approach stressing learning and facilitation rather than prescription.[43] This is much more concerned with building up a long-term relationship and enabling learning than with the one-time solution to a particular problem.

There are several ways in which innovation agents can improve the operation of the innovation process. These include:

- *expert consulting*, providing particular solutions to particular problems;
- *experience-sharing*, transferring what is learned in one context to another. The metaphor here might be that of bees, cross-pollinating as they fly from flower to flower;
- *brokering*, putting different sources and users in contact across a wide range of services and resources. The degree to which the innovation agent is independent is important here as his/her credibility is likely to diminish if there are particular standard solutions being proposed, or particular sources of supply advanced. The metaphor here is that of marriage broker/dating agency trying to set up the most appropriate match between two partners;
- *diagnosis and problem-clarification*, helping users articulate and define the particular needs in innovation. Many user firms lack the capability to understand or prioritise their problems into a strategic framework for action, and outside agencies may be able to assist in this process. A metaphor here would be the general medical practitioner, whose main task is diagnosis and who then prescribes a range of specialist treatments, such as medicines, physiotherapy, diet, surgery, which are then actually delivered by others;
- *benchmarking*, where the process of identifying and focussing on good practice can be established through an intermediary;
- *change agency*, where organizational development can be undertaken with help from a neutral outside perspective.

Growth in the number and range of actors in this area can be seen as part of the wider expansion of the producer services sector.[44] It is particularly important to recognise that traditional consultancies (in finance, engineering or management) are by no means the only source of such support and there has been a rapid expansion in the numbers and range of agencies involved over the past decade.

Over time, this model of using innovation agents as part of a form of consultancy-based innovation support scheme has become widespread, and there was particularly rapid expansion during the 1990s.[45] Table 8.5 summarises some of the key mechanisms whereby consultancy schemes of this kind can augment and extend the reach and effectiveness of innovation policy.

Examples of intermediaries being used in policy support programmes include the Innovation and Technology Counsellors in the United Kingdom, the BUNT programme in Norway and the Tekes Technology Clinic approach in Finland.

Table 8.5
Potential Contribution of Innovation Agents in Capability Building

Characteristic	Illustrative examples
Capability building	Advice and information support via consultancy enabled the development of key management capabilities in identifying needs, exploring and selecting innovations, planning, implementation, project management, etc.
Institution building	Such schemes also offered an opportunity for developing strategic capabilities across the supply side—for example, mobilising a critical mass of technological knowledge and skills in support of particular technologies.
Failure avoiding	Providing targeted advice and direct technical and managerial support offered opportunities to reduce the incidence of costly failures of investments through transferring better innovation management practice - for example, in selecting appropriate applications of new technology and in project management.
Lower cost	Providing innovation support through information and consultancy advisory services was less expensive than direct subsidy, soft loans or other innovation policy options.
Targetting of support	Using consultants as active intermediaries opened up the possibility of reaching user firms more directly than traditional financial support mechanisms, which tended to lack focus and often failed to reach many potential users within a target group.
Decentralised operation	Using consultants enabled a more decentralised mode of operation, involving less monitoring and control by civil servants. Once the broad objectives of a programme were set out it could be largely self-managing, with overall monitoring and quality assurance provided by a small and specialist group—itself sometimes outside of government but contracted with the specific project monitoring role.

Source: J. Bessant and H. Rush, "Building bridges for innovation; the role of consultants in technology transfer." *Research Policy*, 24 (1995), pp. 97-114.

Structural Connections

Most countries have some form of research and technology organisations (RTOs—specialist laboratories, universities, government laboratories, sectoral facilities, etc.)—and these often represent a major investment of public money as well as an accumulation of specialist technological competence over many years.

Potentially these could be of value to SMEs in their technological change activities—but only if there are mechanisms for raising their awareness of the RTOs and for accessing their knowledge and support. Unfortunately for many SMEs there is both an information gap—a lack of knowledge about which RTOs exist and what they might be able to offer—and a perception gap where SMEs see the activities of RTOs as being too advanced and specialised and not applicable to their problems. There is also a perception of high cost which deters SMEs from approaching RTOs.

The problem is not confined solely to the demand side, however; there is also a gap on the supply side. Many RTOs have a tradition of working with larger firms or on major publicly funded contracts, and have not seen SMEs as an important market. There is an understandable pressure inside these organisations to work on intellectually challenging research-based projects rather than on delivering relatively straightforward problem-solving services associated with application and transfer of technology. As a result there is often a gap between the supply of technology transfer service from RTOs and the demand articulated by SMEs.

This pattern is changing, particularly as a response to a more competitive environment for RTOs.[46] Across the world there is a move away from public subsidy and towards greater emphasis on market-related activities; in many cases this has included the privatisation of RTOs. Making the transition to being more market-oriented is not always easy and in a number of cases public policy support has been engaged in trying to help RTOs identify and implement new strategies, many of which stress a more SME-focused role. In the United Kingdom, for example, the Focus Technical Reviews programme involved management consultants helping a selected group of RTOs redefine their roles and operating practices to make them more market-focused, with particular emphasis on their potential work with SMEs.

For example, many traditional suppliers of technology—such as universities and government research laboratories—are coming under increasing pressure to build closer links with prospective customers for their services (the trend towards privatisation has undoubtedly accelerated this process). As they come to realise the limitations of the traditional linear technology push model referred to above, these traditional suppliers are adopting a more proactive stance, often establishing liaison or marketing groups with the aim of getting closer to and understanding the needs of their customers. In the process they are recognising, and developing mechanisms to assist with, the particular problems of technology transfer to different types of user.[47]

Similarly, RTOs organised on a sectoral basis are playing an increasingly important role, not just in generating technology or providing technical services to members, but also in identifying, understanding and articulating user needs and tailoring suitable solutions to these needs. For example, one of the RTOs in Germany is the Verein der Deutscher Giesserei-industrie (VDG), which is a long-established professional technical association for the foundry sector.[48] Working in concert with the two key trade associations (for ferrous and nonferrous castings), the VDG provides a comprehensive network of technology support for an industry characterised by the high proportion of very small firms (around 80% of the firms employ fewer than 50 people). Arguably the strong technological base of the important German foundry industry, which has enabled it to survive despite strong pressures from overseas competitors and from materials substitution, derives in part from the various technology transfer activities of the VDG. These include:

- articulating key common problems through direct contacts with firms, an approach made possible by a decentralised regional representation in each of the Länder;
- educating and raising awareness among small firms of technology-related issues through seminars and visits;
- coordinating key research themes and securing federal and regional government support for this research through R&D grants;
- carrying out such R&D through a network of universities, specialised institutes and in their own subsidiary organization, the Institut für Giessereitechnik (IfG), which employs over 100 specialists covering the whole field of foundry technology and application (significantly, the IfG is jointly owned by the VDG and the trade associations, and occupies the same site);
- disseminating the results of the R&D work via the decentralised network;
- providing on-line consultancy and advice as part of the membership package;
- working with the chambers of industry and commerce to provide detailed input specifications on skills and human resources for the needs of the industry as part of the national vocational training system.

The overall effect of this is to provide even very small firms with a sophisticated but accessible technical backup, and a clear channel through which technology can transfer. (This was exemplified in an interview with a small foundry where, on being asked how such a firm could support a high level of product and process innovation [microelectronics controls, new materials, new product designs], the foundryman picked up the phone to VDG and said "that's my R&D department!".)

Many RTOs of this kind have, often postprivatisation, begun to offer a range of technology transfer support services in addition to their traditional activities. In some cases these services have expanded to become a key part of the overall portfolio—such as the case of PERA International in the United Kingdom. Others—such as IVF in Sweden or the Fraunhofer Institutes in Germany—have traditionally offered a more comprehensive range of services including support for technology transfer.[49]

Transfer of Skills

Another area where significant experimentation is going on concerns the improvement of competence through novel mechanisms designed to help improve skills acquisition. Traditionally the problem for SMEs has often been that buying scarce skills on the labour market is too expensive, but if the SME invests in training it runs the risk of having its key skilled staff poached by richer and larger firms. An accompanying problem is at the graduate level where key skills in technical and managerial areas are in short supply.

This problem has given rise to many policy interventions designed to put in place graduate and other high-level skills and to enable some element of retention over a sustained period of time. For example, in the United Kingdom the Teaching Company programme contact is established via key individuals—Teaching Company Associates—who work full-time on a client site for two years or more, effectively providing the bridge between a university and an industrial user. Over time the relationship matures and traffic begins to flow across this bridge, at first directly related to the project but increasingly covering other aspects and becoming two-way in nature.[50]

CONCLUSIONS

SMEs represent an important element in any economy for at least two reasons. They are responsible for growth in employment and activity, and they are support to larger firms, enabling or constraining their performance. But it is *growth* in SMEs which is the key issue, not just their existence or preservation—and this is fundamentally a challenge of innovation. The issue has been recognised—few national or European Union policies now lack the obligatory SME emphasis—but it can be argued that a gap still exists between what SMEs need to enable effective innovation and the ways in which the NSI is configured to deliver them.

Part of the problem is that we do not know enough about SMEs and their innovative behaviour. For example, we need to look at such questions as, who innovates and how? Looking at particular segments, what are the key constraining problems? Are these the same as those facing large firms—or is there a need for a new research agenda exploring SME-specific barriers to innovation? How can we begin to work with integrated sets of firms within the value stream—and are the innovation dynamics of these large/small firm combinations well understood? And how can we use this knowledge to inform better theory and to improve practical policy mechanisms for enabling SME growth?

These are issues of significance to both policy makers and innovation researchers. Improving competitive performance through innovation is a key challenge for all industrialised societies, and achieving this will depend not only on a handful of world class technologically-based firms, but also on the long tail of SMEs which make up the majority of firms in the economy. This chapter has indicated a few examples of the experiments now being undertaken to try to engage with this problem, and to address some of the issues associated with closing the SME/NSI gap. But while these represent promising additions to the repertoire of policy tools, there is a need for further exploration, development and diffusion of them—and for research to help understand and capture the lessons learned from such experience.

REFERENCES

1. Hoffman, K. et al., Small firms, R&D, technology and innovation in the UK. *Technovation*, 18 (1997), pp. 39-55; Storey, D., *Understanding the small business*. London: Routledge, 1994; OECD, *Small and medium-sized enterprise: Technology and competitiveness*. Paris: Organisation for Economic Cooperation and Development, 1993.

2. Oakey, R., High technology small firms: their potential for rapid industrial growth. *International Small Business Journal*, 9 (1991), pp. 30-42; Rothwell, R. and M. Dodgson, SMEs: their role in industrial and economic change. *International Journal of Technology Management*, 1993 (Special issue on small firms), pp. 8-22.

3. Voss, C., *Made in Europe 3—the small company study*. London: London Business School/ IBM Consulting, 1999.

4. Hoffman, op. cit.

5. Dodgson, M. and J. Bessant, *Effective innovation policy*. London: International Thomson Business Press, 1996.

6. Pavitt, K., Sectoral patterns of technical change: towards a taxonomy and a theory. *Research Policy*, 13 (1984), pp. 343-373; Nelson, R., Why do firms differ and how does it matter? *Strategic Management Journal*, 12 (1991), pp. 61-74.

7. Rothwell, R. and W. Zegveld, *Innovation and the small/ medium sized firm*. London: Pinter, 1982.

8. Dodgson and Bessant, op. cit.

9. Rothwell, R., Small and medium sized firms and technological innovation. *Management Decision*, 16(6), 1978.

10. Lundvall, B., *National systems of innovation: Towards a theory of innovation and interactive learning*. London: Pinter, 1990.

11. Carlsson, B. and S. Jacobsson, Technological systems and economic performance: The diffusion of factory automation in Sweden. In *Technology and the wealth of nations*, D. Foray and C. Freeman, eds., pp. 77-94. London: Pinter, 1993.

12. Skaug, E., Brief description of the BUNT Programme. In *Technology Transfer and Implementation*. London: Science and Engineering Research Council, 1992; Miles, I., ed., *Services, innovation and the knowledge-based economy*. London: Pinter, 1998.

13. Dodgson and Bessant, op. cit.; Vickery, G. and E. Blau, *Government policies and the diffusion of microelectronics*. Paris: Organisation for Economic Cooperation and Development, 1989.

14. Bessant, J., Government support for innovation. In *Innovation*. London: Caspian Publishing, 1997.

15. Piore, M. and C. Sabel, *The second industrial divide*. New York: Basic Books, 1982; Semlinger, K., Public support for firm networking in Baden-Wurttemburg. In *Europe's next step*, R. Kaplinsky et al., eds. London: Frank Cass, 1995; Schmitz, H., *Collective efficiency and increasing returns*. University of Sussex, Institute of Development Studies, 1997.

16. Danish Technological Institute, *Network co-operation—achieving SME competitiveness in a global economy*. Danish Technological Institute, 1991.

17. Carney, M., State development strategies for small enterprises: The role of structural service agencies. *International Journal of Innovation Management*, 1(2) (1997), pp. 151-172.

18. Nadvi, K. and H. Schmitz, *Industrial clusters in less developed countries: Review of experiences and research agenda*. Brighton: Institute of Development Studies, 1994; Schmitz, H., On the clustering of small firms. *IDS Bulletin*, 23(3) (1992); Best, M., *The new competition*. Oxford: Polity Press, 1990.

19. Porter, M., *The competitive advantage of nations*. New York: Free Press, 1990.

20. Murray, R., CITER. In *Background/benchmark study for Venezuelan Institute of Engineering*, H. Rush et al., eds. Brighton: University of Brighton, CENTRIM, 1993.

21. Prahalad, C. and G. Hamel, *Competing for the future*. Cambridge, MA: Harvard University Press, 1994; Kay, J., *Foundations of corporate success: How business strategies add value*. Oxford: Oxford University Press, 1993; Pavitt, K., What we know about the strategic management of technology. *California Management Review*, 32 (1990), pp. 17-26.

22. Senge, P., *The fifth discipline*. New York: Doubleday, 1990; Garvin, D., Building a learning organisation. *Harvard Business Review*, July/August (1993), pp. 78-91; Argyris, C. and D. Schon, *Organizational learning*. Reading, MA: Addison Wesley, 1970.

23. Bessant, H. and D. Francis, Implementing learning networks. *Technovation*, 19(6/7) (1999), pp. 373-383.

24. Kolb, D. and R. Fry, Towards a theory of applied experiential learning. In *Theories of group processes*, C. Cooper, ed. Chichester: John Wiley, 1975.

25. Senge, op. cit.; Garvin, op. cit.

26. Nonaka, I., The knowledge creating company. *Harvard Business Review*, November-December (1991), pp. 96-104.

27. Hedberg, B., How organisations learn and unlearn. In *Handbook of organisation design*, H. Nystrom and W. Starbuck, eds. Oxford: Oxford University Press, 1981.

28. Argyris and Schon, op. cit.; Leonard-Barton, D., The organisation as learning laboratory. *Sloan Management Review*, Fall 1992.

29. Polanyi, M., *The tacit dimension*. London: Routledge and Kegan Paul, 1967.

30. Revans, R., *Action learning*. London: Blond and Briggs, 1980; Revans, R., *Action learning 2*. Buckingham: G. Wills/IMCB, 1983.

31. Keating, M., P. Coughlan, and R. Vail, *Facilitating the move to World Class manufacturing: The design and delivery of a "demonstrator" management learning programme for SMEs*. In British Academy of Management Annual Conference, 1995.

32. Sako, M., Prices, quality and trust: Inter-firm relations in Britain and Japan. *Cambridge Studies in Management*, 18 (1992).

33. Schmitz, H., Collective efficiency: Growth path for small-scale industry. *Journal of Development Studies*, 31(4) (1995), pp. 529-566.

34. Cusumano, M., *The Japanese automobile industry: Technology and management at Nissan and Toyota*. Cambridge, MA: Harvard University Press, 1985.

35. Lamming, R., *Beyond partnership*. London: Prentice-Hall, 1993.

36. Gereffi, G., The organisation of buyer-driven global commodity chains: How U.S. retailers shape overseas production networks. In *Commodity chains and global capitalism*, G. Gereffi and P. Korzeniewicz, eds. London: Praeger, 1994.

37. Kaplinsky, R., J. Bessant, and R. Lamming, *Using supply chains to diffuse "best practice"*. Brighton: Centre for Research in Innovation Management, 1999.

38. CRINE Network website, 1999-01-06

39. First Point Assessment website, 1999-01-06

40.   Rothwell, R., Successful industrial innovation: Critical success factors for the 1990s. *R&D Management*, 22(3) (1992), pp. 221-239; Witte, E., *Organization für Innovationsentscheidungen*. Gottingen: Schwartz, 1973; Allen, T., *Managing the flow of technology*. Cambridge, MA: MIT Press, 1977.

41.   Carlsson and Jacobsson, op. cit.

42.   Bessant, J. and H. Rush, Building bridges for innovation: the role of consultants in technology transfer. *Research Policy*, 24 (1995), pp. 97-114.

43.   Schein, E., *Process consultation, Volume 2: Lessons for managers and consultants*. Reading, MA: Addison Wesley, 1987.

44.   Rush, H., *Producer services*. Brighton: Centre for Business Research, 1988.

45.   Vickery and Blau, op. cit.; Bessant, J. and H. Rush, Innovation agents and technology transfer. In Miles, op. cit.

46.   Rush, H. et al., *Technology institutes: Strategies for best practice*. London: International Thomson Business Press, 1996.

47.   Ibid.

48.   Appleby, C. and J. Bessant, Adapting to decline: Organisational structures and government policy in the UK and West German foundry sectors. In *Government-industry relations in the UK and West Germany*, S. Wilks and M. Wright, eds. Oxford: Oxford University Press, 1986.

49.   Rush, H. et al., Strategies for Best Practice in Research and Technology Institutes: an Overview of a Benchmarking Exercise. *R&D Management*, 25(1) (1995), pp. 17-31.

50.   Senker, J. *An evaluation of the Teaching Company Scheme*. In R&D Management Conference. Manchester: Manchester Business School, 1994.

9

Productivity and Regional Density

Rui Miguel L. N. Baptista

INTRODUCTION

There are significant differences in average labour productivity across different regions: in 1991, industrial output per worker in the most productive region of Great Britain was about 25% larger than in the least productive region; industrial output per worker in the ten most productive counties was twice that of the ten least productive counties. The purpose of this chapter is to examine the significance of localised agglomeration externalities and increasing returns, and how they relate to spatial differences in labour productivity. For this purpose, an empirical model of local productivity is developed, where the spatial density of economic activity is the source of aggregate increasing returns.

Density is here defined as the intensity of labour, human and physical capital relative to physical space. Density is high when there is a large amount of labour and capital per square kilometre. The idea that economic activity accrues advantages from geographical agglomeration has been present in a large and diversified literature. However, density itself does not usually appear to be an explicit element of the theory, and empirical work focusing explicitly on density has been limited.

Density affects productivity in several ways. If there are externalities associated with the physical proximity of production and human capital, such as knowledge spillovers, then density will spur innovation and productivity. Moreover, a higher degree of specialisation is possible within areas of dense activity, establishing a source of increasing returns. Finally, even if technologies have constant returns themselves, but the transportation of products from one stage of production to the next involves costs that rise with distance, then the technology for the production of all goods within a particular geographical area will still experience increasing returns.

The empirical model developed in this chapter is based on the assumption that there are positive externalities resulting from density, following the work by Ciccone and Hall (1996) on labour productivity in the United States. In order to examine the effects of density, a much finer level of geographical detail than a U.K. region, as defined by the Central Statistical Office (CSO) has to be used. Although the matter may not be so serious here as it is for the United States, large areas in the United Kingdom support very little, if any, economic activity. A meaningful measure of density has to use more detailed data by county. The empirical model views the unit of production to be labour, capital and land present in a county.

Data on employment and gross value-added for manufacturing sectors is available for the United Kingdom. This means that, unlike the case of Ciccone and Hall (1996), estimation does not need to involve aggregation from the county to the region (state, in the case of the United States) level. This provides us with an important advantage since, as it will be shown subsequently, it will make estimation of the empirical model developed much easier. The model allows us to create an index of the density of inputs for each county, which depends on the extent of increasing returns.

Estimation of the empirical model of local increasing returns reveals that accounting for the density of economic activity at the county level is essential for explaining geographical differences in productivity. The degree of locally increasing returns has a highly significant effect on local output per worker.

The chapter will discuss the background literature as regards the relationship between spatial density and economic growth, and the measurement of productivity; present the empirical model developed and explain why geographical differences in density and productivity can persist in equilibrium; describe the data used; discuss estimation and summarise the results.

BACKGROUND LITERATURE

Economics of Agglomeration

The economics of agglomeration can be traced in the literature as far back as Marshall (1920), who stressed the importance of localised technological spillovers and pools of specialised human capital. In the urban economics literature, agglomeration externalities are defined as any economies or cost reductions which are possible if a group of firms locates near to each other (Evans 1985). Such effects are extensively discussed by Isard (1956), who identified two different kinds of externalities: location externalities referred to economies derived by firms in the same industry from locating near to each other; urbanisation externalities referred to economies derived by firms in many different industries from locating in the same area.

Vernon Henderson (1974) demonstrated, building on work by Edwin S. Mills (1967), that, in equilibrium, disamenities from agglomeration on the side of households may offset productivity advantages on the side of firms. In later work, Henderson (1986) listed four factors that capture the nature assumed by location externalities in the urban economics literature:

1.   economies of intraindustry specialisation, where greater industry size permits greater specialisation among firms in their detailed functions;
2.   labour market economies, where industry size reduces search costs for firms looking for workers with specific training relevant to that industry;
3.   scale for "communication" among firms, affecting the speed of adoption of innovations;
4.   scale in providing unmeasured public intermediate inputs tailored to the technical needs of a particular industry.

A second branch of the literature on agglomeration hypothesises increasing returns internal to firms. This means that economies of scale lead to noncompetitive local market structures. In his early work on this subject, Mills (1967) assumed a monopolistic market structure.

More recent works use a monopolistically competitive market structure to study agglomeration with increasing returns to scale. Abdel-Rahman (1988), Fujita (1988, 1989) and Rivera-Batiz (1988) employ the formalisation of monopolistic competition originally presented by Spence (1976) and Dixit and Stiglitz (1977) to demonstrate that nontransportable intermediate inputs produced with increasing returns imply agglomeration. Fujita and Rivera-Batiz (1988) point out that the agglomeration of producers results in an agglomeration of intermediate services, while the agglomeration of households raises the variety of consumer services, thereby giving rise to external effects on consumption. Related work by Krugman (1991) demonstrates that agglomeration will result even when transportation costs are small if there are internal scale economies, as long as most workers are mobile.

Fujita and Thisse (1996) provide a review of the main literature on the economics of agglomeration, proposing three main reasons for the formation of economic clusters involving firms and/or households: (1) externalities under perfect competition; (2) increasing returns under monopolistic competition; and (3) spatial competition under strategic interaction. This last stream builds on the seminal works by Hotelling (1929) and Hoover (1937) and focuses mainly on the influence of demand factors, such as competition for geographical markets, in determining agglomeration, and have a limited bearing on productivity.

Spatial Density and Economic Growth

Although traditional development literature, such as Arthur Lewis (1955) emphasised that the spatial distribution of economic activities affects economic progress, it was the more recent wave of new growth theory that provided a resurgence of interest in the role that spatial structure of economic activities plays in advancing (or retarding) economic growth.

Following Romer (1986), economists have generally accepted two facts about growth:

1.   the accumulation of technologies and ideas creates the bulk of economic progress;
2.   the accumulation of ideas depends critically on the flow of knowledge across agents, that is, on knowledge spillovers.

As was stated in the previous subsection, large urban markets provide local economies of scale and urbanisation lowers transport costs. However, urban density also allows for a more rapid spread of knowledge: it helps companies learn what their consumers need and what their competitors are up to, and provides beginners with a wider variety of opportunities and role models.

There are many obvious examples of areas where space seems to have contributed to the spread of technology and the diffusion of ideas. Silicon Valley is surely the most visible example, but similar examples can be found in industries as far apart as pottery (the Sassuolo region in Italy) and financial services (Wall Street and the City of London). Firms have easy access to at least some of the innovations practised by their neighbours, ideas flow in after work conversations between specialised technicians, employees are hired away by competing firms or leave to start up their own businesses, bringing with them the ideas held by their former employers (see Saxenian 1994).

Such facts have given rise to a widespread literature on industrial clusters, following the work by Porter (1990). Clusters can be defined as strong collections of firms, usually concentrated in the same geographical area. These agglomerations are typically linked to the presence of an infrastructure of related and supporting industries and to the proximity of a strong science base: universities and research centres which act as sources of technological knowledge.

It has been argued (see Feldman, 1994; Baptista, 1998) and indeed empirically verified (see Audretsch and Feldman, 1996; Baptista and Swann, 1998; Baptista, 2000) that the geographical concentration of rivals enhances competitiveness and stimulates the creation and diffusion of innovations. Glaeser et al. (1992) have shown that geographically bounded externalities arising from both local competition and diversity have a significant impact on firm growth. The impact of these findings upon productivity and regional economic growth should be very significant.

The amount and quality of human capital plays a determinant role in a region's productivity. Lucas (1988) placed a central role on the accumulation of human capital, emphasising the knowledge possessed by an average member of society, rather than the overall knowledge of the society as a whole. Urban proximity enables people to take advantage of their neighbours' expertise more easily, and can also be useful for the production of more human capital. The large range and density of experiences to be had in a large city can directly raise the level of skill accumulation. Moreover, dense areas provide better conditions for coordination and matching of skills. Becker and Murphy (1992) stressed the role of coordination and specialisation in creating increasing returns through the rapid accumulation of human capital with the same skills. Ideas travel better over small distances and technological knowledge, particularly at the early stages of its life-cycle, is essentially tacit and, therefore, easier to transmit through personal contact (Lundvall, 1988).

Measuring Productivity and Agglomeration

Empirical studies of productivity and agglomeration have focused on city (population) and industry size (employment) as determinants of productivity, and on technological spillovers as determinants of agglomeration economies. Shefer (1973) found effects of local industry size to be unstable over time. Sveikauskas (1975) and Moomaw (1981) used total local population as a measure of size and found this variable to be positively correlated with labour productivity. Segal (1976) developed capital stock data for urban areas and concluded that total factor productivity (and not just labour productivity) was significantly higher in large metropolitan areas than in small ones.

Moomaw (1985) conducted empirical work to evaluate whether population measures urban agglomeration economies (positive externalities) or diseconomies (such as congestion costs), finding that total population provided a significant net effect from positive agglomeration externalities. Moreover, his study confirmed the existence of significant interregional differences in productivity across the United States. Henderson's study (1986) of U.S. and Brazilian data on cities found that the productivity of firms increases with the size of the industry as measured by industry employment.

However, these studies seem to be flawed by the focus on return to city or industry size, since this seems to make results highly conditional on the type of data used and on the particular specification of the production function. In fact, using data focusing exclusively on the food-processing industry, Sveikauskas et al. found significant effects from city size on productivity, but not from industry size. However, replicating Henderson's (1986) methodology on the same set of data, the same authors found significant positive effects on productivity from industry size, and significant negative effects from city size.

Ciccone and Hall (1996) believe that urban density, rather than size, is a more accurate determinant of the amount of agglomeration externalities. This chapter also follows this approach. Moreover, following studies such as Carlino and Voith (1992), human capital differences among regions are considered on the basis of different levels of education.

An important branch of empirical literature on agglomeration has studied wage differentials across regions, instead of focusing directly on productivity. Wages are higher in cities and other dense areas. However, differences in wages may reflect differences in regional and company policies regarding the split between cash and noncash compensation. Figure 9.1 plots weekly manufacturing wages against labour productivity for the counties of Great Britain. Although there is a positive correlation, variability is very high. It seems, therefore, that wages should not be considered a proxy variable for labour productivity, at least in the present specific set of data.

Figure 9.1
Wages and Labour Productivity for U.K. Counties, 1991

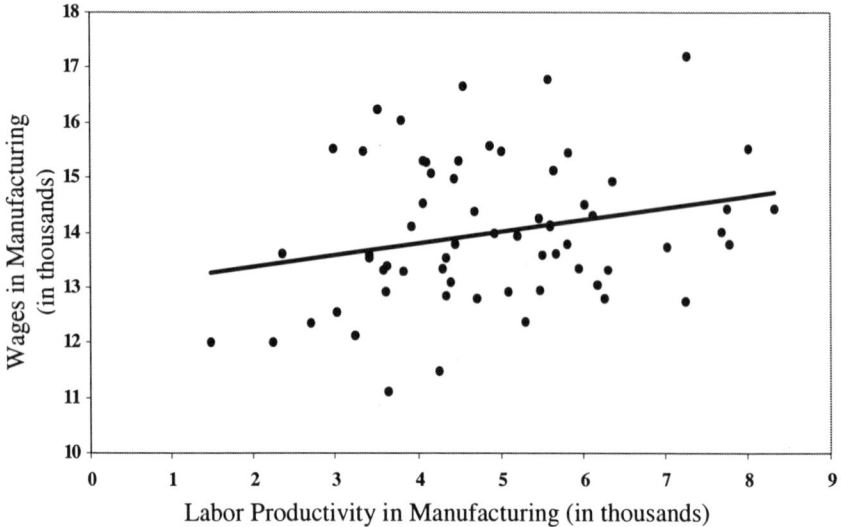

AN EMPIRICAL MODEL OF LOCAL PRODUCTIVITY

A model representing total factor productivity (TFP) at the county level can be built to include increasing returns arising from agglomeration externalities[1]. Let us consider land, labour and capital as the factors of production. Let $f(n, k, q, a)$ represent the production function describing the output produced in a unit (square kilometre = km^2) of space by employing n workers and k units of capital (all space is deemed equivalent). The unit of space is embedded in a larger area (a county, in the present empirical work) with total output q and area equal to a square kilometres. The last two variables are used to introduce the agglomeration externality associated with density.

Following Ciccone and Hall (1996), the present model assumes that the externality depends multiplicatively on a particular measure of density, namely output per square kilometre (q / a). The elasticity of output with respect to density is a constant, $(\lambda - 1) / \lambda$. Capital and labour are used according to a Cobb-Douglas production function yielding constant returns, β being the labour elasticity. Since the empirical model is being designed to test the effects of local density on productivity, the amount of labour is weighted by a variable e, describing labour efficiency. If one expects better qualified labour to be more productive, then this weighting variable should depend on the average level of education of workers. The production function will then be:

$$f = A_c \cdot \left[\left(e_c \cdot n_c \right)^{\beta} \cdot k_c^{1-\beta} \right]^{\alpha} \cdot \left(\frac{q_c}{a_c} \right)^{\lambda - \frac{1}{\lambda}} \tag{1}$$

The subscript c is used to denote the county to which the square kilometre belongs. The constant α is the production elasticity to total factor (labour plus capital) and should be less than one by the amount of the share of land in factor payments. A_c is a Hicks-neutral technology multiplier for county c, reflecting technological differences between counties.

The amounts of labour and capital employed in a county, n_c and k_c, are assumed to be equally distributed among all the square kilometres in the county. Thus, total output in county c is:

$$q_c = a_c \cdot A_c \cdot \left[\left(\frac{e_c \cdot n_c}{a_c} \right)^{\beta} \cdot \left(\frac{k_c}{a_c} \right)^{1-\beta} \right]^{\alpha} \cdot \left(\frac{q_c}{a_c} \right)^{\lambda - \frac{1}{\lambda}} \tag{2}$$

Solving (2) for output per square kilometre yields:

$$\frac{q_c}{a_c} = A_c^{\lambda} \cdot \left[\left(\frac{e_c \cdot n_c}{a_c} \right)^{\beta} \cdot \left(\frac{k_c}{a_c} \right)^{1-\beta} \right]^{\gamma} \tag{3}$$

where $\gamma = \alpha \cdot \lambda$.

Here, γ is the product of the production elasticity to total factor, α, and the elasticity associated with the agglomeration externality, λ. Only γ can be obtained from model estimation. However, as it was stated previously, α is always less than one, while the elasticity from the externality, λ, should be greater than one. Thus, if the estimated value of γ is significantly greater than one, increasing returns arising from agglomeration externalities will dominate decreasing returns resulting from congestion (higher share of land in factor payments).

Capital stock data are not generally available at the county, or even regional level. To deal with capital, it is assumed that its cost, r, is the same everywhere. This allows us to derive a demand function for capital, thereby substituting the factor price for the factor quantity in the production function. If r equals the marginal productivity of capital, k_c, we have:

$$\frac{k_c}{a_c} = \frac{\alpha \cdot (1-\beta)}{r} \cdot \left(\frac{q_c}{a_c}\right) \tag{4}$$

Thus, the county technology described in (3) becomes:

$$\frac{q_c}{a_c} = A_c^{\omega} \cdot \phi \cdot \left(\frac{e_c \cdot n_c}{a_c}\right)^{\theta} \tag{5}$$

where ϕ is a constant that depends on the interest rate. The elasticities for the county technology multiplier, ω, and for the efficiency-weighted labour used in the county, θ, are:

$$\omega = \frac{\lambda}{1-(1-\beta)\cdot\gamma} \tag{6}$$

$$\theta = \frac{\beta \cdot \gamma}{1-(1-\beta)\cdot\gamma} \tag{7}$$

Let us now turn to the labour efficiency variable. The present model assumes that labour efficiency depends log-linearly on the percentage of active workers who have completed a university degree or equivalent[2]. This statistic is only available at the regional level, and not for each county. Thus, the labour efficiency factor weights the labour input (county employment) equally for all counties within a region, acting as a kind of regional fixed effect. Defining h_s as the percentage of active workers with a university degree or equivalent in region s, and η as the elasticity of education, we have:

$$e_s = h_s^{\eta} \tag{8}$$

Using this relationship in equation (5), we have a description of county-level productivity:

$$\frac{q_c}{a_c} = \phi \cdot A_c^{\omega} \cdot D_c(\theta, \eta) \tag{9}$$

where
$$D_c(\theta, \eta) = \left(\frac{h_s^{\eta} \cdot n_c}{a_c}\right)^{\theta} \tag{10}$$

$D_c(\theta, \eta)$ is a density index for county c, reflecting the intensity of economic activity within the county and, thus, the extent of agglomeration externalities. If we assume that most agglomeration externalities take the form of pools of specialised labour and of knowledge spillovers, geographically bounded and mostly conveyed through workers, the county density index should be positively correlated with county productivity.

The functional form presented in equation (9) allows us to estimate θ, and not γ. However, it is easy to verify from equation (7) that, if θ is greater than one, then γ will always be greater than one, regardless of the value assumed by β^3. This means that one is still able to determine if positive agglomeration externalities outweigh negative congestion effects.

A stochastic specification for the county productivity function can then be presented. Let us assume that the county technology multiplier A_c is distributed log-normally around a nationwide underlying level. Let us also assume that measurement error has a log-normal distribution with zero mean. Hence, taking logarithms, equation (9) yields:

$$\log\left(\frac{q_c}{a_c}\right) = \log\phi + \log D_c(\theta, \eta) + u_c \tag{11}$$

where the stochastic error, u_c, is the sum of the measurement error plus ω times the deviation of county productivity from the underlying level in the nation.

Given the form of the density index, D_c, presented in equation (10), and knowing that the education level h_s is only available at the regional level, equation (11) becomes:

$$\log\left(\frac{q_c}{a_c}\right) = \log\phi + \psi \cdot \log h_s + \theta \cdot \log\left(\frac{n_c}{a_c}\right) + u_c \tag{12}$$

where $\psi = \eta \cdot \theta$.

This stochastic form allows for the estimation of the model parameters (θ, ψ and the constant $\log \phi$) through linear regression. The parameter γ can be obtained from equation (7), given the value of β.

If θ exceeds one, γ will also be greater than one, meaning that there are increasing returns to county density. Under neoclassical assumptions, density should be equal everywhere. The marginal product of labour should be higher in areas with density below average, thus attracting workers from other areas. With increasing returns to density, however, a worker becomes more productive when moved to a denser area, which would eventually lead employment to concentrate in a single area. Both alternatives would be possible only if factor mobility was perfect, which is not the case of labour.

The explanation, in any case, is that workers are willing to accept lower wages to remain in the same area, reflecting different preferences as regards density. A stable equilibrium can be achieved where some areas, or counties, remain denser than others. The marginal cost of production is equalised across all counties as the decrease in marginal cost associated with higher density and productivity is offset by higher efficiency wages and higher land prices.

DATA USED IN THE STUDY

The data used in the study concern employment and output (gross value added) in the manufacturing sector for 64 counties of Great Britain[4] (United Kingdom excluding Northern Ireland). Complete data for all counties is available intermittently from the Central Statistical Office and is published in *Regional Trends*. Data sets for the 64 counties (see Figure 9.2) were collected for the years 1987, 1989 and 1991. Data include also the percentage of active workers with a university degree or equivalent, obtained from the same source.

When a company is located in a denser county, there is likely to be less vertical integration. In dense areas there is greater availability of business services and intermediate goods, so a firm is less likely to produce in-house than it would be in a less dense county. Furthermore, a subsidiary company located in a denser county is more likely to be close to its parent company's headquarters, therefore receiving more transferred services from it.

One would expect, therefore, data on gross value added to include a bias against denser counties. However, this bias acts against the agglomeration effects being tested, so it is not a concern.

ESTIMATION AND RESULTS

For estimation, it is assumed that the random element of output per worker is uncorrelated with density and education. This assumption means that it is supposed that density and education are measured with little error and do not respond to the random element of productivity. If the determinants of productivity are the same across counties, this is not a strong assumption.

Since taking logarithms of the county productivity function derived previously yields a linear function, the parameters ϕ, θ and ψ can be estimated by ordinary least squares (OLS) regression. Since both the explained and the explanatory variables are logarithms, heteroskedasticity did not affect the regression.

Estimations were made using each of the three data sets collected. Results are similar for 1991, 1989 and 1987, which seems to confirm their robustness. Table 9.1 presents the estimation results for the constant term (log ϕ) and for the parameters θ and ψ. The value of η can easily be obtained from the other estimates, while the values presented for γ assume that $\beta = 0.7$.[5]

Figure 9.2
Counties of the United Kingdom

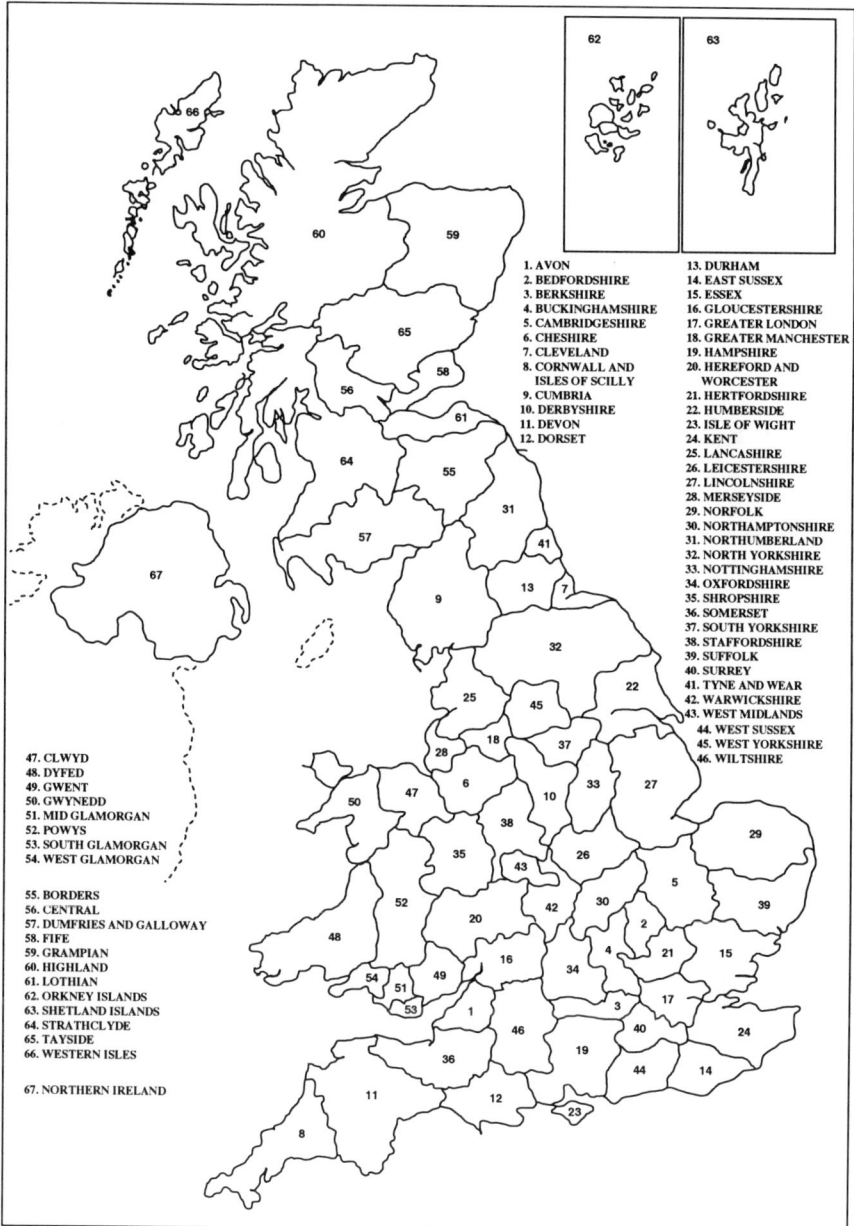

1. AVON
2. BEDFORDSHIRE
3. BERKSHIRE
4. BUCKINGHAMSHIRE
5. CAMBRIDGESHIRE
6. CHESHIRE
7. CLEVELAND
8. CORNWALL AND
   ISLES OF SCILLY
9. CUMBRIA
10. DERBYSHIRE
11. DEVON
12. DORSET
13. DURHAM
14. EAST SUSSEX
15. ESSEX
16. GLOUCESTERSHIRE
17. GREATER LONDON
18. GREATER MANCHESTER
19. HAMPSHIRE
20. HEREFORD AND
    WORCESTER
21. HERTFORDSHIRE
22. HUMBERSIDE
23. ISLE OF WIGHT
24. KENT
25. LANCASHIRE
26. LEICESTERSHIRE
27. LINCOLNSHIRE
28. MERSEYSIDE
29. NORFOLK
30. NORTHAMPTONSHIRE
31. NORTHUMBERLAND
32. NORTH YORKSHIRE
33. NOTTINGHAMSHIRE
34. OXFORDSHIRE
35. SHROPSHIRE
36. SOMERSET
37. SOUTH YORKSHIRE
38. STAFFORDSHIRE
39. SUFFOLK
40. SURREY
41. TYNE AND WEAR
42. WARWICKSHIRE
43. WEST MIDLANDS
44. WEST SUSSEX
45. WEST YORKSHIRE
46. WILTSHIRE

47. CLWYD
48. DYFED
49. GWENT
50. GWYNEDD
51. MID GLAMORGAN
52. POWYS
53. SOUTH GLAMORGAN
54. WEST GLAMORGAN

55. BORDERS
56. CENTRAL
57. DUMFRIES AND GALLOWAY
58. FIFE
59. GRAMPIAN
60. HIGHLAND
61. LOTHIAN
62. ORKNEY ISLANDS
63. SHETLAND ISLANDS
64. STRATHCLYDE
65. TAYSIDE
66. WESTERN ISLES

67. NORTHERN IRELAND

Table 9.1
Estimation Results

Year	1991*	1989*	1987*
Number of Observations	64	64	64
Constant (Ln ϕ)	**4.8090** (1.0359)	2.6793 (0.9662)	3.5329 (1.0173)
ψ	**0.9793** (0.3361)	**0.4484** (0.0351)	**0.7308** (0.3746)
θ	**1.1021** (0.0289)	**1.0268** (0.0671)	**1.0720** (0.0318)
R^2	0,2695	0,2842	0,2086
ADJ R^2	0,24551	0,2499	0,1794
γ ($\beta = 0.7$)	1,0694	1,0119	1,0493
$\eta = \psi / \theta$	0,8885	0,4367	0,6817

* Values in parentheses are standard errors. Values in bold are significant at the 5% confidence level.

Since results are similar for the three years considered, let us focus on the results for 1991. The OLS estimate for θ is 1.102, with a standard error of 0.029[6]. The R^2 or the regression is relatively small (27%), reflecting the importance of county-specific technological factors. Still, density has a significant impact on local productivity, being responsible for more than a fourth of its variability across counties.

The elasticity of labour efficiency with respect to education, η, is 0.889. Since θ is significantly greater than one, so will be γ. This means that positive agglomeration externalities outweigh negative congestion effects. Assuming $\beta = 0.7$, the value for γ would be 1.069, which means that doubling employment density in a county would increase total factor productivity by almost 7%[7]. The value for θ means that, doubling employment density in county, labour productivity would increase by about 10.2%.

Tables 9.2 and 9.3 show the factor density index D_c for the top ten and bottom ten counties in terms of density, evaluated at $\theta = 1.102$ and $\eta = 0.889$, while also showing gross state product per worker (labour productivity). Not surprisingly, Greater London, with its extreme concentration of employment, provides the highest value of the density measure. The second and third counties in terms of density are West Midlands (Birmingham) and Greater Manchester. Although it shows a density index which almost doubles that of Greater Manchester, Greater London presents only a small advantage in terms of productivity. It should be pointed out, however, that, concentrating most companies' headquarters, a lot of employment in Greater London which is classified as "manufacturing" may in fact refer to service activities.

Table 9.2
Density Index and Productivity, Top 10 Counties (θ = 1.102; η = 0.89)

	Density Index	**Productivity**
Greater London	2,337	7,792
West Midlands	1,403	7,020
Greater Manchester	1,178	7,746
Humberside	0,931	5,666
Tyne & Wear	0,913	5,467
Merseyside	0,860	4,435
West Yorkshire	0,458	5,475
South Glamorgan	0,411	4,058
Cleveland	0,386	7,780
South Yorkshire	0,378	6,352
Top 10 Average	**0,925**	**6,179**

Some cases deserve further attention. For instance, Leicestershire shows a labour productivity level ranking it fourth in Great Britain, while its density index is below average. On the other side of the spectrum, Merseyside and South Glamorgan, in Wales, rank in the top ten in terms of density, but show productivity levels below average. The fact remains, however, that average productivity for the top ten counties in terms of density about doubles that of the ten bottom counties.

Table 9.3
Density Index and Productivity, Bottom 10 Counties (θ = 1.102; η = 0.89)

	Density Index	Productivity
Central	0,038	3,021
Islands	0,037	1,486
Lincolnshire	0,034	3,028
North Yorkshire	0,030	3,576
Grampian	0,023	3,408
Highland	0,022	3,514
Northumberland	0,020	3,605
Tayside	0,018	4,288
Gwynedd	0,016	2,238
Dyfed	0,015	2,700
Bottom 10 Average	**0,025**	**3,086**

The positive correlation between density and labour productivity is apparent from Figure 9.3. This figure plots the percentage deviations of the density index from its mean against percentage deviations of labour productivity from its mean. Outliers are mostly counties in Scotland, where density is much smaller than in the rest of Great Britain, and average area (in square kilometres) is much larger than average. One striking feature of the data as plotted is the perception of an upper bound to the effect of density on productivity.

Figure 9.4 plots density against productivity, where employment has been weighted by the efficiency factor computed from the education variable (again, in terms of percentage deviations from the mean). No major changes are found as regards Figure 9.3 and all outliers remain. This is possibly because the data on education can only identify differences at the regional level, and not for individual counties. Figure 9.5 plots density against county wages. The pattern is similar to that of productivity. Again, there seems to be a clear upper bound to the effect of density, which is probably associated with congestion effects.

Figure 9.3
Density and Labour Productivity by County, 1991

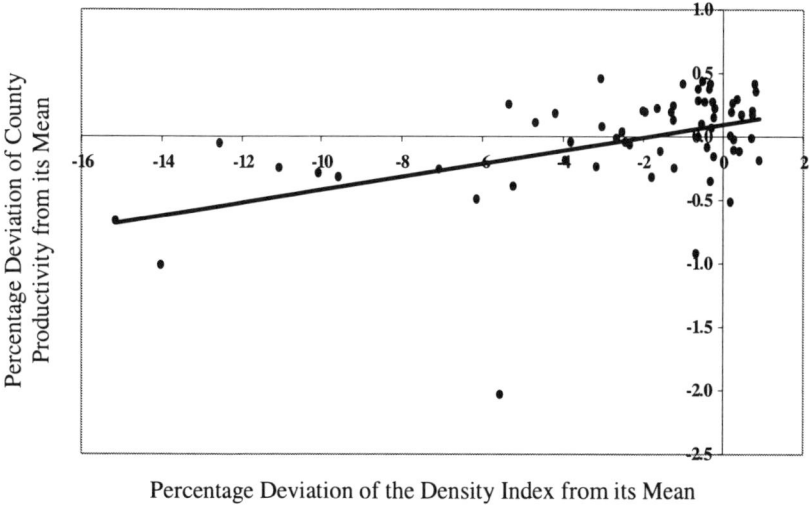

Percentage Deviation of the Density Index from its Mean

Figure 9.4
Density and Labour Productivity by County Adjusted for Education, 1991

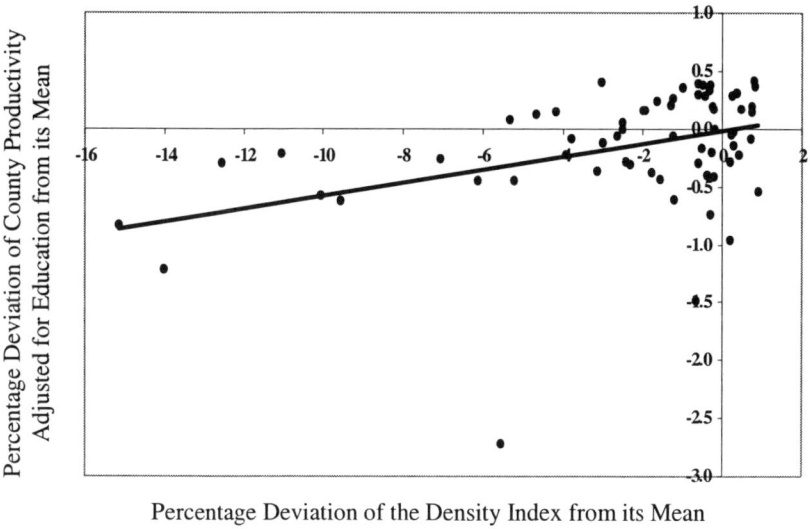

Percentage Deviation of the Density Index from its Mean

Figure 9.5
Wages and Density, 1991

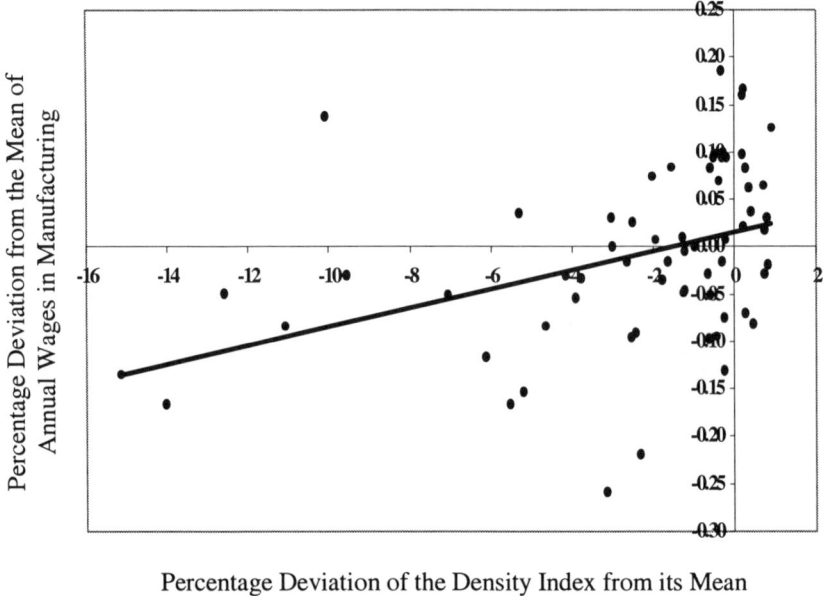

Percentage Deviation of the Density Index from its Mean

CONCLUDING REMARKS

Locally bounded increasing returns play a significant role in explaining productivity differences across U.K. counties. This chapter estimates that doubling employment density in a county would increase average labour productivity by almost 7%. The degree of locally increasing returns can explain more than 25% of the variation in labour productivity across counties, after the level of education of the work force is controlled for.

The chapter defines a county-level density index which can be matched against labour productivity. A striking result from this is the fact that there seems to be an upper bound to the effect of local density on productivity. This is possibly due to congestion effects that limit the level of density that can be achieved at the county level of geographical aggregation.

This work can be extended in several ways. One obvious research course is to compare increasing returns to density with increasing returns to size, which is the main variable used by the literature to examine the pervasiveness of agglomeration economies. One other related research subject is the distinction between agglomeration effects arising from the proximity of firms in the same industry—localisation economies—and agglomeration effects occurring across industries, that is, arising from the general diversity and scale of urban areas—

urbanisation effects. For this purpose, one needs to have data available by industry at the local (county) level.

The availability of a proper time series of data may also allow us to estimate the significance of dynamic spillovers arising from localised increasing returns on productivity. However, as it has been pointed out, data at this level of geographical aggregation has been available only intermittently.

Finally, it should be possible to introduce new elements into the local production function, in order to increase the amount of variability on local productivity that is explained by the model. Appropriate measures of local infrastructure and public capital, as well as of local innovative performance, in terms of the introduction and diffusion of significant, general-purpose, process innovations might increase the explanatory power of the model. However, one should be aware that the highest levels of public infrastructure, and also on innovative activity, are usually associated with larger and denser areas.

Ultimately, this stream of literature should play an important role in the measurement of the part of total economic growth that can be associated to localised technological spillovers and with increasing returns and rising density.

NOTES

1. The basis for the model presented here can be found in Ciccone and Hall (1996). Their models were devised to allow for aggregation up to the state level, since no productivity data at the county level was available. The data assembled for this chapter presents no such problem, which allows for considerable simplification of both models and estimation procedures.

2. Ciccone and Hall (1996) use average years of education to account for the level of education of the labour force. That measure was not available when building the present data set, so the percentage of university-educated active workers is used as an alternative.

3. Ciccone and Hall find that, for $\beta = 0.7$, the relation between γ and θ is such that the overstatement of g associated with the treatment of capital is small.

4. Great Britain has, in fact, 66 counties. The Orkney and Shetland Islands were excluded due to the lack of manufacturing output data.

5. This is the same assumption made by Ciccone and Hall (1996).

6. One striking feature of the results is the small size of the standard error for θ, making it not just significantly different from zero, but also significantly greater than one (except for 1989). Ciccone and Hall suggest that this is because, departing from $\theta = 1$, small changes in θ correspond to large changes in the density index D_c. Furthermore, given the form of the density index presented here, $\theta = 0$ would make log $D_c = 0$, which would result on density having no impact on productivity.

7. This effect will, of course, depend on the elasticity of output to total factor, β. For $\beta = 0.5$, the increase would be about 4.8%, while for $\beta = 0.3$ it would be about 2.8%.

REFERENCES

Abdel-Rahman, H. M. (1988). Product differentiation, monopolistic competition and city size. *Regional Science and Urban Economics*, 18(1), pp. 69-86.

Audretsch, D. B. and Feldman, M. P. (1996). Knowledge spillovers and the geography of innovation and production. *American Economic Review*, 86, pp. 630-640.

Baptista, R. (1998). Clusters, innovation and growth: a survey of the literature. In *The Dynamics of Industrial Clustering: International Comparisons in Computers and Biotechnology* (eds. G. M. P. Swann, M. Prevezer and D. Stout). Oxford: Oxford University Press.

Baptista, R. (2000). Do innovations diffuse faster within geographical clusters? *International Journal of Industrial Organisation*, 18(3), pp 515-535.

Baptista, R. and Swann, P. (1998). Do firms in clusters innovate more? *Research Policy*, 27(6), pp. 527-542.

Becker, G. and Murphy, K. (1992). The division of labour, co-ordination costs and knowledge. *Quarterly Journal of Economics*, 106, pp. 407-444.

Carlino, G. A. and Voith, R. (1992). Accounting for differences in aggregate state productivity. *Regional Science and Urban Economics*, 22(4), pp. 597-617.

Ciccone, A. and Hall, R. E. (1996). Productivity and the density of economic activity. *American Economic Review*, 86(1), pp. 55-70.

Dixit, A. and Stiglitz, J. E. (1977). Monopolistic competition and optimum product diversity. *American Economic Review*, 67(3), pp. 297-308.

Evans, A. W. (1985). *Urban Economics: An Introduction*. Oxford: Basil Blackwell.

Feldman, M. P. (1994). *The Geography of Innovation*. Dordrecht: Kluwer Academic Publishers.

Fujita, M. (1988). A monopolistic competition model of spatial agglomeration: differentiated product approach. *Regional Science and Urban Economics*, 18(1), pp. 87-124.

Fujita, M. (1989). *Urban Economic Theory: Land Use and City Size*. Cambridge, MA: Cambridge University Press.

Fujita, M. and Rivera-Batiz, F. L. (1988). Agglomeration and heterogeneity in space. *Regional Science and Urban Economics*, 18(1), pp. 1-5.

Fujita, M. and Thisse, J.-F. (1996). Economics of agglomeration. *Journal of the Japanese and International Economics*, 10, pp. 339-378.

Glaeser, E. L., Kallal, H. D., Scheinkman, J. and Shleifer, A. (1992). Growth in Cities. *Journal of Political Economy*, 100, pp. 1126-1152.

Henderson, J. V. (1974). The sizes and types of cities. *American Economic Review*, 64(4), pp. 55-70.

Henderson, J. V. (1986). Efficiency of resource usage and city size. *Journal of Urban Economics*, 19(1), pp. 47-70.

Hoover, E. M. (1937). Spatial price discrimination. *Review of Economic Studies*, 4, pp. 182-191.

Hotelling, H. (1929). Stability in competition. *Economic Journal*, 39, pp. 41-57.

Isard, W. (1956). *Location and Space-Economy*. Cambridge, MA: The MIT Press.

Krugman, P. (1991). Increasing returns and economic geography. *Journal of Political Economy*, 99(3), pp. 483-499.

Lewis, W. A. (1955). *The Theory of Economic Growth*. London: Allen and Unwin.

Lucas, R. E. Jr. (1988). On the mechanics of economic development. *Journal of Monetary Economics*, 22, pp. 3-42.

Lundvall, B.-A. (1988). Innovation as an interactive process: from user-producer interaction to the national system of innovation. In *Technical Change and Economic Theory* (eds. G. Dosi, C. Freeman, R. Nelson, G. Silverberg and L. Soete). London: Pinter Publishers.

Marshall, A. (1920). *Principles of Economics*. London: Macmillan.

Mills, E. S. (1967). An aggregative model of resource allocation in a metropolitan area. *American Economic Review*, 57(2), pp. 197-210.

Moomaw, R. L. (1981). Productivity and city size: a critique of the evidence. *Quarterly Journal of Economics*, 96(4), pp. 675-688.

Moomaw, R. L. (1985). Firm location and city size: reduced productivity advantages as a factor in the decline of manufacturing in urban areas. *Journal of Urban Economics*, 17(1), pp. 73-89.

Porter, M. (1990). *The Competitive Advantage of Nations*. London: Macmillan.

Rivera-Batiz, F. L. (1988). Increasing returns, monopolistic competition and agglomeration economies in consumption and production. *Regional Science and Urban Economics*, 18(1), pp. 125-153.

Romer, P. (1986). Increasing returns and long-run growth. *Journal of Political Economy*, 94, pp. 1002-1037.

Saxenian, A. (1994). *Regional Advantage: Culture and Competition in Silicon Valley and Route 128*. Cambridge, MA: Harvard University Press.

Segal, D. (1976). Are there returns to scale in city size? *Review of Economics and Statistics*, 58(3), pp. 339-350.

Shefer, D. (1973). Localisation economies in SMSAs: a production function analysis. *Journal of Regional Science*, 13, pp. 55-64.

Spence, A. M. (1976). Product selection, fixed costs and monopolistic competition. *Review of Economic Studies*, 43(2), pp. 217-235.

Sveikauskas, L. A. (1975). The productivity of cities. *Quarterly Journal of Economics*, 89(3), pp. 393-413.

10

From Technology Policy for Regions to Regional Technology Policy: Towards a New Policy Strategy in the EU

Michael Guth

INTRODUCTION

The chapter starts with a short analysis of EU regional policy conception, its objectives and results with regard to regional cohesion and innovation. Although financial resources dedicated to regional research, technological development and innovation (RTDI) activities increased tremendously from 2 billion EUR in 1989-1993 to some 5 billion EUR in 1994-1999, regional technology disparities have not decreased. On the contrary, the technology gap, measured as GERD/GDP, is about 5:1 while the cohesion gap between the poorest and the richest Member States is less than 2:1.

The reasons for the modest performance of regional RTDI policy in Europe are multifold. On the one hand the chapter describes some structural deficits of the European regional policy. A lack of regional strategy formulation, a policy which is to a large extent bound to the administration of the programmes and to the commitment of the resources rather than to the envisaged objectives and a predominance of supply-side oriented actions are the main deficits discussed here. It is argued that the technology and innovation scene in the Member States, which can be seen as closed-shop interest groups which still follow a linear vision of innovation, can be regarded as another decisive factor for the failure of the technology policy for regions in the current setting.

Against the background of this analysis the chapter outlines some basic elements for a new rationale of European RTDI policies. A starting point here is a new definition for the role of the regions. A new innovation policy in and for Europe needs to combine three elements: the European dimension, the national states and the regions. At the European level the chapter asks for a more integrated implementation of both the regional and the R&D policy lines. The new role the regions can play in R&D and innovations needs also to be appreciated by the Member States and the national R&D establishments in particular. At the regional level the chapter proposes three guiding principles for a revised regional technology policy in Europe:

1. a territorialised R&D policy approach, which takes the regional socioeconomic situation sufficiently into account;
2. strategical backing rather than ad-hoc approaches for regional innovation strategies;
3. at the fringe of the information society, regional innovation strategies must take information and communication technologies (ICT) and in particular Internet aspects on board.

COHESION GAP, TECHNOLOGY GAP AND EU REGIONAL POLICY

Economic and social cohesion has been an implicit goal of European integration. The so-called Single European Act (1986), however, for the first time codified and further defined this objective: in addition to the former EEC treaty the new article 130a delegated the political responsibility for an enhanced economic and social cohesion to the Union. And the Union was obliged to develop and conduct a policy which fosters a coherent economic development in all Member States. The new and explicitly formulated goal was to reduce income and development gaps between the regions.

In a European perspective and according to the historical experiences in Europe, large income disparities coincide with political conflicts and political radicalisation of society. It is because of this reason that regional income divergences are tolerated in Europe, but in far smaller levels than in the United States. Against this background the cohesion goal can be interpreted as the attempt to avoid critical income gaps within the European regions. With the Single European Act the Union also formulated the policy tools which should ensure a coherent economic and social development in Europe: (1) coordination of the policies in the Member States, (2) the single-market programme, (3) the European structural funds, (4) the European Investment Bank (EIB) and (5) other financial instruments. This chapter will concentrate mainly on the structural funds interventions.

The integration of the explicitly formulated cohesion goal into the treaty is evidently only one side of the coin. The economic reality in the Member States and in the regions is different. Official statistics of Eurostat and the European Commission show the existence of a still tremendous economic or cohesion gap.

In its latest report on the economic and social situation in the regions,[1] the European Commission presents the first indications for a coherent economic development in the Union. According to the Commission the per capita GDP of the ten (25) poorest regions increased from 41% (52%) of EU average to 50% (59%) between 1986 and 1996 (see Table 10.1). In the poorest Member States (cohesion four countries Ireland, Portugal, Greece and Spain) per capita GDP increased from 65% in 1986 to 76.5% in 1996. On the other hand, unemployment rate in the 25 regions suffering from the highest unemployment in Europe went up from 20.1% in 1987 to 23.7% in 1997, whereas the unemployment rate in the most prosperous 25 regions remained relatively stable (increase from 3.1% to 4.2 %).

However, if the poorest regions are catching up or not, the existent gap is still enormous. The city of Hamburg (Germany), for example, is the richest region in Europe. Its per capita GDP measured in purchase power standards (PPS) was about twice (192%) the European average (in 1996). The per capita GDP of the ten poorest regions on the other hand reached between 40% (Guadeloupe, overseas territory of France) and 55% (Dessau, Germany) of the European average in 1996.

Alongside the theoretical formulation and discussion of the new growth theory, the creation and the accumulation of knowledge was regarded both by economists and practitioners as one core precondition for economic development and the catching-up of regions lagging economically behind. That was the basis for the idea that technological development—and education and training—would stimulate regional economic growth and thus foster coherent economic developments among the regions. Therefore, beginning with the 1989 EU structural funds' programmes but with a growing significance in the programmes starting in 1994, research, technological development and innovation-oriented measures (RTDI measures) were endowed with increasing financial resources in the regional aid schemes of the EU.

However, despite increasing financial efforts dedicated to technological development, regional technology disparities have not decreased. The technology gap between the cohesion four countries and the other EU Member States (measured with indicators like number of patent application or R&D expenditures) is still larger than the economic gap. The 1995 basic indicators as presented in Table 10.2 show the gap between the poorest four Member States and the other 11 states with regard to financial input (GERD as % of GDP) as well as concerning the structure of the national innovation systems (BERD as % GERD) and concerning innovation output.

Table 10.1
GDP Per Capita, Richest and Poorest Regions of the European Union, 1986 and 1996

1986			1996		
Region	GDP*	Rank	Region	GDP*	Rank
Hamburg (D)	185	1	Hamburg (D)	192	1
Reg. Bruxelles-Cap./Brussel Hfdst. Gew. (B)	163	2	Reg. Bruxelles-Cap./Brussel Hfdst. Gew. (B)	173	2
Île de France (F)	162	3	Darmstadt (D)	171	3
Darmstadt (D)	152	4	Luxembourg (Grand-Duché)(L)	169	4
Wien (A)	148	5	Wien (A)	167	5
Greater London (UK)	148	6	Île de France (F)	160	6
Bremen (D)	144	7	Oberbayern (D)	156	7
Stuttgart (D)	143	8	Bremen (D)	149	8
Oberbauyern (D)	141	9	Greater London (UK)	140	9
Luxemborg (Grand-Duché)(L)	137	10	Antwerpen (B)	137	10
Upper 10	153		Upper 10	158	
Stockholm (S)	133	11	Stuttgart (D)	135	11
Ahvenanmaa/Aland (FIN)	132	12	Groningen (NL)	134	12
Lombardia (I)	132	13	Emilia-Romagna (I)	133	13
Uusimaa (FIN	129	14	Lombardia (I)	132	14
Valle d'Aosta (I)	129	15	Valle d'Aosta (I)	131	15
Berlin (D)	128	16	Uusimaa (FIN)	129	16
Emilia-Romagna (I)	125	17	Trntino-Alto Adige (I)	128	17
Mittelfranken (D)	124	18	Grampian (UK)	128	18
Antwerpen (B)	124	19	Friuli-Venezia Giulia (I)	126	19
Karlsruhe (D)	123	20	Karlsruhe (D)	126	20
Düsseldorf (D)	122	21	Veneto (I)	126	21
Grampian (UK)	122	22	Berkshire, Buckinghamshire, Oxfordshire (UK)	124	22
Noord-Holland (NL)	117	23	Mittelfranken (D)	124	23
Köln (D)	117	24	Stockholm (S)	123	24
Piemonte (I)	117	25	Salzburg (A)	123	25
				121	

Table 10.1 (continued)

Upper 25	138	
Guyane (F)	37	1
Gualdeloupe (F)	37	2
Alentejo (P)	37	3
Acores (P)	40	4
Madeira (P)	40	5
Réunion (F)	40	6
Centro	42	7
Voreio Aigaio (EL)	44	8
Extremadura (EL)	44	9
Algarve (P)	44	10
Lower 10	**41**	
Ipeiros (EL)	47	11
Martinique (F)	49	12
Dytiki Ellada (EL)	49	13
Norte (P)	51	14
Ionia Nisia (EL)	52	15
Andalucía (E)	53	16
Castilla-La Mancha (E)	54	17
Galicia (E)	55	18
Thessalia (EL)	55	19
Anatoliki Makedonia, Thraki (EL)	56	20
Kriti (EL)	57	21
Dytiki Makedonia (EL)	58	22
Kentriki Makedonia (EL)	58	23
Calabria (I)	59	24
Peloponnisos (EL)	61	25
Lower 25	**52**	

Upper 25	143	
Guadeloupe (F)	40	1
Ipeiros (EL)	44	2
Réunion (F)	46	3
Guyane (F)	48	4
Acores (P)	50	5
Voreio Aigaio (EL)	52	6
Martinique (F)	54	7
Madeira (P)	54	8
Extremadura (EL)	55	9
Dessau (D)	55	10
Lower 10	**50**	
Andalucia (E)	57	11
Dytiki Ellada (EL)	58	12
Magdeburg (D)	58	13
Peloponnisos (EL)	58	14
Calabria (I)	59	15
Alentejo (P)	60	16
Centro	61	17
Anatoliki Makedonia, Thraki (EL)	61	18
Thüringen (D)	61	19
Mecklenburg-Vorpommern (D)	61	20
Dytiki Makedonia (EL)	62	21
Ionia Nisia (EL)	62	22
Norte (P)	62	23
Thessalia (EL)	63	24
Galicia (E)	63	25
Lower 25	**59**	

* GDP in purchase power standards (PPS), with reference to EUR15=100.   *Source*: European Commission (1999a), p. 200, table 2.

Table 10.2
EU Technology Gap: Basic Indicators at Member State Level, 1995

	GERD as % of GDP	BERD as % of GERD	Patent application per Mio. population
EU 11	2.05	64%	108
EU 4	0.82	45%	11

Source: European Commission (1998), table 1.

The four poorest Member States enjoyed higher GDP growth rates during the 1990s than the other Member States (EU11). As a consequence the GDP gap decreased by 14 index points between 1986 and 1996.2 With regard to the input indicator, R&D expenditures per GDP in Germany and Sweden for 1989 allocated 2.87% and 2.94% of their GDP to R&D expenditures. At the same time Greece brought up only 0.38% of its GDP to R&D activities. The gap between Greece and the two countries on the top was about 7:1. According to Eurostat estimations Germany fell to 2.14% in 1997 whereas Sweden—standing at the top of all Member States—allocated some 3.61% of its GDP to R&D expenditures. By that time Greece was reaching some 0.57%.[3] The gap between Sweden and Greece was thus still about 6:1 for R&D input. With regard to another input indicator (R&D personnel as a percentage of labour force) the gap seems to be less significant. According to Eurostat, which unfortunately does not provide the data for Sweden and Germany, in Denmark (1997 estimations) an estimated 1.84% of the total labour force were either employed directly in R&D or supplied direct service to R&D. In Greece this figure reached 1%. And for Spain and Ireland Eurostat estimated 0.94% and 1.21% respectively. For Portugal, standing at the end of the European range, Eurostat estimated 0.61%.[4] The European Commission in its second report on science and technology indicators argued in 1997 that the data available confirm "that the technology gap, measured as GERD/GDP, is about 5:1 while the cohesion gap between the poorest and the richest Member States is less than 2:1".[5]

The technology disparities are even larger when we look at regional levels within the Member States. In terms of GERD/GDP in 1995 the 25 least developed regions dedicated only 0.5% to R&D while the EU average was about 2%. In terms of high-technology employment, in the 25 most advanced regions high technology accounts for some 14.6% of total employment whereas the 25 least developed regions reach only 4%.[6]

We do not have to talk only about an intra-European technology gap, we also have to observe a structural gap between the European Union and the United States and Japan as well (see figure 10.1).

Figure 10.1
Technology Gap: Europe, United States and Japan, 1998

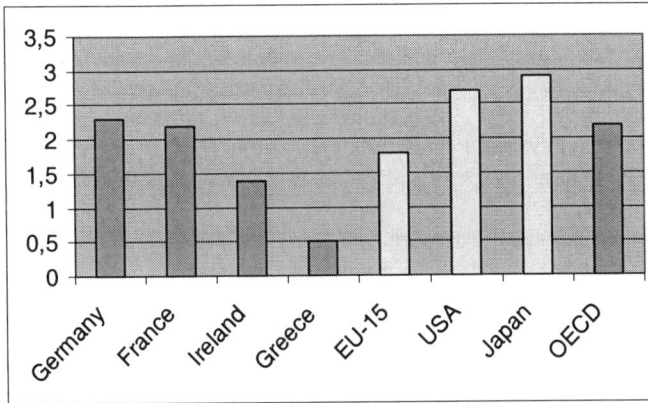

Source: OECD (2000), pp. 70ff, author's graphic.

The technology gap in Europe is far more relevant for Europe's future development than the economic disparities. If technological know-how and the pathway of transferring know-how into products determines regional economic development, the technology gap will counteract the first positive cohesion results.

The reason for the relative inefficiency of the funds spent in the field of technological development is not only the still predominant linear policy vision and practice in the regions and in the Member States. This has a transmission model in which new knowledge and basic research is being transferred via applied research processes into the technological development and implementation of new products and new production processes. The chapter attempts to show that a second reason for the failure can be seen both in the implementation processes as well as in the specific delivery mechanisms of R&D funds in the Member States.

Against this background the next session will present some basic structural elements of the regional aid policy in Europe. Special attention will be given to R&D and innovation orientated measures.

ANALYSIS OF INNOVATION POLICY IN
THE FRAMEWORK OF EU REGIONAL POLICY

EU structural funds interventions are based on cooperative programmes which are borne both by the European Commission and the Member States. The regulations differentiate between programmes initiated by the Member States and those initiated by the European Commission. The first are programmes which follow the so-called priority objectives defined by the European Council. And the latter are the so-called Community Initiatives which are based on guidelines published by the Commission. Community Initiatives supplement the programmes initiated by the Member States and concentrate on issues and strategies of particular interest for the Union as a whole. In both cases the actual intervention programmes must be formulated by the Member States. These two intervention procedures represent the vast majority of the available funds. In the current period about 5.35% are earmarked for Community Initiatives and 94% for the programmes initiated by Member States. A third intervention procedure is the so-called innovative measures (see Figure 10.2). With these financial resources the Commission can start its own initiatives, pilot actions, studies and so on.

RTDI-oriented measures were supported in all three intervention categories. According to estimations of the European Commission about 2 billion EUR were spent in the field of RTDI between 1989 and 1993.[7] The spending philosophy was for the most part based on the linear innovation process and thus primarily supply-side-driven. As a result in particular technology and technology transfer infrastructures (research institutes, research and technology centres, technology transfer centres) emerged both in the less favoured regions (so-called objective 1 regions) and in the old industrialised regions facing severe structural change and industrial unemployment (so-called objective 2 regions).

As a result of the new theoretical wave of emphasising that innovation cannot simply be explained in a linear model, the European Commission in its guidelines for programming attempted to shift RTDI policy measures from supply-side orientation to more process-oriented measures which support the interaction between the firms, the knowledge producer, the technology organisations, the policy makers and other relevant stakeholders. In short, the European Commission intended to integrate the new concept of innovation systems into the regional policy programmes of the Member States, in particular into objective 1 and 2 programmes.

Figure 10.2
Intervention Procedures and Relative Resources of EU Structural Funds

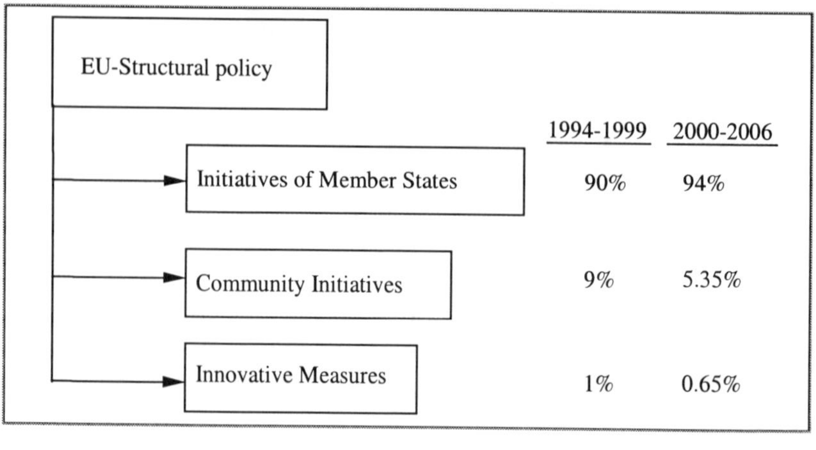

		1994-1999	2000-2006
EU-Structural policy			
	Initiatives of Member States	90%	94%
	Community Initiatives	9%	5.35%
	Innovative Measures	1%	0.65%

As a result the budget allocated to RTDI measures increased from 2 billion EUR (1989–1993) to 5 billion EUR in 1994-1999. A first analytical look into that major spending block and its results provided two thematic evaluations which were conducted on behalf of the European Commission (DG Regio).[8] The main results of these two evaluation works are:

- Until the end of the 1994-1999 intervention period, in objective 1 programmes about 50% of RTDI budget was dedicated to technology infrastructure and public research institutes.
- In objective 2 regions there was a clear shift to be observed in 1994. Generally the importance of RTDI measures grew with the new generation of programmes during 1994–1996. Also, the 1994-1996 period represents a major change in the strategic conception of RTDI in the programme documents. From 1994 onwards one can observe an increasing shift in favour of innovation which became the most important category of action by 1999. However, infrastructural measures (e.g., support for business innovation centres which represent an infrastructural innovation measure) have remained predominant.
- Both studies criticise the absence of appropriate theoretical and strategical foundations of the RTDI measures implemented in the different regions.
- Actual results of the interventions were rather modest. At least a systematic review of the results could not be provided in objective 1 or objective 2 evaluations. Both studies, however, display interesting cases of success.

As far as this author is concerned, the results show that although both the theoretical thinking of the economists and the political intentions of the European Commission have shifted from the linear model of innovation to an innovation system approach, Member States' implementation policy obviously still sticks in the linear model.[9] The lack of strategic backing also indicates an implicit thinking which follows the linear model.

Here it is argued that the reasons for the modest performance of RTDI measures within the structural funds programmes lie for the most part in the technology and innovation systems in the different Member States. Also, there is the risk that the partnership approach of structural funds in the given status, which requests the integration of all relevant socioeconomic regional actors into the programming and implementation phase of the programmes, and which at first glance might be regarded as a sensible and quality-assuring requirement, has counterproductive effects as well. The RTDI systems and the subsequent programmes in the Member States are for the most part horizontally organised. Additionally it holds that in particular in the major countries national RTDI programmes follow the rule of supporting excellence in terms of technological development or research. A national regional cohesion goal (if at all) comes only third in the objective pyramid.

These developments have created a specific structure of organisations, institutions and other stakeholders (in short, a specific R&D and delivery system) which has encompassed both the horizontal R&D strategies and the linear innovation model. In the programming phase the R&D establishment defends its particular interest and hinders new approaches being integrated into the structural funds programmes. From a theoretical point of view the RTDI establishment can be regarded as a cartel or an Olson-type rent-seeking group.

A further point which also explains modest results and shows a lack of adequate strategy is the dominant absorption orientation of EU Member States. As a matter of fact, in particular net contributors but also the other Member States intend to receive as much structural funds resources as possible. Also, the Member States and especially the national programme managers intend to spend all resources allocated to them. Research and technology development (RTD) infrastructural measures generally are more expensive and are far easier to manage than measures which stimulate regional demand for R&D, innovation or innovative products. Thus, programme managers systematically prefer infrastructure measures even if RTD infrastructure in specific regions is already well developed or even beyond saturation point.

Both points lead to rather obsolete supply-side oriented programmes. At the same time the situation makes it difficult for the European Commission to break up the existing cartels in the countries and to foster a more "innovation system" oriented RTDI approach.

Against this background, the next section will outline some basic elements for a new rationale of European RTDI policies.

A NEW RATIONALE FOR REGIONAL INNOVATION POLICY IN THE EU

The question should be raised, why is the regional dimension of technology policy and innovation important? Earlier in this chapter the significance of the technology gap in Europe was discussed. It became evident that there is a huge divergence to be observed at the level of the Member States and—even more significant—at the regional level. Here it is argued that at the fringe of both the knowledge economy and the information society regions need to innovate. That is a must in particular for regions lagging economically behind but also for those facing technological or industrial changes. In its typology of innovative regions the Commission identified 13 regions (the whole of Greece, apart from Athens, Galicia in Spain, Calabria in South Italy and major parts of Portugal) as sleeping birds with high dependence on agriculture, low economic growth and very limited technological level. And 33 regions were classified as question marks. The question-mark regions are somewhat rural, enjoy considerably high economic growth rates, show some technological activity but face the highest average unemployment rate.[10] The regions in these two clusters need innovation in order to develop comparative advantages and eventually to catch up in economic terms. Obviously this argument is also credible for the so-called cash-cow regions and for the eight European star regions. To sum up, it is the regional level where the disparities with regard to both economic indicators and R&D are most significant. Innovation strategies must therefore have a real regional dimension.[11]

In its communication to the European Council "Towards a European Research Area" the Commission formulated policies totally in line with the views presented here: "the conditions must be studied and put in place for a real territorialisation of research policies (adaptation to the geographical socio-economic context), and a better understanding and strengthening is needed of the role that the regions can play, in addition to the Member States and the Union, in the establishment of a more dynamic European research area on the international scene."[12]

This little paragraph points out two major issues: First, the prevalent RTDI policy strategies at the national level do not take the regional economic situation properly into account. That provides room for policy strategy formulation and policy action at the regional level. These regional technology policy strategies can be tailor-made to the specific regional environment and to the regional demands.[13] To be sure, the regional activities are to supplement national and EU activities and not to substitute them. Second, internationalisation processes are heavily interlinked to regional activities. R&D activities or investments into a new technology always take place at a given location in a specific region. Furthermore, foreign direct investment settlements are also bound to a region in the target country.

The new policy vision of the European Commission is in line with the views presented here. However, the reality in Europe is somewhat different. This chapter has shown two issues: (1) the horizontal orientation of the RTDI policy scene in the European Member States, not taking the regional situation sufficiently into account, and (2) the deficits of RTDI actions in the framework of structural funds interventions (lack of strategy, supply-side driven, absorption oriented). The European Union, entering the information society and at the edge of the enlargement towards the countries in Eastern Europe, cannot afford to neglect the potential and the role of the regions. This holds in particular with regard to innovation policy. A new innovation policy in and for Europe needs to combine three elements: the European dimension, the national states and the regions (see Figure 10.3).

At the European level the so-called framework programmes on R&D are the most important policy instruments. The main objective of EU R&D policy is to provide impulses for welfare increasing progress in important technological fields and to stimulate the dissemination of new technologies.[14]

The budget of the current 5th framework programme (1998-2002) amounts to 14.96 billion EUR. The relative financial capacity of the common R&D programmes is with some 5.4% of total public R&D expenditures (in civil sectors) rather marginal. According to the Treaty (Art. 130h) the Union and the Member States should coordinate their R&D activities. The Commission very much emphasises the coordination of the national programmes. Furthermore, the Commission argues that "the non-existence of a European research area is due to the compartmentalisation of public research systems and to the lack of coordination."[15] A second financial resource with regard to RTDI is given in the structural funds programmes. Within the structural funds a budget of some 5 billion EUR was allocated to RTDI for 1994-1999. The synergy between the cohesion-oriented structural funds and the R&D framework programmes, however, still needs to be better exploited.

Figure 10.3
Three Elements of a New Innovation Strategy in Europe

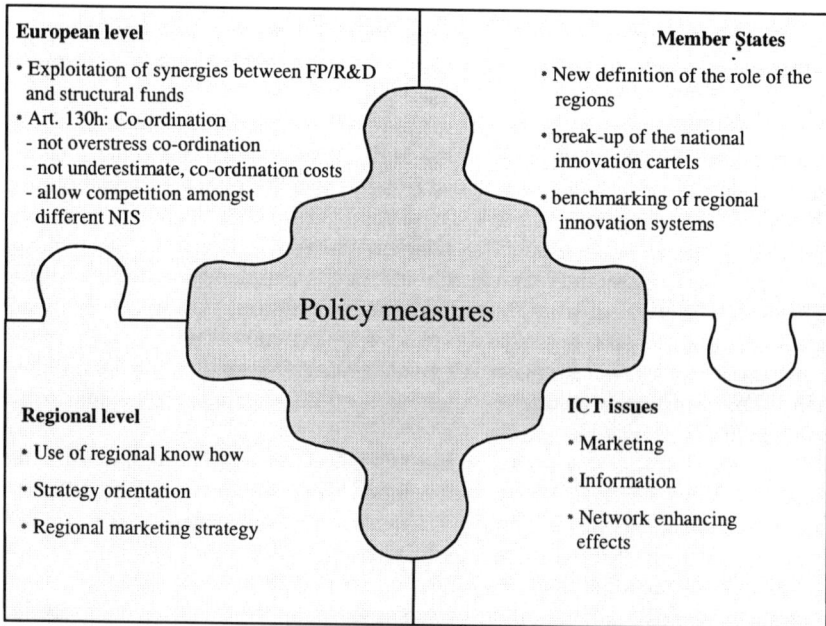

European level
* Exploitation of synergies between FP/R&D and structural funds
* Art. 130h: Co-ordination
  - not overstress co-ordination
  - not underestimate, co-ordination costs
  - allow competition amongst different NIS

Member States
* New definition of the role of the regions
* break-up of the national innovation cartels
* benchmarking of regional innovation systems

Policy measures

Regional level
* Use of regional know how
* Strategy orientation
* Regional marketing strategy

ICT issues
* Marketing
* Information
* Network enhancing effects

At the level of the Member States we are also arguing for a new definition of both the regional dimension of R&D policy and the role the regions can play in R&D policy in order to better exploit the European innovation potential. With regard to the coordination between EU and Member States' policies the coordination efforts should not be underestimated. A balance between necessary coordination on the one hand and competition in the field of R&D and innovation-oriented policy conceptions on the other would be more pragmatic, less bureaucratic and thus eventually more efficient. This analysis of R&D policies at the European level and in the Member States is obviously rather superficial. An in-depth analysis would however go far beyond the scope of this chapter, which is concentrating on regional issues.

Coming back to the regional level, a core recommendation is the formulation of regional RTDI strategies based on a comprehensive analysis of the situation. The key for more success is a regional approach which is based on the know-how and experiences of the regional innovation system. A second point which inevitably coincides with regionalisation is a requirement for a better monitoring of the implementation of the measures. With regard to this respect the author argues for specific regional innovation monitoring systems which must also include all relevant RTDI and political actors as well as the enterprise sector in a region.

At the fringe of the realisation of the information society in Europe regional innovation strategies must take ICT and in particular Internet aspects on board.

On the one hand the World Wide Web provides access to knowledge and know-how available in Europe and elsewhere. On the other hand regional innovation networks must not neglect the network-enhancing effects of the Internet. The Irish case of outstanding growth and rapid structural economic change shows that the new knowledge-based economy coincides with the need for revised innovation strategies. Traditionally the Irish economy had its competitive advantages in the food and transport equipment sectors. In its innovation strategy the government, however, did not concentrate on the formerly strong sectors of the economy. On the contrary, Ireland has concentrated its innovation strategy on sectors in which almost no indigenous potential used to be available, namely soft-ware and electronic equipment. Mainly multinational investors provided know-how and/or research capacity in this field. Based on a geographical (Krugman) perspective it is obvious that light weighted (or weightless) technology-intensive products are well suited for peripheral regions provided there is an adequate endow-ment of human capital.[16] Products and services which can be digitised thus may provide an advantage for peripheral regions even if these products do not belong to the traditional regional industrial portfolio. Regional innovation strategies should take that into account. This seems to be of particular importance for the accessing countries in Eastern Europe as well.

Innovative measures in structural funds would be able to provide room for experimentation and piloting of new policy approaches. However, as Figure 10.2 showed, the relative budget earmarked for innovative measures diminished remarkably from 1% in 1994-1999 to 0.65% for the period 2000-2006. Further-more, with regard to the European Regional Development Fund (the main financial instrument in regional policy) the Commission plans to address the new generation of innovative actions directly to the regional governments. Additionally it is planned to base the interventions on specific programmes (rather than on a project base as before). Against the background of the situation in the Member States the new approach, which has a charming intellectual setting, is too straightforward. As RTDI policy currently is dominated by the national governments, one would have to assume that the regions do not have their own money resources to cofinance regional innovation programmes. If the money comes from the national level (or from the level which is currently running R&D programmes; in Germany, e.g., that could also be the Länder level) there is the risk that the innovation budget will be absorbed by traditional policy actions.

Innovation is the only strategy for the European regions to gain or to main-tain competitive advantages. The European R&D policy as well as the regional programmes are important policy tools which can support this strategy. Support for Eastern European accession countries could be difficult if the economic and technological disparities in the EU remain resistant. A new vision of the role of the regions, a territorialised R&D policy approach, which takes the regional socioeconomic situation into account, and a strategic backing of regional actions could be guiding principles for a revised regional technology policy in Europe.

NOTES

1. See European Commission (1999a).
2. See ibid, p. 199, table 1.
3. See Eurostat (1997), p. 75, table 24.
4. See ibid, table 23.
5. See European Commission (1997), p. 340.
6. Eurostat data. Cited from Crauser (2000).
7. see European Commission (1999b), p. 126.
8. For objective 2 see Zenit GMBH, Ade S.A. and Enterprise PLC (1999). And for objectives 1 & 6 see Circa Group et al. (1999). Both documents downloaded in March 2000: http://www.inforegio.org/wbdoc/docgener/-evaluation/rado_en.htm.
9. For an introduction to the systemic approach of innovation see Lundvall (1992). A more recent and empirical-based work on national innovation systems can be found in OECD (1999).
10. See European Commission, (1997), p. 360f.
11. See Braczyk, Cooke, Heidenreich, et al. (1998).
12. European Commission (2000), p. 26.
13. That also implies that not every region has to find its future in high technology. The regional situation may make other strategies seem to be more successful. This can be an adaptation-oriented approach but also an innovation in services. In any case regional innovation policy strategies would have to take the strengths and weaknesses, the opportunities and threats of the given territory into account. This chapter however focuses on technology-oriented innovation strategies.
14. For more details and a discussion, see Welfens, Audretsch, et al. (1998), p. 141ff.
15. European Commission (2000), p. 13.
16. See Gorg and Ruane (1999).

REFERENCES

Besse, S. and Guth, M. (2000), Europäische territoriale Beschäftigungspakte—Darstellung, Beispiele und erste Ergebnisse, in Friedrich-Ebert-Stiftung, Ed. (2000) *Europäisch denken—vor Ort handeln, Perspektiven lokaler Beschäftigungspolitik*, Department Arbeit und Soziales, No. 93, Bonn, May 2000.

Braczyk, H.-J., Cooke, P., Heidenreich, M. et al., Eds. (1998), *Regional Innovation Systems*. London: UCL Press, 1998.

Circa Group et al. (1999), *Thematic Evaluation of the Impacts of Structural Funds (1994/99) on Research, Technology Development and Innovation (RTDI) in Objective 1 and 6 regions*, 1999. Available at http://europa.eu.int/comm/regional_ policy/ sources/docgener/evaluation/rdti_en.htm.

Crauser, G. (2000), *Regional Innovation Policy under the new Structural Funds*. Speech of the Director General of DG Regio of the European Commission. Available at http://europa.eu.int/comm/regional_policy/index_en.htm.

Deilmann, B. (1995), Wissens—und Technologietransfer als regionaler Innovationsfaktor, Duisburger geographische Arbeiten, Duisburg 1995.

European Commission (1997), *Second Report on Science and Technology Indicators*. Luxembourg, 1997.

European Commission (1998), *Reinforcing Cohesion and Competitiveness through Research, Technological Development and Innovation*, COM(1998)275 final.

European Commission (1999a), *Sechster periodischer Bericht über die sozio-ökonomische Lage und Entwicklung der Regionen in der Europäischen Union* (Sixth Periodic

Report on the social and economic situation and development of the regions of the European Union). Luxemburg, 1999. Available at http://europa.eu.int/comm/ regional_policy/sources/docoffic/official/reports/radi_en.htm.

European Commission (1999b), *Die Strukturfonds in 1998. Zehnter Jahresbericht* (The Structural Funds in 1998. Tenth Annual Report). Luxemburg, 1999.

European Commission (2000), *Towards a European Research Area.* Communication from the Commission to the Council, the European Parliament, the Economic and Social Committee and the Committee of the Regions, COM(2000)06 final.

Eurostat (1997), Research and Development, *Annual statistics*. Luxembourg, 1997.

Görg, H. and Ruane, F. (1999), European Integration and Peripherality: Are There Lessons from Ireland? *Trinity Economic Papers Series*, Paper No. 99/10.

Lundvall, B.-Å., Ed. (1992), *National Systems of Innovation: Towards a Theory of Innovation and Interactive Learning*, London: Pinter, 1992.

OECD (1999), *Managing National Innovation Systems*. Paris, 1999.

OECD (2000), *OECD in Figures*. Paris, 2000.

Welfens, P.J.J., Audretsch, D. et al. (1998), *Technological Competition, Employment and Innovation Policies in OECD Countries*. Heidelberg: Springer, 1998.

Welfens, P.J.J. and Guth, M. (1996), *EU-Strukturpolitik in Deutschland: Entwicklung, Effizienzüberlegungen und Reformoptionen* (EU-Structural policy in Germany: Development, efficiency considerations and reform options), Discussion Paper No. 21, University of Potsdam, European Economy and International Economic Relations, Potsdam, 1996.

Zenit GMBH, Ade S.A. and Enterprise PLC (1999), *Evaluation of Research, Technological Development and Innovation Related Actions under Structural Funds (Objective 2)*, 1999. Available at http://europa.eu.int/comm/regional_policy/sources/docgener/evaluation/rafon_en.htm.

11

Regional Innovation Policy at the Community Level: Evidence from the RITTS Programme to Promote Regional Innovation Systems

Fabienne Corvers

OBJECT AND OBJECTIVES OF THE CHAPTER

This chapter will describe, analyse and evaluate a European programme abbreviated RITTS (Regional Innovation and Technology Transfer Strategies and Infrastructures) financed by the European Commission in the area of regional innovation policy. The RITTS scheme supports regional actors in designing a regional innovation strategy in order to promote regional development which is demand-led and consensus-based. The RITTS initiative can be seen as a strategic planning tool intended to promote regional innovation systems throughout the European Union and beyond. Between 1994 and 1999, three generations of RITTS projects have seen the light, covering over 70 regions throughout the European Union, Norway and Iceland[1].

The role of the European Commission is that of facilitator—providing European regions with a methodological framework to facilitate the design of regional innovation strategies. RITTS projects share the conviction that regional actors, regional governments in particular, should and can play an important role as a catalyst, a facilitator and a broker in the articulation of the regional innova-

tion system (Landabaso et al., 1999). Based on the philosophy of the programme, this chapter will assess whether RITTS can contribute to the promotion
of regional innovation systems.

In order to be able to assess the merits of the RITTS scheme, this chapter
will start with a short overview of the literature on national and regional systems
of innovation and will describe why this innovation systems approach is appealing to policymakers. Besides new academic insights, a series of policy events
influenced the way of thinking in EU policy circles of which the RITTS
programme was one of the results. A short overview of these policy events and
policy documents will be given, and the RITTS programme itself will be
described and analysed in terms of results achieved. The merits of the RITTS
scheme will be evaluated in light of being a strategic planning tool to promote
regional innovation systems. This will be followed by some recommendations
for future policy intervention at the Community level in the area of regional
innovation policy based on the evaluation results. Some sections rely heavily on
the recent RITTS evaluation study carried out by Charles et al. (2000). This
chapter will end with a conclusion and an overview of the references used.

THEORETICAL OVERVIEW OF SYSTEMS OF INNOVATION

National Systems of Innovation

The concept of a national system of innovation was first introduced by
Bengt-Åke Lundvall in the book he edited, *National Systems of Innovation—
Towards a Theory of Innovation and Interactive Learning*, published in 1992.
The idea, however, can be found in the work of Friedrich List called *The
National System of Political Economy* published in 1841 which, according to
Chris Freeman (1995) could also have been titled *The National System of
Innovation*. Another book that draws attention to the national systems of inno-
vation concept was the one edited by Richard Nelson, *National Innovation
Systems—A Comparative Analysis*, published in 1993. Since it was launched in
the early 1990s, the concept's appeal to both academics and policymakers has
increased. One explanation for its success could be that the concept of national
systems of innovation could be regarded as a tool for analysing economic
development and economic growth (Lundvall, 1996:24). Given its assumptions,
this concept seems to offer better insights in economic phenomena and can adapt
economic theories, policies and institutions to what is going on in the real world
(Lundvall, 1996:17). Archibugi and Michie (1995) conclude that, despite
increasing economic globalisation, "national systems of innovation take on more
rather than less importance". Also policymakers have been charmed and inspired
by the concept, such as the European Commission when designing the RITTS
programme.

Within the national systems of innovation (NSI) concept, the evolutionary,
systemic view on innovation underlines the social process of innovation, unlike
the neoclassical, linear research-to-market model of innovation. Freeman (1995)
defines a national system of innovation as "the network of institutions in the

public and private sector whose activities and interactions initiate, import, modify and diffuse new technologies" (see also OECD, 1992). Soete and Arundel (1993) speak of the capacity of firms to create, diffuse, apply and adapt technological knowledge whereby firms do not operate in a vacuum, but within a knowledge system. The ways in which the elements of creation, diffusion, application and adaptation of this knowledge system interact as well as their interplay with social institutions (such as laws, values, norms) determine the overall innovative performance of an economy. According to Freeman (1995: 14) variations in national systems can be contributed to the variety of national institutions which affect the relative rates of technical change and hence of economic growth.

Nelson and Rosenberg (1993:4) study the NSI concept through "an orientation [that] is not limited to the behaviour of firms at the forefront of world's technology, or to institutions doing the most advanced scientific research, ... but is more broadly on the factors influencing national technological capabilities". According to them, the NSI concept has been inspired by a combination of "a strong belief that the technological capabilities of a nation's firms are a key source of their competitive prowess, with a belief that these capabilities are in a sense national, and can be built by national action" (Nelson & Rosenberg, 1993:3). Differences between national systems of innovation are considered to explain differences in national economic performance.

According to Lundvall (1992:2) "a system of innovation is constituted by elements and relationships which interact in the production, diffusion and use of new, and economically useful, knowledge and that a national system encompasses elements and relationships, either located within or rooted inside the borders of a nation state[2]".

The Usability of the Innovation Systems Approach in Public Policy

The fact that evolutionary economics place innovation at the heart of industrial dynamics and economic growth of both nations and regions has triggered of a lot of policy interest in the topic. Lundvall (1992:5) gives three reasons why the NSI concept may be useful for the design of public policies in the area of research, technological development (RTD) and innovation (RTDI). First, the NSI concept might be helpful in analysing the specific systemic context in which national government intervenes and making it understood. In other words: which elements and relationships exist and how do they interact in the production, diffusion and use of new, and economically useful, knowledge? Second, the NSI concept might help to understand how other national systems of innovation work. And third, once the workings of the respective national systems of innovation are understood, one can start to learn from successes as well as failures. In other words: which elements and/or relationships have been beneficial for or detrimental to the production, diffusion and use of new, and economically useful, knowledge?

In order to be able to design innovation policy and to assign proper roles to the public and private sectors in enhancing technological capabilities, a better understanding of national systems of innovation is required (OECD, 1992). An equal argument can be used for the design of innovation policy at the regional level—that is, a better understanding of the regional system of innovation.

The fact that innovation is considered to be at the heart of economic growth of both nations and regions, combined with the fact that innovation processes do not take place in isolation, but within a knowledge system, has triggered a lot of policy interest in the question: How do we design a national/regional system of innovation? According to Cooke et al. (1997:478) this gives rise to a serious problem of what a national/regional system of innovation actually is: a conceptual or an operational system?

Some authors, such as Carlsson (1995) argue that a national/regional system of innovation is the objective of technology policy to design. Others such as Nelson and Rosenberg (1993) argue that a national/regional system of innovation is not purposively designed, but has been shaped by the history of industrialisation of the country, the laws of a nation, the existence of a common language and a shared culture, the national science, education and training system, the style of politics, the scale of public policies and programmes at micro, meso and macro levels, the competitiveness and export orientation of firms and so on.

Nelson (1993:517-520) sees a number of difficulties using the NSI concept in policy circles. First, the innovation capabilities of a country depend on a wide range of factors, including labour markets, financial systems and monetary, fiscal and trade policies. It is therefore not possible, when designing innovation policy, to focus only on the institutions and mechanisms that deal with innovation in the narrow sense, thereby neglecting the wider perspective or "Umfeld" in which firms operate. Second, national systems of innovation perceived as a conceptual system suggest more uniformity and connectedness than exists in real life (Nelson, 1993:518). What about intrasystem conflict (Cooke et al., 1997:478)? Third, increasing globalisation and regionalisation of the economy might question the importance of national systems of innovation for innovation processes.

Yet the NSI concept has already entered the vocabulary of policymakers at the regional, national and international levels since the early 1990s (Lundvall, 1993:5)[3]. This has led Nelson (1993:520) to point out two dangers of NSI thinking among policymakers. In the first place, policymakers have shown a tendency to copy what they perceive as successful in other NSIs without systematic analysis of the underlying success factors (e.g., Silicon Valley as a blueprint). In the second place, policymakers have not refrained from accusing other national systems of innovation of illegitimate aspects which create unfair competition (e.g., MITI support to key Japanese industries). In other words, the reality of the NSI concept among policymakers seems to be to either copy or condemn other national systems of innovation.

Regional Systems of Innovation

Based on the concept of national systems of innovation, a new, yet related concept called regional systems of innovation (RSI) has emerged (Morgan & Nauwelaers, 2000; Braczyk et al., 1998; Cooke et al., 1997; Edquist, 1997). Both concepts stress the relationship, if not dependence, of the individual firm on its surroundings (the business environment) to become and stay competitive, whereby the firm is no longer seen as an isolated actor (Hassink, 2000; Kilper & Latniak, 1996). If national variations in economic growth could be contributed to differences in national systems of innovation, so could regional growth differentials be contributed to differences in regional systems of innovation.

According to Arcangeli (1993) it should not be forgotten that regional systems of innovation and economies of agglomeration have always underpinned national systems of innovation from the beginnings of the industrial revolution. Also Lundvall (1992:4) agrees that more than hundred years ago, when the rapid industrialisation and modernisation of European countries started, the international specialisation was often reflected in a regional specialisation within the countries. Authors such as Storper (1991), Camagni (1990) and Porter (1990) have all assumed that globalisation and international specialisation have their roots in the strengthening of specialised technological districts and regional networks (Lundvall, 1992:3).

Research done on high-tech regions (e.g., Saxenian (1994), on Silicon Valley) have underlined the importance of geographical scale for innovation processes. As knowledge plays a crucial role in the evolutionary view on innovation, so does the regional level play a crucial role in providing access to knowledge and favouring learning processes among regional firms based on proximity. As Cooke and Morgan (1994) suggest, "innovation is first and foremost a collective social endeavour, a collaborative process in which the firm, especially the small firm, depends on the expertise of a wider constituency than is often imagined (workforce, suppliers, customers, technical institutes, training bodies, etc)." The organisational capacities of these networks of relationships become a crucial determinant of the performance of what can be labelled as the regional system of innovation (Nauwelaers & Reid, 1995:17). According to Nauwelaers and Reid (1995), the notion of regional systems of innovation stresses the importance of both collaborative networks and the openness of the local environment to internal and external factors influencing the innovation process. They have presented their regional system of innovation in Figure 11.1.

Figure 11.1
Regional System of Innovation

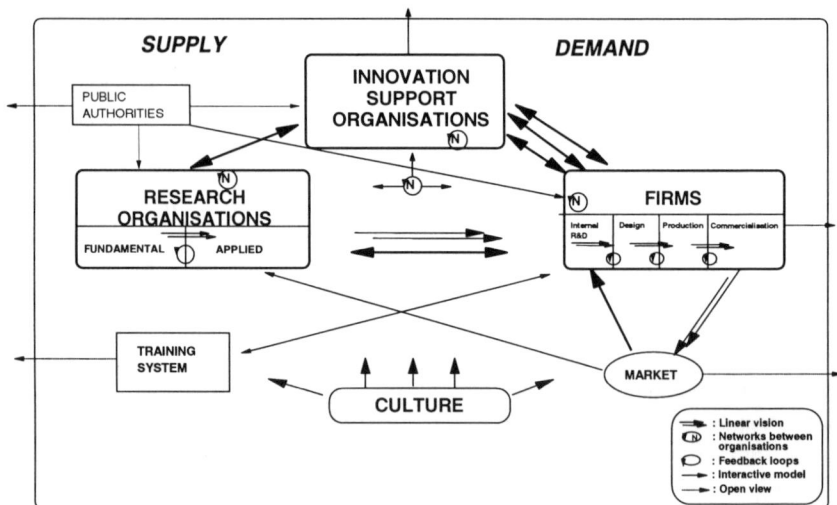

Elements of a Regional System of Innovation

On the basis of Figure 11.1, a supply side can be distinguished, involved in the creation of knowledge and materialised in organisations dealing with science, education and training such as research organisations, universities, training institutes as well as a demand side. The demand side refers to firms which are using the scientific and technological output of the supply side, but also of other firms in the demand side. Any policy which claims to be demand-led should take the needs of firms (financial, personnel, managerial, organisational, technological, etc.) as a starting point for policy design. In between are the innovation support organisations which bridge the gap between supply and demand. Depending on their assignment(s) these organisations can range from technical centres, technology broker organisations, service companies in the technology field, chambers of commerce, regional development organisations, business innovation centres, interface units of universities and research organisations to venture capital firms, innovation finance providers and (regional) banks.

Nauwelaers and Reid (1995:20-21) distinguish a fourth element which they call "the environment of the regional system of innovation" consisting of:

- the market—main stimulus, but also final beneficiary of innovation;
- training system—producer of skilled manpower necessary to undertake research in the supply side and to implement innovation processes in the demand side;
- public authorities—their policies affect both supply and demand side as well as innovation support organisations;

- culture—materialises in the prevailing entrepreneurial culture, the level of regional identity, the extent of collective confidence in regional agents, and so on.

Finally, Nauwelaers and Reid put a lot of emphasis on the degree of openness of the regional system of innovation. The arrows in Figure 11.1 depict the network of relations between the elements within the RSI and between the elements and the outside world (such as other regional systems of innovation, the national system of innovation, other national systems of innovation, other regional systems of innovation in other national systems of innovation).

POLICY EVENTS AND DOCUMENTS

The European Commission is certainly one of the policymaking organisations where the systems of innovation concept has influenced the design of policy programmes, of which the RITTS programme is a clear illustration. Within RITTS, the innovation systems approach is used as a tool, first to analyse the regional system of innovation in a systematic manner, and second to support the design of a regional innovation strategy based on the results from the analysis.

The European Single Market Initiative

However, the new academic insights of the late 1980s and early 1990s are only partly responsible for shaping new policy ideas in the area of regional and technology policy. The design of the RITTS programme was also influenced by the main policy event of the 1980s, the launch of the European Single Market initiative in 1986. The Single Market would be the next logical step in the economic integration process of the European Economic Community, facilitating the free movement of goods, services, labour and capital between the then 12 Member States. As the Single Market would abolish the remaining nontariff trade barriers among the Member States, national industries would be exposed to European-wide competition (Corvers, 1994). The European Single Market promoted greater concern with the competitiveness and productivity of Europe's firms. For policies and programmes designed by the European Commission, competitiveness became the keyword and their focus shifted accordingly.

The European Community's regional policy, for example, focused increasingly on assisting the restructuring of regional production systems in order to make regions more competitive. RTD and, more generally, the capacity to innovate and upgrade, particularly in products and processes, were considered to be vital, although not the only, components of regional competitiveness (CEC, 1994a).

The European Community's technology policy was increasingly considered to be most effective in making Europe more competitive and cohesive at the regional level, given the importance of small and medium-sized enterprises (SMEs). This change in focus was influenced by two important policy documents: the White Paper on Growth, Competitiveness and Employment from 1994 (CEC, 1994b) and the Green Paper on Innovation from 1995 (CEC, 1995) further detailed in its subsequent First Action Plan for Innovation of 1996 (CEC, 1996).

White Paper on Growth, Competitiveness and Employment

The White Paper on Growth, Competitiveness and Employment identified a number of weaknesses from which Europe suffered, including unduly low levels of RTD investment, a lack of coordination at various RTD levels and a comparatively limited capacity to convert scientific breakthroughs and technological achievements into industrial and commercial success (CEC, 1994b). A similar list of weaknesses emerged from the Green Paper on Innovation (CEC, 1995). The White Paper identified the need to define a global strategy bringing together the public authorities, research bodies and the various sectors of society concerned, while the Green Paper stressed the importance of the regional level in the formulation and implementation of such a strategy.

Green Paper on Innovation

In 1996 the European Commission launched a big debate on innovation in Europe. The basis for discussion was the aforementioned Green Paper on Innovation, adopted by the European Commission in December 1995. The Green Paper analysed in detail the climate for innovation in Europe, concluded that improvement was essential and presented a comprehensive set of proposals. The main question of the debate was "what are the factors that encourage—and discourage—innovation in Europe?" Despite Europe's excellent scientific performance, the technological and commercial performance in high-technology sectors such as electronics and information technologies has deteriorated over the last 15 years. Four main obstacles were identified:

1. The financing of innovation: Europe did not seem to have the financial mechanisms to fulfil the needs of innovative growth firms to the same extent as its rivals.
2. The protection of innovation: the patent system in Europe was more costly, less understood and therefore less used than in the United States or Japan.
3. The administrative environment was more complicated than it need be, particularly for small and medium-sized enterprises.
4. The research effort was insufficient, particularly when measured in number of R&D scientists and engineers in the workforce (EU 4.5 per thousand compared with 7.6 in the United States and 8.0 in Japan).

First Action Plan for Innovation

In order to remedy the obstacles, an Action Plan was prepared and adopted by the European Commission in December 1996. The Action Plan addresses three key questions:

1. how to foster a real innovation culture in Europe (focus on education and training, mobility, innovation management in companies);
2. how to make sure that the environment in Europe allows innovation to thrive (focus on establishing a favourable regulatory and administrative framework, intellectual property rights);
3. how to improve the link between Europe's research capabilities and innovation (focus on start-up of technology-based companies, university spin-offs).

The 17 action lines suggested for implementation—at regional, national and Community level—fall into these three areas.

Regional Innovation Policy at the Community Level

Shaped by the aforementioned policy events and documents, regional innovation policy emerged at the Community level, early in the 1990s, in a partly cooperative, partly competitive liaison between two directorates-general: DG XIII (technology policy) and DG XVI (regional policy). It has been argued (Corvers & Nijkamp, 2000; Hassink, 1992) that regional innovation policy can be seen as the result of a gradual convergence of two previously distinct policy areas, namely regional and technology policy, which have both undergone a change in policy ideas. The RITTS programme can be considered a clear illustration of this change in underlying policy ideas.

THE RITTS PROGRAMME IN MORE DETAIL

The RITTS programme provides regional policymakers with a tool to support the development of a regional innovation strategy that is based on the identification of innovation needs of regional firms—hence demand-led—and on a quality assessment of the innovation services provided by the regional innovation support infrastructure to firms—also demand-led instead of supply-push. Innovation needs are those that should be fulfilled in order to be able to (successfully) introduce a new product or process into the marketplace and can be technical, managerial, commercial and/or financial (RIS, 1996:5).

Purpose of RITTS

The explicit purpose of the RITTS scheme is (RITTS, 1994:35-36):

*   To "provide local and regional governments and/or development organisations with support in the analysis and/or development of their innovation and technology transfer support infrastructure by offering them access to advice from experienced Community experts".
*   In order to achieve that, the EC provides financial support to "local or regional governments and/or development organisations willing to set up a transnational team of experts to review the design, impact and effectiveness of technology diffusion organisations and services that constitute the regional technology transfer and innovation support infrastructure and the interaction among them, to develop

strategies aimed at improving this infrastructure, and to share experience in this area".

Particular emphasis is laid on:

1. the analysis of expressed and latent needs of firms, and in particular of the smaller companies along with those which do not usually implement innovation projects;
2. the necessary work for maximum coherence in the assignments, the goals and the modes of intervention of transfer structures and local, regional and national actors, which act as the sponsors or financiers to these assignments.

The development of this regional innovation strategy should be the outcome of a process that involves all the regional actors related to RTD, innovation and associated business support activities—hence consensus-based—such as local and regional governments, local and regional economic development organisations, regional representatives of national agencies in charge of innovation, technology, science, economic and/or regional policy, central government ministries in those areas, research organisations, higher education institutes, technology transfer organisations, innovation support organisations, large businesses, R&D laboratories, business associations and trade unions.

A regional innovation strategy developed in the framework of RITTS should reflect:

- A *bottom-up* approach: it should be demand-driven, based on strengthened dialogue between firms, particularly SMEs, regionally based research and technology transfer organisations and the public sector in order to assess the needs (expressed and latent) of regional firms and to aim at meeting these needs effectively.[4]
- A *regional* approach: there should be a specific territorial dimension which takes full account of the national and international context; the starting point should be the strengths and weaknesses of the regional economy. Perhaps more important, RITTS should build a consensus at the regional level on the priorities for action between the principal actors involved.
- A *strategic* approach should be applied to regional development in the fields of technological progress and innovation. They should plan for short- and medium-term actions that fit with the long-term objectives and priorities defined by the region. RITTS does involve not only the completion of a study, but above all requires the production of an action plan.
- An *integrated* approach: the efforts of the public sector (local, regional, national and European) and the private sector should be linked towards the common goal of increasing regional productivity and competitiveness. They should try to maximise the economic impact of regional, national and European programmes.
- An *international* approach: a RITTS should adopt an international perspective in terms of the analysis of global economic trends as well as on the need to co-operate nationally and internationally to be more effective in the field of RTD and innovation. In more practical terms, the use of a Reference Panel (composed of experienced European experts or other regional practitioners) to which the results of the work are presented at crucial stages of the project is a means to ensure this external viewpoint on the project.

The organisations that can apply to RITTS are:

- regional authorities, representing either assisted regions (at NUTS II level) where the majority of the population live in ERDF eligible areas (European Regional Development Fund) or nonassisted regions;
- organisations which are not formally regional authorities, but have a formal mission regarding technology-based regional development and which can demonstrate commitment and backing from regional authorities;
- regional authorities and organisations of the European Economic Area (EEA = Norway, Iceland, Liechtenstein).

Financial Framework and Policy Rationale of RITTS

The financial framework of the RITTS programme was the SPRINT Programme (RITTS, 1994 Call for Proposals) and its successor the INNOVATION Programme (RITTS, 1995; and RITTS 1997 Call for Proposals). Both programmes are part of the Fourth Framework Programme for Research and Technological Development (1994-1998) which brings together all EU programmes and resources dealing with RTD activities at the Community level. One of the objectives of the INNOVATION Programme was to promote an environment favouring innovation and the absorption of technologies, especially by interacting with the network of authorities, structures and bodies which, at local or regional levels, contribute to the optimisation of the innovation system and to the definition and implementation of policies to promote innovation and technological development.

The rationale given by the European Commission focused on firms, particularly SMEs, which face a number of weaknesses with respect to innovation such as finance, human resources, design, marketing, organisation and management issues. As innovation is increasingly considered to be a crucial factor for the firm's ability to compete, it is important that the environment of firms provides access to the know-how and information needed. Particularly the regional level fulfils an important role as this is the level closest and most natural to firms.

Regional and local authorities have acquired through decentralisation and devolution larger roles in devising and implementing strategies and concrete policy actions aimed at supporting the innovation process in firms. Investments have been made in developing a variety of structures such as science parks, innovation centres, university-industry liaison offices, technology demonstration centres, technology diffusion networks and the like.

RITTS aims at enhancing the operating efficiency of the regional innovation and technology transfer support infrastructures and policies towards satisfying firms' needs, particularly of SMEs. Although new schemes or structures could be an outcome of the RITTS exercise, the emphasis lies rather on optimising the existing regional infrastructures and policies than on increasing the level of regional resources dedicated to supporting RTD, technology transfer and innovation. In addition, it examines the efficiency of policies directed at these issues, and the allocation of resources and tasks within the region's SME support infrastructure directed at innovation, technology diffusion and exploitation.

Community funding for RITTS projects will not exceed 50% of the cost of a RITTS project. Maximum contribution will be limited to € 250.000 in ERDF[5] assisted areas, and € 175.000 in nonassisted areas.

Methodological Framework

As said earlier, the influence of the innovation systems concept as a tool for systematic analysis and policy development, is striking in the methodological approach of RITTS. The work to be undertaken within a RITTS exercise has to comprise five interdependent themes—core specifications—in order to ensure the successful implementation of the initiative. These five interdependent themes are:

1. Building a *regional consensus*. This includes a communication strategy to raise awareness within the region with regard to the exercise, involving the main stakeholders concerned, keeping them informed of the progress of the exercise and seeking their opinion and feedback. A steering committee should be set up as a major tool to build this consensus.
2. The *identification of regional firms' needs* (expressed and latent). This work should take into account, among other things, the impact of global market and technology trends on the regional economy.
3. An *analysis of the regional supply* in terms of innovation and technology transfer support services as well as of the pure RTD resources. This part also includes an analysis of the strategies of the main regional actors.
4. Based on the work undertaken with regard to the previous themes (in particular the strengths and weaknesses analysis of regional firms, of an assessment of the regional technology and innovation support demand, an identification of the gaps and duplications in the technology and innovation support supply), the steering committee and external experts will define a *strategic framework* and agree upon *priority actions*.
5. The last theme is the initial *implementation* of actions defined and the definition and setting up of a *monitoring system*, that will help to follow and evaluate the actions undertaken.

The work to be done on the five interdependent themes is divided into three successive stages (stage 0, 1, 2). It is expected that a RITTS project lasts up to 18 months in total. Between the three stages some loop back can take place.

Stage 0: this is the definition stage and can last up to three months. It should be used to prepare the rest of the exercise and finalise the various parts of the work programme. It can include:

*   the setting up of the steering committee (informing and winning commitments from various members);
*   a work programme with clear milestones, timetable, budget and a description of the various studies, surveys to be carried out;

- the exact composition of the project team (management unit, staff and external consultants, to be involved in the project), together with the exact role of each of them.

Stage 1: this is the information-gathering and assessment phase, including themes such as the analysis of regional firms' needs and the analysis of the regional innovation and technology transfer support supply. The information should identify the structure and relevance of a region's innovation system with respect to demand, international linkages/orientation, potential obstacles to a regional consensus and so on. The purpose of this stage—which lasts on average between six and nine months—is to:

- provide the basis for a decision as whether or not to proceed to the next detailed stage;
- help solicit support for the initiative;
- provide a basis for developing a plan of action.

Stage 2: this stage is concerned with establishing regional priorities as a result of a regional debate and the validation of stage 1 results, as well as to start implementing priority actions and setting up an evaluation and monitoring system. Its normal duration should be about five or six months. The purpose of Stage 2 is to provide a blueprint for the development and launch of the regional innovation strategy, defining the role of each party involved in the implementation and operation. It should, therefore, include:

- the presentation of priority actions (including those with an international dimension) and of their coherence as a whole;
- the detailed presentation of the leading stake holders and of their role in the defined priority actions.

RESULTS OF THE RITTS PROGRAMME

The first RITTS programme was launched in 1994 and between 1994 and 2000 three generations of RITTS projects have been implemented. The first generation, 1994-1996 (emerging from the 1994 Call for Proposals), encompasses 22 selected projects. The second generation, 1996-1998 (emerging from the 1995 Call for Proposals), encompasses 21 projects. The third generation, 1998-2000 (emerging from the 1997 Call for Proposals), encompasses 29 selected projects. This brings the total number of RITTS regions to 72 throughout the whole European Union and European Economic Area[6].

Type of Regions

The difference between assisted and nonassisted areas in the European Union was already briefly mentioned. It relates to the possibility for regions whether or not to receive money from the Structural Funds (such as the European Regional Development Fund). Experience with RITTS shows that the programme is particularly appealing to regions which do not have the status of assisted area. Although these regions are not considered problem areas in EU terms, they themselves perceive the innovation capabilities of their region as not up to standard and in need of revision. As far as geography is concerned, there are no limits to where RITTS regions can be found. They range from the very north of Sweden to the island of Crete, from the industrial heart of England to the tourism-based economy of the Canary Islands, from the agricultural environment of Neubrandenburg to metropolitan areas such as Madrid, Milan, Berlin, Bremen, Hamburg, London, Lisbon, Dublin, Copenhagen, Rotterdam and Stockholm.

Type of Project Promoters

Given that RITTS is an initiative explicitly intended for regions, three assumptions can be made:

1. Given that RITTS is an initiative explicitly intended for regions, most of the selected RITTS proposals will come from regional authorities.
2. Given the differences in governance structure among the participating European countries ranging from federalism to unitary state forms, those countries with a strong regional government tier will be better represented among the RITTS regions.
3. Given the formal policymaking capacity (which is regionwide and sometimes includes financial competencies to raise taxes) of regional authorities, regional governments will be more successful in formulating a demand-led and consensus-based regional innovation strategy if the contractor is a regional government instead of an innovation-relevant organisation.

To start with the first assumption, looking at the statistical distribution of selected RITTS proposals over type of organisation, there is no evidence that most of the RITTS proposals will come from regional authorities. The distribution is more even: 57% (41 proposals) originate from regional authorities whereas 43% (31 proposals) come from other, so-called innovation-relevant organisations such as universities, regional development organisations, technology transfer agencies, regional offices of national agencies and science and technology parks.

Second, 72 selected RITTS projects are divided over 16 European countries (Luxembourg did not hand in any proposal during three Calls for Proposals) of which only three have a true federal governance structure: Austria, Belgium, Germany. These three countries represent 19.4% (14 proposals) of the total selected proposals. Countries well known for their weak regional government tier such as France, Portugal and the United Kingdom represent 22.2% (16 proposals). Intermediary cases such as Greece, Italy and the Netherlands represent 25% (18 proposals). In other words, the statistical distribution of selected proposals over participating countries shows that countries with a strong regional government tier are not better represented in the RITTS programme nor countries with a weak regional government tier to be worse represented.

And third, it could be expected that regional authorities are better equipped to design a regional innovation strategy as they have a regionwide competence base and are policymakers per excellence. It could be assumed that their commitment to the project would be stronger given their assignments and competencies (sometimes including financial competencies to raise taxes). Again, this does not seem to be the case. The success of a RITTS project depends, besides formal assignments, also on the project promoter and his or her ability to motivate people, to acquire political support and to build consensus on a common regional innovation strategy. Both the scope of manoeuvre and the motives for participation influence this ability, in combination with personal qualities. The regional situation is a factor that induces motives for participation—what has to be changed to optimise the current innovation support system?—but also the scope of manoeuvre of the project promoter—can changes be conducted by my own organisation?

Duration of a RITTS Project

Although the official duration was estimated at 18 months, reality shows that all regions needed more time to complete the RITTS project: at least two years (24 months) and sometimes even more. Various aspects explain for this longer duration.

First, the initiative was new and unknown to the regional partners. This programme was not about getting EU money to build a road, a regional airport or set up a new university department, but it was about getting money to engage in strategic thinking about the future development of the region. The idea was appealing in conceptual terms, but not always easy to explain in operational terms. Making the RITTS programme understood and, even more, making the region's participation in this initiative understood was a time-consuming task for the project promoters.

Second, affirming the role of project promoter was not always self-evident. Some regional partners would dispute why this particular organisation should be entitled to run the RITTS project instead of them. Running the project itself turned out to use up more staff, time and budget than was originally envisaged.

Third, selecting and guiding the team of international and national consultants was equally less self-evident. Project promoters had to invest a lot of time in explaining to the consultants what was expected of them. In many cases, the consultants invested more time in the analyses of demand and supply than was budgeted for, causing problems for the time and budget allocation to Stage 2.

Fourth, setting up the Steering Committee was time-consuming, but even more so to make it work as a true steering committee. Some regions had 30 members on the committee, others only 3. Obviously, it takes more time and trouble to reach consensus among 30 than among 3 organisations. In some regions, members of the Steering Committee would start to question the methodological approach of the research in Stage 2 when they should actually have been discussing the regional innovation strategy, setting priorities and defining policy options to implement the developed strategy.

RITTS as a Consensus Catalyst

What was appreciated by the participating RITTS regions is the fact that RITTS provided them with an instrument to bring the regional partners together. The fact that RITTS had this European label triggered off interest from regional actors who otherwise would not have been interested in starting collaborations with other regional partners. Bringing people and organisations together is a first step to consensus.

Participation in RITTS was project-based. This allowed for an allocation of tasks, time, deadlines, deliverables, responsibilities and budget. This, in turn, created a spirit of belonging to a privileged group designing the future of the region.

Systematic Approach to RSI

To some regions, the systematic approach to the regional system of innovation was most appealing. The methodological framework given by the European Commission showed them how to assess the RSI in their region. Approaching the strengths and weaknesses of the region in this fashion was never done before and generated a lot of enthusiasm. The systematic approach to the regional system of innovation also generated new insights, new data, new indicators and new ideas.

Reflection Followed by Action

The RITTS scheme is action oriented. Although analyses of and reflection on the regional technology transfer and R&D infrastructures are part of a RITTS project, the goal is to change and optimise the existing situation. A RITTS project should result in a strategy that provides the framework for optimising innovation policies and infrastructures at the regional level especially with regard to their relevance for the needs of small and medium-sized companies. Many regions appreciated this action orientation, even though they complained that the European Commission did not provide for additional money to fund the implementation of actions itself.

RITTS Advocates Regional Openness

Finally, the RITTS programme emphasises the importance of regarding the region in its relation to the rest of the world. Instead of regarding the region as an island, it is important to assess the global economic and technological trends and how they might affect the region. In order to provide for a more outward outlook, the European Commission finances accompanying measures to organise exchange of experience by means of seminars, workshops, conferences, exchange of staff, use of international consultants, reference panels, and the like.

RESULTS FROM THE RITTS EVALUATION

The RITTS programme has recently been evaluated by Charles et al. (2000). In this section, the key evaluation results will be presented, distinguishing between the project level and the programme level of RITTS.

RITTS at the Project Level

At the project level, it is clear that, although there is a common methodological framework, every RITTS project has different characteristics. The regions which were more successful in designing their regional innovation strategy took as a starting point for the analyses: what exists in the region instead of let's copy Silicon Valley. Experiences with RITTS also show that the commitment to the project's success is not necessarily stronger in case the project promoter is a regional authority. The scope of manoeuvre as well as the motives for participation of the organisation in charge seem to be equally important driving forces, in combination with personal qualities of the person representing this organisation.

RITTS at the Programme Level

At the programme level,[7] the overall conclusion of the evaluation team is that the RITTS programme has achieved its objective to support regional policymakers in upgrading the regional innovation and technology transfer infrastructure for a better response to SME needs. The implementation of RITTS showed an additional achievement of the programme, less clearly planned, concerning capacity building for innovation policy in the regions. However, RITTS is an ambitious programme, containing several complementary objectives which, taken as a complete menu, have proved to be beyond the reach of most regions. Also, the implementation of RITTS has taken place in a much more complex manner that the sequential model: analysis-strategy-action-evaluation suggested in the programme design.

Therefore, the success in RITTS needs to be appraised with regard to the starting conditions in the region: achieving significant changes in a number of dimensions of RITTS is the key criterion to be applied rather than comparing results with some best-practice model. A very strong message emerging from the Charles et al. (2000) evaluation is that RITTS is a (key) moment in a long-term policy-building process.

In the evaluation a distinction is made between regional capacity and project driving force (see Figure 11.2).

Figure 11.2
Underlying Factors Behind RITTS' Successes and Failures

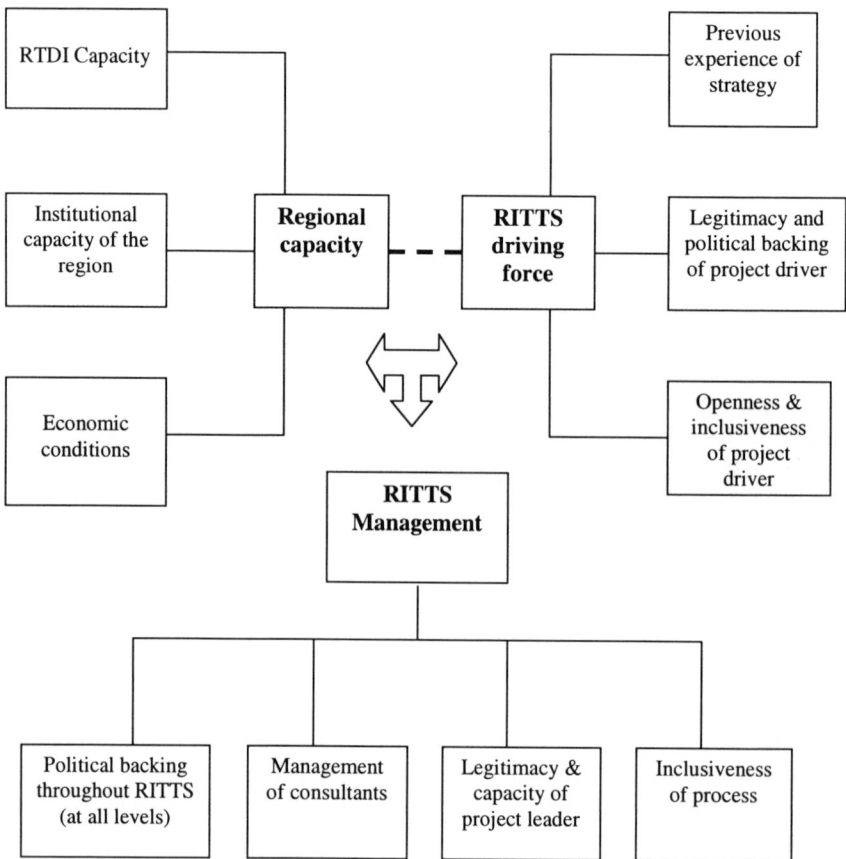

Regional capacity encompasses the institutional capacity of the region, its political autonomy, maturity and competence in developing innovation policies and innovation support, openness to learn from outside, along with its socio-economic conditions in terms of economic development and innovative capacity. *Project driving force* which encompasses the legitimacy, political backing and strength of the project driver, the experiences of the partnership in working together and developing strategy, and the openness and inclusiveness of the partnership.

The preconditions for a RITTS project are a complex combination of these various attributes of both the regional capacity and the project driving team, and so simple typologies as explanation for outcomes are impossible to identify. Overall, the evaluation showed that, while preconditions to RITTS in the form of regional capacity and project driving force were important determinants of the

RITTS achievements, even more important were the choices made on how to run the project. Failures to achieve significant results in RITTS were mainly due to weaknesses in project management and wrong management decisions rather than to unfavourable starting conditions.

The crucial management decisions for achieving the best results in the RITTS are:

- political backing at early stage and throughout the project,
- sound management of consultants,
- choice of project leaders with high legitimacy and capacity,
- intense and bottom-up involvement of regional innovation stakeholders in the project.

Overall Appreciation of the RITTS Programme

The RITTS programme had positive impacts on the following aspects:

1. It brought in a much-needed *move towards strategic thinking* for innovation-oriented regional development.
2. It offered mechanisms and incentives to *create a regional dialogue* in fragmented regions (in geographic, institutional and cultural senses).
3. It helped to develop a *broader concept for innovation*, different from technology transfer, and put this higher on the policy agenda.
4. It supported many regions to *clarify the scene of innovation support infrastructure* and to develop *actions* to rationalise, better define and augment the visibility of this infrastructure.

The degree of success achieved by the various RITTS has been shown to depend not only on the prevailing starting situation in the regions, but also, and more importantly so, on the appropriateness of management practices in the RITTS process itself.

RECOMMENDATIONS FOR FUTURE RITTS PROJECTS

The RITTS evaluation team (Charles et al., 2000:xii-xiii) proposes several suggestions for further implementation of projects under the RITTS programme.

Recommendation 1: The RITTS decision maker should be either regional policymakers or a neutral agency enjoying lasting and solid policy support from the regional authority.

Recommendation 2: The leadership of the RITTS by a regional authority should go along with an inclusiveness of the exercise, where all actors in regional innovation and technology transfer infrastructure have a voice in the process.

Recommendation 3: Steering Committees should be chaired by a senior person who can command respect and ensure the participation of key figures from other organisations.

Recommendation 4: The area definition of the RITTS should correspond to a meaningful area in policy terms.

Recommendation 5: RITTS projects should devote resources to a permanent, experienced, full-time management team for the duration of the project.

Recommendation 6: Practical training on the job for innovation policy designers in the form of exchanges of experience, visits, and such, should be made available at national and international levels.

Recommendation 7: RITTS should include a more explicit concern for the legacy of consultants in the regions in terms of methods and skills. Consultants should be well embedded in the RITTS process, rather than seen as neutral evaluators.

Recommendation 8: RITTS should start with a solid Stage 0 where the detailed objectives and questions to be solved by RITTS are discussed and detailed. This could even take the shape of a feasibility stage for the whole project.

Recommendation 9: There is a role for the Commission to exploit the pool of knowledge gathered by the set of consultants involved in RITTS. The best way to do this would be in frames facilitating the exchange of tacit knowledge rather than codified pieces (available in RITTS reports), through, for example, seminars for experts or collaborative studies of those experts.

Recommendation 10: Interregional activities should be enlarged beyond the project coordinator, and involve key organisations and firms in the region.

Recommendation 11: Exchanges of experience between projects in the same country, through national seminars, creation of a network, visits and exchanges, should be favoured.

Recommendation 12: Benchmarking methods and data should be made available to regions wishing to go further in exploiting the diversity of European regions in innovation policy, freed from a limited best-practice approach which mostly leads to marketing presentations of nontransferable instruments.

Recommendation 13: A more systematic informing of national administrations about RITTS-like exercises need to be organised, not only at the level of each project, but also by the European Commission.

Policy Recommendation Beyond RITTS

The RITTS programme seems to reinforce a linear view of innovation give its supply-demand focus of the methodology in Stage 1 of RITTS. The vast majority of regions have followed that model in practice. Accordingly, the main thrust of the analysis stage of many RITTS projects is expressed as follows: What do regional SMEs need in terms of support for innovation, and what does the regional supply organisation offer? How do the demand and supply match and where are the gaps between the two?

PROBLEM 1: Preventing the recognition of innovation as an interactive process. The main problem with such a linear approach to innovation is that it is in contradiction with the current understanding of the real nature of innovation. Both current academic research and innovation enquiries (including those undertaken in RITTS projects) emphasise a nonlinear view on innovation: innovation is a firm process, mainly driven by market considerations, and the main issue for innovation is how firms organise to be more innovative. Thus, the main problem is not how to match a set of supply services with firms' needs, but how to improve innovation management processes in firms. The focus is on an innovation system with firms at the core as they are the source of innovation, rather than on an innovation support system with suppliers and intermediaries at the core. Then the policy question shifts to how styles of innovative behaviour and capabilities for innovation in firms can be improved. Therefore, there is a risk that the results of supply-oriented RITTS, whatever their intrinsic value, will fall short in tackling the more open question, how can innovativeness in firms be improved?

PROBLEM 2: A closed view on the regional innovation system. The demand-supply approach in RITTS implicitly favours a closed view of the regional innovation system: in the overwhelming majority of regions involved in RITTS, the above question has been treated with a limited focus on regional firms and regional supply organisations, as if the objective would be necessary to match the two. In an open, globalised world, where the knowledge base necessary for the development of firms is growing in diversity and complexity, such a closed approach is unlikely to be successful. If firms are put at the core of the exercise, then it does not matter if the support comes from within or outside the region, the important point is how firms search for, absorb, engineer and use this support for their own innovation goals.

PROBLEM 3: Overlooking the importance of policy aspects. The supply-demand approach in RITTS prevented sufficient attention being paid to policy involvement in the implementation of RITTS, at least in the first years of implementation of the programme. Under the RITTS approach, it is implicitly assumed that fine-tuning the supply and intermediary organisations in better accordance to SMEs needs could be achieved, if those suppliers and intermediaries are put together and agree on the directions of the changes to be made. The RITTS issue is thus mainly one for technology transfer agencies or intermediaries, and relates to the optimisation of their own strategy. However, such an approach neglects the fact that various policy levels need to be involved in decisions, because in the majority of cases they will hold the power to decide on and finance the desired changes. Furthermore, the role of policymakers is essential in taking a broader view of the various instruments to be put in synergy for better support to firms, and to ensure that specific actions taken are for the benefit of the society. This underrecognition of the role of policy to reach RITTS objectives has also led to a neglect of the involvement of the national level in these exercises.

Policy Suggestions for the Future

Thus, if regional governments have the goal to promote more innovative-ness in their business sector, there is more to do than a RITTS, which has a more restricted goal of improving the regional support infrastructure. Even if the two goals may overlap in some respect, the first goal needs another approach, focused on the innovation process itself.

The targets of such a future scheme should shift towards and concentrate on:

- Opening windows of opportunities for firms to favour innovation, both through a variety of types of knowledge useful for their activities, and through access to foreign sources of knowledge.
- Developing innovation management capacities in firms.
- Putting more accent on human resources as main carriers of knowledge to support innovation moves.
- Transforming the underlying best-practice framework for interregional exchange activities into favouring exchange in a learning perspective.
- Increasing absorptive capacities of regional policymakers and policy implementers towards new ideas and new practices in innovation policy.

The evaluation points to the relevance of building a new European scheme, on the basis of the RITTS programme, which would:

- incorporate a more up-to-date notion of innovation, less linear in nature,
- focus on an open vision of the regional innovation system,
- develop the tools proposed beyond a simple supply-demand approach, and
- integrate the policy dimension at the core of the scheme.

Such a new scheme would differ from its parent RITTS in the following ways:

1. It should not include (only) a demand analysis, in the sense that the expectations and rate of satisfaction of the firms towards the regional support infrastructure are investigated, but, rather, it should be founded on an analysis of the behaviour of firms with regard to innovation.
2. The supply analysis should not be foreseen at the start, but, rather, after the firms' analysis, there should be a decision on what are the key drivers and barriers for innovation in the firms located in the region.
3. The consensus-building idea should be abandoned and, rather, the inclusiveness criterion should be put at the forefront.
4. The international dimension, mainly brought in by the consultants in the RITTS implementation, should be kept as it was one of the main successful elements of this scheme. Both the use of consultants and the shape of interregional networks should be made more specific, moving away from the more generalist tools that were developed in the early RITTS.
5. The driving role of the regional authorities as masters of the whole process should be rooted in the scheme.

CONCLUSION

This chapter has described, analysed and evaluated a Community initiative in the area of regional innovation policy, namely RITTS (Regional Innovation and Technology Transfer Strategies and Infrastructures). By means of RITTS, the European Commission has provided European regions with financial resources and a methodological framework to facilitate the design of regional innovation strategies based on a thorough analysis of their regional system of innovation.

Although the RITTS initiative can be seen as a strategic planning tool to promote regional innovation systems, the actual results are more modest. First, the innovation systems approach as developed in academic circles has inspired the RITTS programme design, but was adjusted more in accordance with an innovation support systems approach. As a result, the evolutionary, systemic view on firms' innovation capabilities diminished and was overtaken by a more linear, neoclassical view on demand and supply of innovation needs and in-novation services.

Second, policy documents such as the White Paper on Growth, Com-petitiveness and Employment and the Green Paper on Innovation stressed the importance of formulating and implementing a regional innovation strategy with all innovation-relevant stakeholders concerned. This consensus-based approach, however, has proven to be more difficult to manage in practice than in theory. A more fruitful approach might be to install mechanisms where both regional firms and innovation-relevant organisations are involved and consulted on a regular basis.

Further, experience from RITTS teaches us that all regional actors, not just the official regional authorities, can play a significant role in the region's innovation system. The success of a RITTS project depends, besides formal assignments, also on the project promoter and his or her ability to motivate people, to acquire political support, to detect areas for consensus and to secure funds for the implementation of a regional innovation strategy. Also, various policy levels need to be involved besides the regional level, because in the majority of cases these policymakers will hold the power to decide on and fi-nance the desired changes.

Finally, the policy area dealing with regional innovation is new, conceptually, which has wide-ranging effects on the way policy is formulated, implemented and evaluated. Regional innovation policy can only be successful if the region's strengths and weaknesses, opportunities and threats, characteris-tics and constraints are assessed in a broader reference framework benchmarking beyond the region. As such, there exists no generally applicable blueprint on how to promote regional innovation systems. Each attempt to do so has to be tailor-made for the region in question.

NOTES

1. Members of the European Economic Area—Norway, Iceland and Liechtenstein—were allowed to participate in the RITTS programme.

2. This implies, for example, that a foreign-owned firm will be part of two different national systems—its home country and its host country.

3. Lundvall mentions the example of the OECD which launched the Technology/Economy Programme (TEP) in 1988, an ambitious effort to understand the importance of technology for economic change. When the outcome of this programme was presented in 1991 in Montréal, the NSI concept was given a prominent place in the conclusions.

4. Needs should not be limited to technological issues, but include managerial, financial, commercial, training and organisational issues (RITTS Information Package, 1995:4).

5. European Regional Development Fund regions assigned as Objective 1, 2, 5b or 6 areas were eligible for a maximum cofinancing amount of €250.000 within the framework of RITTS.

6. The total number of 72 RITTS projects is based on calculating the number of project proposals that were selected in the evaluation stage and that materialised once the contract between the European Commission and the region was signed.

7. This section is a short version of the RITTS Final Evaluation Report's Executive Summary (Charles et al., 2000:vii-xv).

REFERENCES

Arcangeli, F., 1993, Local and global features of the learning process. In Humbert, M. (ed.), *The Impact of Globalisation on Europe's Firms and Industries*. London: Pinter.

Archibugi, D. and J. Michie, 1995, The globalisation of technology: a new taxonomy. *Cambridge Journal of Economics*, 19(1), pp. 121-140.

Braczyk, H-J., Ph. Cooke and M. Heidenreich (eds.), 1998, *Regional Innovation Systems—The Role of Governances in a Globalized World*. London: UCL Press Ltd.

Camagni, R. (ed.), 1990, *Innovation Networks: The Spatial Perspective*. London: Belhaven-Pinter.

Carlsson, B. (ed.), 1995, *Technological Systems and Economic Performance: The Case of Factory Automation*. Dordrecht: Kluwer.

CEC, Commission of the European Communities, 1994a, *Competitiveness and cohesion: trends in the regions*. Fifth Periodic Report on the social and economic situation and development of the regions in the Community. Luxembourg: Office for Official Publications of the European Communities.

CEC, Commission of the European Communities, 1994b, *White Paper on Growth, Competitiveness, Employment—The Challenges and Ways Forward into the 21st Century*. Luxembourg: Office for Official Publications of the European Communities.

CEC, Commission of the European Communities, 1995, *Green Paper on Innovation*. Luxembourg: Office for Official Publications of the European Communities.

CEC, Commission of the European Communities, 1996, *First Action Plan for Innovation*. Luxembourg: Office for Official Publications of the European Communities.

Charles, D., C. Nauwelaers, B. Mouton and D. Bradley, 2000, *Assessment of the Regional Innovation and Technology Transfer Strategies and Infrastructures (RITTS) Scheme*

—*Final Evaluation Report.* Luxembourg: Office for Official Publications of the European Communities.

Cooke, Ph., M. Gomez Uranga and G. Etxebarria, 1997, Regional innovation systems: Institutional and organisational dimensions. *Research Policy,* 26, pp. 475-491.

Cooke, Ph. and K. Morgan, 1994, The Creative Milieu: A Regional Perspective on Innovation. In Dodgson, M. and R. Rothwell (eds.), *The Handbook of Industrial Innovation.* London: Edward Elgar.

Corvers, F., 1994, *Economic Integration the European Way: Stronger Firms, Stronger Regions.* Maastricht: MERIT Research Memorandum 2/94-040.

Corvers, F. and P. Nijkamp, 2000, Regional Development and Interorganizational Policy Making in a Pan-European Context. In Kohno, H., P. Nijkamp and J. Poot (eds.), *Regional Cohesion and Competition in the Age of Globalization.* Cheltenham: Edward Elgar.

Edquist, Ch. (ed.), 1997, *Systems of Innovation—Technologies, Institutions and Organizations.* London: Pinter.

Freeman, C., 1995, The "National System of Innovation" in historical perspective. *Cambridge Journal of Economics,* 19, pp. 5-24.

Hassink, R., 1992, *Regional Innovation Policy: Case-Studies from the Ruhr Area, Baden-Württemberg and the North East of England.* Utrecht: KNAG/Faculteit Ruimtelijke Wetenschappen, Rijksuniversiteit Utrecht.

Hassink, R., 2000, Political Lock-Ins in Old Industrial Areas—A Comparative Study on Shipbuilding and Textile Regions in Germany and South Korea. Internal Paper. Geographical Institute of the University of Bonn.

Kilper, K. and E. Latniak, 1996, Einflußfaktoren betrieblicher Innovationsprozess—Zur Rolle des regionalen Umfeldes. In Brödner, P., U. Pekruhl and D. Rehfelf (eds.), *Arbeitsteilung ohne Ende? Von den Schwierigkeiten inner- und überbetrieblicher Zusammenarbeit.* Munich: Rainer Hampp Verlag, pp. 217-240.

Landabaso, M., C. Oughton and K. Morgan, 1999, Learning Regions in Europe: Theory, Policy and Practice through the RIS Experience. Paper prepared for the 3rd International Conference on "Technology and Innovation Policy: Assessment, Commercialisation and Application of Science and Technology and the Management of Knowledge", Austin, TX, 30 August-2 September.

List, F., 1841, *Das Nationale System der Politischen Oekonomie.* Basel: Kyklos Verlag. English edition, 1904, *The National System of Political Economy.* London: Longman.

Lundvall, B.-Å. (ed.), 1992, *National Systems of Innovation—Towards a Theory of Innovation and Interactive Learning.* London and New York: Pinter.

Lundvall, B.-Å., 1993, User-producer Relationships, National Systems of Innovation and Internationalisation. In Foray, D. and C. Freeman (eds.), *Technology and the Wealth of Nations.* London: Pinter.

Lundvall, B.-Å., 1996, Reflections on How to Analyse National Systems of Innovation. In Kuusi, O. (ed.), *Innovation Systems and Competitiveness.* Helsinki: ETLA and VATT, pp. 17-25.

Morgan, K. and C. Nauwelaers (eds.), 2000, *Regional Innovation Strategies: The Challenge for Less Favoured Regions.* London: Jessica Kingsley Publishers.

Nauwelaers, C. and A. Reid, 1995, *Innovative Regions? A Comparative Review of Methods of Evaluating Regional Innovation Potential.* Louvain-La-Neuve: RIDER and Luxembourg: Office for Official Publications of the European Communities.

Nelson, R. (ed.), 1993, *National Innovation Systems—A Comparative Analysis.* New York and Oxford: Oxford University Press.

Nelson, R. and N. Rosenberg, 1993, Technical Innovation and National Systems. In *National Innovation Systems—A Comparative Analysis*. New York and Oxford: Oxford University Press, pp. 3-21.

OECD, 1992, *Technology and the Economy: The Key Relationships*. Technology/ Economy Programme (TEP). Paris: OECD.

Porter, M., 1990, *The Competitive Advantage of Nations*. London: Macmillan.

RIS *Guidebook*, 1996. Brussels: Commission of the European Communities, DG XVI.

RITTS, 1994, Regional Innovation and Technology Transfer Strategies and Infrastructures, Call for Proposals 93C - 328/13. Science Park Consultancy Scheme RITTS. Core Specifications and Form of Application. Luxembourg: Commission of the European Communities, DG XIII.

RITTS, 1995, Regional Innovation and Technology Transfer Strategies and Infrastructures, DG XIII & DG XVI Call for Proposals for RITTS and RIS. Information package containing guidelines for submitting a proposal. Luxembourg: Commission of the European Communities, DG XIII.

RITTS, 1997, Regional Innovation and Technology Transfer Strategies and Infrastructures, Call for Proposals. Information Package and Proposal Forms. The INNOVATION Programme. Luxembourg: Commission of the European Communities, DG XIII.

Saxenian, A., 1994, *Regional Advantage—Culture and Competition in Silicon Valley and Route 128*. Cambridge, MA and London: Harvard University Press.

Soete, L. and A. Arundel (eds.), 1993, *An Integrated Approach to European Innovation and Technology Diffusion Policy*. A Maastricht Memorandum. Luxembourg: Office for Official Publications of the European Communities.

Storper, M., 1991, *Production Organisation, Technological Learning and International Trade*. Mimeo.

12

Information and Communication Technologies and Economic Development of Peripherally Located Regions: Experiences in the Netherlands

Marina van Geenhuizen

INTRODUCTION

There is an increased attention for the use of information and communication technologies (ICTs) as an instrument in regional economic policy. This is based on the assumption that a widespread application of these technologies is beneficial for regional economies in peripherally located areas, due to what might be called "shrinking" distances or even the "the death of distance". In the latter situation each place is connected with places elsewhere, simultaneously, without distance as a substantial cost component. Under such circumstances firms would be "footloose", meaning having a larger freedom in their selection of a location. In this chapter, we address a differentiated approach to the benefits of ICTs in economies in peripherally located regions, based on a survey of the literature and a field study in the Netherlands. Like many other new technologies, ICTs can flourish only if there is a fertile seedbed. This means that the production environment should satisfy particular needs, such as concerning the labor market and entrepreneurial spirit among the local population. The focus of the chapter is on the potentials of ICTs in regional development in the

Northern part of the Netherlands, located outside the economic core of the country. The results achieved in this analysis lead to various policy recommendations, particularly the design of an ICT policy embedded in a broader learning policy.

ICTs AND REGIONAL ECONOMIC GROWTH: EUPHORIA

It is a widely held view that ICTs will overcome traditional barriers of time and space, and therefore will enhance the competitiveness of economies in remote areas. ICTs cause various advantages unknown from traditional infrastructures. First, different from other (physical) networks, ICT networks have the advantage of externalities with regard to both access and use (Capello, 1998). The individual user value of access to the network depends on the number of existing subscribers linked with the network. There is not only a benefit for a newcomer in the network but also for existing ones (even unknown ones). Moreover, there is a positive externality in the use of the network. Both parties benefit from using a network connection, whereas often only one of them is charged. In practice, the decision of firms to join an ICT network is not simply related to the number of subscribers already in the network, but to the number of *relevant* groups linked by the network, like suppliers, customers, knowledge organizations and financial institutes. Second, ICT networks cause a larger variety of socioeconomic benefits than other infrastructures. There is not only an increased accessibility of regions but also a number of structural benefits based upon the need for competence to make use of ICTs. Furthermore, ICT networks can be established much cheaper than traditional transport infrastructures, user costs are continuously decreasing and there are virtually no environmental damages.

The rationale for interest in the use of ICTs in regional policy is mainly based on an assumed positive relationship between ICT investment and economic growth (e.g., the World Bank, 1994). There are various ways in which investment in ICTs may contribute to economic growth. First, new jobs arise from the design and construction of the infrastructure at hand. This is, of course, only a small temporary affair. Second, structural growth may be enhanced by an increase in productivity and a reduction of transactions costs of existing firms due to the use of ICT-based innovations, such as new systems of internal and external logistics (e.g., inventory control in supply-chain management). A third impact may be the establishment of entirely new firms based upon ICT products and services, as well as the attraction of firms or (some of their departments) from elsewhere by an "ICT minded" production structure and environment.

An important assumption in the reasoning behind the role of ICTs is that specific shortcomings in the region can be overcome and regional comparative advantage reestablished. What might be caused by the use of ICTs is a shrinking of physical distance between remote areas and the economic core, particularly since the downfall of particular costs of electronic communication (Gillespie and Williams, 1988). Telecommunication may particularly *substitute* certain types of personal interaction. In this context, some authors have put forward the idea of the *death of distance* (Cairncross, 1997). Accordingly, ICT networks would improve the access of remote areas by being potentially ubiquitous. Through ICT each spot on earth would be linked with each other, everywhere simultaneously, with distance no longer as a cost determinant of communication. As a consequence, urban meeting places and markets give way to virtual gathering and exchanges for plugged-in managers and customers (Mitchell, 1995). Urban functions are allowed to spread into a world where space and time converge.

In this situation of declining importance of physical distance and proximity, firms would have a much larger freedom to select their location. Certain parts of firms—along functional lines or otherwise—would avoid a location in the economic core because of high congestion costs and enjoy a location with cheaper land prices and better living conditions for their employees (Amirahmadi and Wallace, 1995; Tacken and van Reisen, 1998). Along functional lines, one may think of order booking, direct marketing and sales promotion. Internet and EDI (Electronic Data Interchange) enable the decentralization of such functions. In addition, PDI (Product Data Interchange) enables a decentralized production with design in the core area, even in the current situation of increased product complexity and increased variation in product types. Moreover, the use of multimedia (a combination of text, electronic images and sound) together with new network technology allows for videoconferencing and teleinstruction in such a way that various departments of firms can be decentralized.

However, various theoretical and empirical observations of the role of ICTs in regional economic development have dimmed the excitement. These observations are concerned with (1) potentials to basically change the economic position of regions, and (2) potentials of firms to enjoy the benefits of ICTs. This will be discussed in the next two sections.

A DIFFERENTIATED VIEW:
DEVELOPMENT POTENTIALS OF REGIONS

Based upon various theories or paradigms, the beneficial role of ICTs in the economic development of remote regions can be questioned. The first limiting factor of the role of ICTs refers to the *seedbed conditions* in the recipient economy and society. There has been a shift from the "Golden Bullet" paradigm to the complexity paradigm, meaning that there is a growing awareness that ICTs cannot simply be transplanted into a remote economy with information directly affecting this economy like a bullet hitting the target (Jipp, 1963). Rather, ICTs need to be seen in connection with the level of complexity in the recipient economy. There is no simple linear cause-effect model of introduction of ICTs in the regional economy. The outcome is the result of a complex interplay between the technologies and the power field in which they develop (Graham, 1998). The impact of ICTs on regional economic growth, for example, varies according to the regional production structure (e.g., the size of local firms and their embeddedness in larger supplier/customer networks).

Regions with many large firms linked with metropolitan areas and the global economy seem to benefit more from ICTs than regions with a dominance of independent small and medium-sized firms (SMEs) (Capello, 1998; Cornford and Gillespie, 1993). In addition, it needs to be emphasized that a certain amount of networking (interaction) between firms must exist before communication can be mediated through ICTs. Policy cannot create networks where none existed before simply by providing an ICT infrastructure (Melody, 1991). The point made here is that the competitive position of regions cannot be changed merely through the introduction of ICTs in a supply-side approach.

The complexity paradigm can easily be linked with general theories of regional economic development, particularly evolutionary ones (e.g., Boschma and Lambooy, 1998). The latter theories say *inter alia* that once emerged, divergences between regions will not simply disappear but tend to remain and often reinforce themselves. In other words, regions that get ahead tend not only to stay ahead but consolidate their position. The basic underlying principle is that of increasing returns on investment (private and public infrastructure) and agglomeration of firms and organizations, reinforcing a long-term concentration of new technologies in core areas (Arthur, 1994). There are increasing returns to scale and to adoption. The former refers to decreasing unit costs related to production increase, whereas the latter refers to the fact that the more new technology is adopted, the more is learned about it and the more incentive arises for further adoptions.

According to evolutionary theories, organizations like firms, governments, and the like tend to base their decisions on a limited rationality due to their perception of the situation they are facing. They act mostly based on routines leading to decisions that keep them relatively close to development paths chosen in the past (van Geenhuizen, 1999; Nelson and Winter, 1982). Thus, they adapt themselves only incrementally to changing conditions (technologies). In terms of spatial behavior of firms the above arguing means that new firms—as spin-offs from large (inert) ones—usually locate close to the parent organization for market and information reasons. Accordingly, economic core areas often grow in adjacent regions, a pattern that complies with a strong stability in regional economic and technological development. Based on the role assigned to chance in evolutionary developments, however, one may expect the emergence of a few *new* technology growth centers in unexpected locations, as an exception to the general pattern of stability. The point made in the previous reasoning is that the advancing of ICT use (like any other policy) cannot change the basic (structural) position of remote regions in the short- to medium-term along incremental development paths.

In the context of regional policy using ICTs, the question of the *magnitude* and *type* of spread (decentralization) from the economic core to remote regions is a relevant issue. In explaining the mechanisms underlying spread, an important role is attached to different types of knowledge (OESO, 1996). Accordingly, high-quality personal interaction and transfer of implicit (tacit) knowledge ask for geographic proximity (concentration) and tend to remain in metropolitan areas. By contrast, contacts based on routine and codified knowledge can be established and maintained over a large distance from the economic core, a situation which enables geographic spread. Thus, activities with a small in- and output of physical goods and routine knowledge contacts may prefer to locate in remote areas, thereby avoiding congestion costs and enjoying a better living climate (Tacken and van Reisen, 1998).

Simultaneously, a certain concentration may take place in metropolitan areas, because of the social (knowledge) networks and up-to-date telecommunication networks there. In fact, large cities are ahead in terms of investment in telecom infrastructures (Mitchell, 1995; Graham and Marvin, 1999; Ministry of Economic Affairs, 1999). Thrift (1997) calls for attention in this respect for the supremacy of the City of London as the world's major telecommunication hub based on a wide variety of connections, such as direct small disk delivery, microwave and five fiber optic rings. But the question remains whether the process of concentration is limited to a few world cities or includes also smaller towns.

To summarize, there is no unambiguous answer from the literature to date concerning the magnitude and scope of spread under the influence of ICTs. Some authors state that much interaction cannot be replaced by telecommunication and will remain in large cities, while routine service activities may spread to remote regions based on cost advantages (back offices) (Graham and Marvin, 1997). For other authors, however, the question is still open whether certain parts of *strategic* functions of firms can move to remote regions.

There is another perspective that sheds some light on the potential for the spread of economic activity and the role of ICTs. In a renewed attention for industry clusters, a strong emphasis is put on advantages of proximity. Proximity of mutually dependent firms would create various important economies, such as based on knowledge transfer, a pool of specialized workers and specific relations of firms with suppliers and customers (Porter, 1998). The emergence of such clusters is connected with a common awareness and understanding between actors in the region, often based on shared values and a common culture (institutional layer) (Grabher, 1993). The relevance of clusters may have two implications. First, disconnecting firms from clusters in the economic core area by moving them to remote regions is not realistic because of various diseconomies (no matter which function). Second, an improved use of ICTs in remote regions needs to be coupled with the fostering of existing or new industry clusters here.

EMPIRICAL STUDIES: INDIVIDUAL FIRMS

Empirical studies of the impact of ICTs on the micro level of firms are still rare. Their scope is also relatively narrow because most of them deal with the use of ICTs in existing firms and improvement of their competitive power, not on the establishment of new firms. The results, no doubt, point to a serious questioning of the positive impacts of ICT use on economic growth of regions (Table 12.1). Disappointing results are true for both the use of (advanced) ICTs by firms and an improved competitive power of these firms through ICTs. Often-cited causes are the absence of an entrepreneurial spirit among the local population (partly related to the absence of competition), the small size of most firms involved and concomitant lack of risk-taking, and the modest embeddedness of remote regions in urban economic networks. The point made here is that there is apparently a need for a demand-side approach, particularly focusing on SMEs.

Table 12.1
Empirical Results on the Micro Level of Firms (Main Examples)

AUTHORS*

Capello, 1994/98 (Southern Italy versus Northern Italy)
Various conditions on the micro-level prevent a beneficial use of network externalities, leading to the absence of a critical mass. Conditions hampering the use of ICTs (and improved economic performance) include risk-avoiding behavior and limited competition due to stable market positions.

Gillespie et al., 1994
Three factors explain regional differences in the use of telecommunication, i.e. entrepreneurial spirit, presence of information professionals, and the embeddedness of rural regions in urban networks.

Clark, Ilbery & Berkeley, 1995 (rural Lancashire and Warwickshire in the United Kingdom)
There is a limited use of advanced telematics by rural firms. Use decreases with firm size. There are various explanations: high costs compared to perceived benefits and risks (rapidly changing technology, inappropriate systems), a shortage of formal training in telematics use, and a lack of awareness of potential applications of advanced telematics.

Richardson & Gillespie, 1996 (Scotland)
There is a limited contribution of telecommunication services to the competitive position of firms.

Newlands & Ward, 1998 (Scotland)
There is a difference in use of (advanced) ICTs between regions, based on culture and habits (rural Scotland versus Aberdeen) and embeddedness of rural areas in urban economic networks (remote versus short distance). There is small evidence of an increased competitive power of rural regions.

* The study areas are within parentheses.

A further type of empirical studies to be discussed here deals with specific needs in the production environment among ICT-based firms. In this respect, a distinction can be made between manufacturing of information hardware (equipment) and information products on one side and producing information and communication services on the other side (Houghton, 1999). The results are derived from a study of ICT firms located in the Northern part of the Netherlands (Huntink et al., 1996). It needs to be stressed that the results give only a partial impression, leaving reasons why firms are *not* located here outside the focus of investigation. Given this limitation, the results can be summarized as follows. The most important location factor is the historical embeddedness in the region. This factor is true for manufacturing firms that have moved to the North from other parts of the country in the past decades based on cost advantages. A second factor is the proximity to customers. This holds mostly for ICT service firms. In addition, the working and living climate as well as the price/quality relation of premises are generally important factors of location of ICT-based firms in the North of the Netherlands. In contrast, there is a low importance for factors like telecommunication infrastructure and knowledge infrastructure. We may understand these research results as follows. Routine manufacturing (hardware) can easily take place in peripherally located regions, alongside ICT services that focus on local customers. The case of knowledge-intensive service firms, however, is much more complex, due to specific labor market requirements.

This can be illustrated with the location of selected call centers. The recent growth of international call centers in the town of Maastricht (in the very South of the country close to Belgium and Germany) for example, is based *inter alia* on the availability of native speakers of the Dutch, French and German languages, as well as a pool of native speakers of other languages connected with the international (European) functions here (van Geenhuizen, 2000). It seems reasonable to assume that most peripheral regions cannot attract this type of call center, simply for the reason of labor market shortages. In addition to this, the establishment of call centers may be influenced by the system of tariff zones for telephone calls and the location of the main concentrations of potential customers in this zone system.

ICT POLICY IN THE NETHERLANDS AND REGIONAL DEVELOPMENT

Since the mid-1990s the government of the Netherlands has set the ambitious goal to achieve a leading position in the information technology (IT) sector in Europe, in terms of economic growth and employment. To this purpose, a distinct policy line deals with the creation of quality ICT firms and the alleviation of financial constraints: the Twinning Centers and Investment Funds initiative. The vision is to advance local communities of innovative ICT firms that are closely linked with universities, large ICT firms and global networks. The direct role of the national government is to sponsor the creation of business incubation centers, the so-called Twinning-Centers (now named Twincubators). These provide a full range of services and contacts, including access to international networks, in order to enable promising start-up businesses to grow (Ministry of Economic Affairs, 1998). In addition, the government is involved in providing investment capital for promising ICT initiatives through a government-sponsored seed/start-up fund, as well as a government-sponsored coinvest-ment fund which facilitates joint private-sector and government investment in expanding early-stage ICT firms. The policy targets can be summarized as follows:

- A direct creation of approximately 600 new quality ICT firms within 10 years.
- A growth of the revenue base of existing ICT firms due to the lifting of investment constraints.
- Enhancement of the capabilities of the venture capital industry to support early-stage ICT companies.
- In the longer term, a shift in conditions for entrepreneurship in ICT through the above-indicated mechanisms, such as an enhanced role of universities and an im-proved brain pool, as well as a cultural shift towards technological entrepreneurship.

The Twinning Centers and Investment Funds initiative forms part of a sector policy. However, from a regional development perspective, the planned location of the Twincubators is relevant. There is a pool of potential cities, from which already Amsterdam and Eindhoven are selected. The pool of cities clearly reflects *existing growth patterns* in that all but two cities are located in or adja-cent to the economic core of the Netherlands. The underlying location factors include the proximity of a top university linked with outstanding regional networks; the presence of large high-tech firms accompanied with a pool of specialized workers; and the availability of facilities similar to university science parks. Whether the attraction of promising ICT start-ups will be successful in Twincubators in the coming years is dependent upon various factors, such as a strict use of criteria in the selection of candidates and the supply of competitive (cheaper) office space in the city at hand.

Given the ambitions, the above national initiative is not intended to serve economic growth in remote areas. One might, therefore, expect a high priority for ICTs in regional development among policy actors at the regional level (provinces) in the North of the Netherlands. If we consider the most recent policy intentions of the three Northern provinces, however, it appears that ICT has not yet moved up the policy agenda, whereas there is an overwhelming attention for measures concerning traditional infrastructures, like rail, roads and waterways (SNN, 1999). The latter measures serve an improved physical con-nection of the economic core (Randstad) with the three provinces in the North in establishing an international transport corridor with Northern Germany and Scandinavia. The only policy in which explicit attention is given to ICTs refers to the labor market (training and retraining of labor force). In a future elaboration of the policy, attention will most probably also be given to premises and the supply of specific ICT services here.

Most recently, the provinces have reflected on their role in improving ICT use in regional development. They have identified the following roles: to increase awareness of the potentials of ICTs among local actors, and to improve access of these actors to telecommunication networks. It needs to be emphasized that access to these networks is relatively poor for large parts of the Northern Netherlands (including Emmen), because there is no connection with a part of the vital telecommunication infrastructure. In the Netherlands, the construction of glass fiber cables and the like is considered to be a matter for the market (Ministry of Economic Affairs, 1999). Accordingly, provinces can only play an important role in intermediating for finance (e.g., to help to establish private consortia and public-private partnerships that are willing to invest in improving telecommunication infrastructure connections). The Emmen region as a part of the Northern Netherlands will be discussed in detail in the next section.

THE REGION OF EMMEN

The region of Emmen (Southeast Drenthe) is located in the Northeast of the Netherlands adjacent to the German border (Figure 12.1). The town of Emmen is approximately 190 km from Amsterdam. The distance to the seaport of Rotterdam is 270 km. Bridging the distance to Amsterdam by train takes approximately two hours and to Rotterdam two and a half hours. Such distances and travel times are generally not a disadvantage, but according to the *perceptions* in a small country like the Netherlands these are quite substantial. Seen from the Western part of the economic core area (the Randstad) Emmen is quite far away. In addition, the town is outside the major transport corridors running from Amsterdam and Rotterdam to Germany.

Figure 12.1
The Netherlands

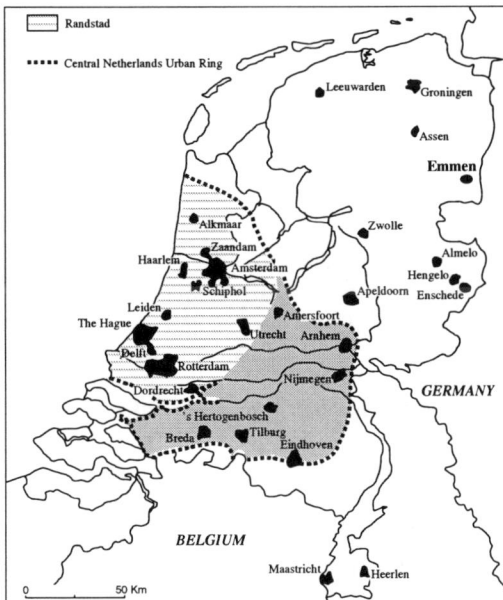

With almost 95,000 inhabitants (municipal level) in 1997, Emmen ranks 25th in the city system of the Netherlands (Table 12.2). The population in the region is relatively small as a share of the country's total (almost 1%) and this is reflected in a relatively thin population. In terms of regional product the region lags behind the country as a whole (89 versus 100) but performs better than other regions in Drenthe (a level of 76). If we consider the production structure of the region, three important anomalies come to light. First, there is a large size for mining (16% of value added), due to oil production and administrative quarters of the NAM (Nederlandse Aardolie Maatschappij) in the region.

Table 12.2
Main Characteristics of Emmen and the Emmen Region

Indicators	Emmen	Emmen Region[a]	National
Population (1997)			
Size	94,528	152,799	15,6 million
Density (per km2)	304	172[b]	446
Regional Economy (1996)			
Gross Regional Product per Capita[c]	n.a.	89	100
Selected Economic Sectors (% share):	n.a.		
- Agriculture (Value Added)		5.6%	2.9%
(Employment)		3.0%	1.7%
- Mining (Value Added)		15.6%	2.8%
(Employment)		1.9%	0.2%
- Manufacturing (Value added)		18.7%	17.4%
(Employment)		24.9%	16.5%
- Transport (Value Added)		4.0%	6.6%
(Employment)		5.2%	6.9%
- Producer Services[d] (Value Added)		18.4%	25.5%
(Employment)		13.3%	18.7%
Labor Market (1995/97)			
Net Participation Rate	56%	57%	59%
Registered Unemployment	8%	7%	6%

a.   Southeast-Drenthe (Corop region).
b.   Province of Drenthe.
c.   Index (The Netherlands is 100).
d.   Financial services and remaining business services

Source: Various statistics of the Netherlands Central Bureau of Statistics.

Second, the regional economy faces a less developed service sector (financial services and other business services), particularly regarding value added (18% versus a national average of 26%). In addition, there is a relatively large manufacturing sector, particularly regarding employment size (25% against a national average of 17%). The latter complies with the presence of various large routine manufacturing subsidiaries in the region. To mention a few examples (1998): AKZO Nobel Fibers (1,870 employees), NPBI (production laboratory for blood products) (930 employees), Honeywell Combustion Control (650 employees), NAM (Shell, Exxon) (530 employees), DSM Engineering Plastics (480 employees) and Ericsson Radio Systems (340 employees).

The economic structure can be qualified as weak in two respects. There is a relatively large dependency on production subsidiaries with headquarters located somewhere else, and the prevailing economic sectors are partly weak with strong price competition and/or a large surplus in the market (e.g., synthetic fibers production and metal-electronic industries). Furthermore, based on the literature concerning industrial districts one may assume that a strong presence of production subsidiaries has influenced industrial relations and the underlying institutions in the region in two ways, namely a weak development of entrepreneurship among the local population and a weak development of flexible cooperation between regional firms based on subcontracting and supplier relationships (e.g, Grabher, 1993).

To date, the region of Emmen has not attracted many ICT-based firms (Jacquet, 1998). The only relatively large one is Ericsson Radio Systems (hardware), but this firm is about to close the manufacturing department here whereas a part of the R&D will be maintained. As far as ICT-services is concerned, there is a small call center. In addition, a medium-sized software company plans to open its help-desk activities in Emmen.

Despite a relatively weak economic structure, the labor market as it is today faces no particular anomalies. The participation rate in the region and registered unemployment are close to the national average. The same is true for the town of Emmen. Further, the level of education in the region is also close to the national average (van Geenhuizen, 2000). The region suffers, however, from a phenomenon that is clearly related to a weak economic structure—a brain drain. Thus, students from the Higher Education Institute (Hogeschool Drenthe) leave the region after graduation in search for jobs elsewhere (Jacquet, 1998). The weak economic structure of the region of Emmen has been acknowledged by the Dutch government for a long time. Like most of the North, it has a special status for assistance in regional economic development (IPR region). This means that investment projects receive a subsidy of 25 to 35%. In addition, in the European regional policy it has received the status of an Objective 2 Region.

ICT INITIATIVES AND STRENGTHS AND WEAKNESSES

Various initiatives were taken in the region in order to stimulate ICT activities. To date, however, these initiatives have not yet led to successful action. What is striking is the low level of investment (private and public) in telecommunication services in the municipality and adjacent municipalities. With an amount not exceeding €23,000 per km^2 per year, investments are less than 10% of those in large cities and various medium-sized towns in the economic core (DDV, 1999). A number of reasons can be advanced in explanation with regard to local policymaking. First, there is a lack of a clear vision about the potentials of ICT use, such as the link with economic development. A second factor is the way in which initiatives come into being. These often emerge as a result of brainstorming sessions, with follow-up elaboration only upon optional action. A further reason is that ICT initiatives are derived from other existing policy fields, leading to a fragmented approach and (sometimes) overlap. What is missing is a coherent policy vision that underpins the initiatives and guarantees a systematic follow-up, including the assignment of priorities (van Geenhuizen and Jacquet, 1999; Jacquet, 1998). Notwith-standing this situation, the municipality of Emmen together with the Hogeschool Drenthe, a private actor and an intermediary organization (Syntens) has been successful in realizing an initiative called "learning paths". This initiative intends to generate positive impacts both on the supply side of the labor market and the supply of innovative software products.

It needs to be realized that the above indicated absence of a formalized ICT policy is not unique. A study in the United Kingdom (Gibbs and Tanner, 1997) shows that only a minority of local authorities have a current or planned ICT policy as a part of economic development measures (35%). This does not mean that local authorities without an ICT policy are not involved in ICT initiatives. Local authorities do not always develop initiatives based on a prior established policy. Such initiatives can develop in an *ad hoc* fashion, often led by a scant availability of funding, rather than systematic strategic objectives. In the United Kingdom, the most-cited difficulty faced by local authorities in ICT policy involvement is lack of finance (Gibbs and Tanner, 1997).

Based on extensive interviews with regional and local actors, various strong and weak points with regard to an increased use of ICTs could be revealed (Jacquet, 1998). The analysis included ICT use in existing firms, the attraction of firms from elsewhere, new firm formation based on ICTs and the competitive power of firms in the region due to the use of ICTs (Table 12.3).

The strong points can be summarized as follows:

- locally available knowledge from the Hogeschool Drenthe and from various large manufacturing companies,
- availability of relatively cheap premises,
- a positive labor mentality of the local population,
- bilangualism in the labor market,
- good living conditions in the region,
- a potentially large customer market,
- size and central place function of Emmen.

The weaknesses are mainly concerned with qualifications in the labor market concerning ICT use, the supply of ICT education, ambitions of the local population in terms of entrepreneurship, availability of specific ICT knowledge and the use of available knowledge. Furthermore, a remarkable shortcoming in the view of regional actors is the persistent disadvantage of the distance to the

Randstad. Apparently, there is no shrinking of distance to the core area. When considering the type of policy needed to avoid or mitigate these shortcomings, one can arrive at the conclusion that ICT policies need to be embedded in a larger comprehensive development policy for the region itself, with a particular emphasis on learning.

Table 12.3
Strong and Weak Points for ICT Use in the Emmen Region and Province of Drenthe

Aspect	Weak Points	Strong Points
Increase of ICT use by firms	Shortage of knowledge in firms and among employees about the potentials of ICT use. Limited availability of ICT services and products in the region. A shortage of (focused) enhancing policies by the government.	Knowledge at the Hogeschool Drenthe
Attraction of firms from elsewhere	An image of a routine manufacturing area. Remaining disadvantages of perceived distance to the core (Randstad). A limited availability of employees that match ICT requirements.	Availability of premises (favorable price-quality ratio). An excellent living climate. Positive mentality of labor force: strong loyalty with employer and strong orientation to customer services. Availability of native speakers of Dutch and German.
Formation of new ICT-based firms	Absence of ICT education and ICT schools. Lack of entrepreneurial (risk-taking) mentality among locals. A shortage of (focused) enhancing policies by the government.	Abundance of cheap space for growth. Presence of large firms as customers in the region. Availability of native speakers of Dutch and German.
Improving the competitive power of the region	Unknown and under-used knowledge potentials in the region. A shortage of pro-active and integrated policy. A shortage of co-operation within the region.	Knowledge at the Hogeschool Drenthe and large firms. Emmen is a medium-sized town serving altogether 250,000 inhabitants in the region (a critical mass).

Source: Jacquet (1998); Huntink et al. (1996).

POLICY IMPLICATIONS

Aside from an infrastructure-led approach, there is now a wide recognition of the necessity of an *institutional* approach to regional economic growth (e.g., Maskell et al., 1997; Morgan, 1997). The latter approach addresses the importance of institution building, such as coherent sets of expectations, built into conventions, permitting the actors involved to develop and coordinate necessary resources. Improving the learning capability of actors in the region in a bottom-up and interactive way is a major aim in this approach. The learning capability of a region refers to a range of critical conditions under which regional actors can use the available knowledge (technology) in an efficient way, in order to improve their economic performance (Camagni, 1992; van Geenhuizen and Nijkamp, 1998; 1999; Morgan, 1997). These conditions encompass *inter alia* the following:

- A sufficiently large *support* among regional actors such as local governments, SMEs and large firms' representatives, knowledge institutes, Chambers of Commerce for learning as a broad socioeconomic strategy. In addition, *trust* among regional actors and reciprocity are important conditions.
- An active participation of regional actors in *learning networks* (both formal and informal ones) in view of acquiring and applying new knowledge. Innovations emerge as a result of complex processes of interaction between firms (such as in buyer and supplier relationships) and between firms and knowledge institutes.
- An active effort of regional actors to become *learning organizations*, meaning an open information circulation from outside and from inside the organization, and an effective translation of information into strategies.
- A continuous effort in *transforming (translating) new knowledge* in order to avoid barriers stemming from different disciplinary backgrounds (policy fields) and differences in educational level of regional actors.
- A continuous effort in *improving human capital*. The enhancement of education and training, including skills in how to learn in an effective way, is concerned with the local residential population, employees of local firms, as well as local policy makers and politicians.
- At a higher level: a continuous effort to *improve the self-organizing power* of regional actors with regard to coordination and renewal of the above activities. The latter includes *inter alia* the monitoring of developments with a focus on the achievements of goals set in the region and on competing policies and developments in other regions.

To summarize, policy efforts need to concentrate on institutional arrangements that foster learning as a self-understanding attitude and activity. Such arrangements need to be coupled with a targeted policy for the use of ICTs. Based on experiences in the United Kingdom and the in-depth case study of the Emmen Region in the Netherlands, the following recommendations can be addressed to local and regional authorities:

- To give more specific attention to ICTs as a field relevant *in itself*, not as an issue derived from other policy fields.

- To design a more transparent vision on ICT applications and links with promising economic clusters in the region, given various weak points that cannot be changed overnight. This serves an improved integration of ICT policies in regional policy and planning.
- To *create a better support and commitment* for such a vision (policy), including priorities, necessary for follow-up in specific actions. In this way coordination and steering can improve in order to reach the goals that have been set.
- To *demonstrate more often in a pro-active way* the potential use of up-to-date ICTs to firms on an ongoing basis, and to initiate SME networking in demonstration projects which can be mediated through ICTs in a subsequent stage.
- To *further improve education and training in ICTs*, both as a broad subject and focused on practical applications in particular branches and firms.

It can be concluded that the "death of distance" is not (yet) reality for the region of Emmen and many other peripherally located regions. Therefore, there is a need to improve the use of opportunities available in the region itself. This would mean, first, to respond to a major prerequisite for success of ICTs use (i.e., to embed ICT policies in a learning policy in order to improve the demand side). Second, it seems wise to focus ICT initiatives on (potentially) strong sectors (clusters) in the region. In the case of Emmen, these may include horticulture, transport and value-added logistics, manufacturing in strong sectors as well as specific city functions of (inter)national importance, such as tourist, cultural and sporting events (SNN, 1999). A further important aim is to create sufficient support in order to finance improvements of the connection of Emmen with the major telecommunication transport infrastructures in the country, enabling a further attraction of knowledge-intensive firms that exchange large amounts of data.

At the same time, it must be admitted that systematic empirical knowledge to underpin ICT policies is still rather scarce. Particularly, the influence of physical distance to the economic core on regional business dynamics related to ICT is not fully understood. Accordingly, with regard to policy-oriented research we would give high priority to a systematic comparative study of firm dynamics in regions outside economic cores in various European countries (large and small ones). Such a study would uncover those positive dynamics that are assumed to be linked with the use of ICTs—innovation in existing firms, new firm formation based on ICTs and relocation (decentralization) of departments of firms. Aside from an investigation of the magnitude of these dynamics, the focus needs to be on the conditions that foster them, including regional policy and local initiatives.

Further, a more fundamental research would focus on *virtual network relationships*, that is, on the way in which various regional places outside economic core areas (adjacent, remote, etc.) are getting interwoven with complex media infrastructures and media activities linking these places with other spaces. The study of the region of Emmen has brought to light the phenomenon of *dependency* of the regional economy from decision centers elsewhere, based on ownership relations in the firms at hand. The question is whether current dependency in place-based interaction in material space is maintained (reinforced) or replaced by other (new) relationships in electronically mediated interaction in virtual space. Although access in virtual space is not determined by distance and economic functions are not filtered by land-rent gradients, evolutionary notions say that the mechanisms at work in creating barriers and thresholds in networks in virtual space are strongly connected with the historical ones in material space. Thus, an important question is concerned with the conditions under which dependency relationships in virtual networks may be avoided or the disadvantages mitigated. A follow-up question would be concerned with the role of regional policy and local initiatives in shaping those conditions.

REFERENCES

Arthur, W.B. (1994). *Increasing Returns and Path Dependence in the Economy*. Ann Arbor: University of Michigan Press.

Amirahmadi, H. and Wallace, C. (1995). "Information technology, the organization of production and regional development". *Environment and Planning A*, 27: 1745-1775.

Boschma, R. and Lambooy, J. (1998). "Economic evolution and the adjustment of the spatial matrix of regions". In Van Dijk, J. and Boekema, F. (eds.), *Innovation in Firms and Regions* (in Dutch). Assen: Van Gorcum & Comp.

Cairncross, F. (1997). *The Death of Distance. How the Communications Revolution Will Change Our Lives*. Boston: Harvard Business School Press.

Camagni, R. (ed.) (1992). *Innovation Networks: Spatial Perspectives*. London: Belhaven Press.

Capello, R. (1998). "Telecommunication network externalities and regional development policy implications". In Capineri, C. & Rietveld, P. (eds.), *Network Interconnectivity*. Ashgate: Avebury.

Clark, D., Ilbery, D. and N. Berkeley (1995). "Telematics and rural businesses: an evaluation of uses, potentials and policy implications." *Regional Studies*, 29(2): 171-180.

Cornford, J. and Gillespie, A. (1993). "Cable systems, telephony and local economic development in the UK." *Telecommunications Policy*, 17(8): 589-603.

Cronin, F.J., McGovern, P.M., Miller, M.R. and Parker, E.B. (1995). "The rural economic development implications of telecommunications. Evidence from Pennsylvania". *Telecommunications Policy*, 19(7): 545-559.

DDV (1999). *Spatial Disparities in Telecommunication Infrastructure* (in Dutch). De Meern: DDV.

Geenhuizen, M. van (1999). "An evolutionary approach to firm dynamics: adaptation and path dependency". In Rietveld, P. and Shefer, D. (eds.), *Regional Development in an Age of Structural Economic Change*. Ashgate: Aldershot.

Geenhuizen, M. van (2000). "Regional economic development and ICT: how about the distance to the Randstad". In Bouwman, H. (ed.), *Silicon Alleys, Cyber-cities and Digital Valleys. Multi-media and Telecommunication as Impulse for Local and Regional Development* (in Dutch). Utrecht: Lemma.

Geenhuizen, M. van and Jacquet, M. (1999). "ICT and regional development" (in Dutch). *Informatie & Informatiebeleid*, 17(2): 65-71.

Geenhuizen, M. van and Nijkamp, P. (1998). "Improving the knowledge capability of cities: the case of Mainport Rotterdam." *International Journal of Technology Management*, 15(6/7): 691-709.

Geenhuizen, M. van and Nijkamp, P. (1999). "The learning capability of regions: conceptual patterns and policies". In Rutten, R., Bakkers, S., Morgan, K. and Boekema, F. (eds.), *Knowledge, Innovation and Economic Growth: Theory and Practice of the Learning Region*. London: Edward Elgar, pp. 38-56.

Gibbs, D. and Tanner, K. (1997). "Information and communication technologies and local economic development policies: the British case". *Regional Studies*, 31(8): 765-774.

Gillespie, A., Coombes, M. and Reybould, S. (1994). "Contribution of telecommunications to rural economic development: variations on a theme?" *Entrepreneurship and Regional Development*, 6: 201-217.

Gillespie, A. and Williams, H. (1988). "Telecommunications and the reconstruction of regional comparative advantage". *Environment and Planning A*, 20: 1311-1321.

Grabher, G. (ed). (1993). *The Embedded Firm. On the Socio-economics of Industrial Networks*. London: Routledge.

Graham, S. (1998). "The end of geography or the explosion of space? Conceptualizing space, place and information technology". *Progress in Human Geography*, 22(2): 165-185.

Graham, S. and Marvin, S. (1997). *Telecommunications and the City. Electronic Spaces, Urban Places*. London: Routledge.

Graham, S. and Marvin, S. (1999). "Planning cybercities? Integrating telecommunications into urban planning." *Town & Country Planning Review*, 70(1): 89-114.

Houghton, J. (1999). "Mapping information industries and markets". *Telecommunications Policy*, 23(9): 689-699.

Huntink, W., van der Weele, E. and Weijers, T. (1996). *IT&T-Sector in the North of the Netherlands* (in Dutch). Apeldoorn: TNO.

Jacquet, M.E. (1998). "Telecommunication and stimulation of regional economic development. A focus on the Emmen region." Delft: Faculty of Policy, Technology and Management (doctoral thesis).

Jipp, A. (1963). "Wealth of nations and telephone density". *Telecommunications Journal*, July: 199-201.

Maskell, P., Eskelinen, H., Hannibalson, I., Malmberg, A. and Vatne, E. (1997). *Competitiveness, Localised Learning and Regional Development—Specialisation and Prosperity in Small Open Economies*. London: Routledge.

Melody, W. (1991). "The information society: the transnational economic context and its implications". In Sussman, G. and Lent, J. (eds.), *Transnational Communications: Wiring the Third World*. London: Sage, pp. 27-41.

Ministry of Economic Affairs (with Booz-Allen & Hamilton) (1998). *Netherlands' ICT Twinning Centers and Investment Funds. Building the Mind-set and the Skill Base for the Information Society*. The Hague: Ministry of Economic Affairs.

Ministry of Economic Affairs (1999). *Document for Spatial-Economic Policy* (in Dutch). The Hague: Ministry of Economic Affairs.

Mitchell, W.J. (1995). *City of Bits: Space, Place, and the Infobahn*. Cambridge: The MIT Press.

Morgan, K. (1997). "The learning region: institutions, innovation and regional renewal". *Regional Studies*, 31(5): 491-503.

Nelson, R.R. and Winter, S.G. (1982). *An Evolutionary Theory of Economic Change*. Cambridge, MA: Harvard University Press.

Newlands, D. and Ward, M. (1998). "Telecommunications infrastructures and policies as factors in regional competitive advantage and disadvantage". Paper for the 38th Congress of the European Regional Science Association, 28 August-1 September, Vienna.

OESO (1996). *The Knowledge-based Economy*. Paris: OESO.

Porter, M.E. (1998). "Clusters and the new economics of competition". *Harvard Business Review*, 76(6): 77-90.

Richardson, R. and Gillespie, A. (1996). "Advanced communications and employment creation in rural and peripheral regions: a case study of the Highlands and Islands of Scotland". *The Annals of Regional Science*, 30: 91-110.

SNN (1999). *A Compass for the North. Spatial-Economic Development Programme for the North of the Netherlands 2000–2006 (Concept)* (in Dutch). Assen: Samenwerkingsverband Noord-Nederland.

Tacken, M.H.H.K. and Reisen, A.J.J. van (1998). *Developments in Science and Technology. ICT in Relation to Mobility and Location Behaviour in Rural Areas* (in Dutch). NRLO-Report no. 98/14. The Hague: NRLO.

Thrift, N. (1997). "New urban areas and old technological fears". In Leyson, A. and Thrift, N. *Money/Space: Geographies of Monetary Transformation*. London: Routledge, pp. 323-354.

World Bank (1994). *World Development Report 1994*. Washington DC: World Bank.

13

Network Building Between Research Institutions and Small and Medium Enterprises: Dynamics of Innovation Network Building and Implications for a Policy Option

Junmo Kim

INTRODUCTION

As the 21st century begins, salience of technology development gathers greater attention, inheriting the full thrust from the 1990s. In designing a strategy for science and technology development, with its intellectual tradition mainly from the 1980s, one of the clear trends has been network building (Debresson & Amesse, 1991; Levy & Samuels, 1991; Drucker, 1998). Confronted with similar external and internal changes, advanced countries in the Western Hemisphere have pursued a policy trend of network building, materialized as science and technology networks and the National Innovation System (Nelson, 1993; OECD, 1997a; Lundvall, 1992; Sandholtz, 1992; EC, 1995). Rationale behind the network comes from a phenomenon called a productivity paradox that these countries commonly experienced through the 1970s (Perez & Freeman, 1988; Krugman, 1996). This phenomenon, which benefits from science and technology investment increases at unproportionally lower rates with a unit increase of science and technology investment, has become a serious social issue as the

245

economies become more reliant on the development of science and technology (Nelson & Romer, 1996; Rosenberg, 1982, 1996; Conceição, Heitor and Oliveira, 1998; Dosi, 1988). Networking was proposed as a way to cope with the problem.

While this mentioned dynamic has been the major trend in advanced nations, developing countries, such as Korea, although following a different development track, seem to share the necessity of science and technology policy network building as a way to cope with new environments. With this context, this chapter focuses on the feasibility of research network building between government-funded research institutions and small and medium enterprises (SMEs) by surveying research on both sides. The aim of the survey can be articulated in two ways. First, by asking highly experienced researchers at government labs who have already accumulated cooperative research experience with SMEs in Korea on the feasibility of research network building in a more formal way than they had experienced, this research tried to deduce policy implications of research network building in Korea with eventual aim of providing a generalizable argument. Second, since this chapter utilized existing theories on research networks, the utilization endowed this research with an opportunity to test theoretical arguments in the Korean contexts for policy implications.

LITERATURE REVIEW

Global Context of Network Building

In understanding the salience of research networks, it is reasonable to track several key developments that have been shared among advanced economies since the postwar economic boom, which are closely interconnected to deduce the research network. These key developments include the relative decline of Fordism (Berger, 1981), the productivity paradox and the gradual reduction of enrollment in engineering schools.

Fordism, or the mass-production system that has been the pillar of economic growth since the postwar period, had its system requirements including a stable macroeconomic management by government, namely the Keynesian Demand Management policy, which became more difficult to maintain (Boyer, 1988; Boltho, 1982). While several factors have contributed to the relative decline, from a technology perspective, the decline implied that the production system of the future would require either an input of more intensive use of technology or a paradigm change into "flexible specialization" (Piore & Sabel, 1984; Shimada, 1991). The course taken was to adopt both. Flexible specialization inspired mass-production systems to be flexible as exemplified by such systems as Just-In-Time (JIT). More important, however, the other path—intensive use of technology—naturally called the importance science and technology as the main engine for growth (Nelson & Romer, 1996; Freeman, 1987; Rosenberg, 1982, 1996; Dosi, 1988; Sandholtz, 1992). This emphasis brought a dilemma, since from the 1970s on advanced economies began experiencing a phenomenon called the "productivity paradox" (Perez & Freeman, 1988; Krugman, 1996).

Furthermore, an important pillar of science and technology promotion—the supply of science and engineering students—has decreased in advanced economies, which indirectly shows that salaries in science and technology research positions have been lower than their private-sector equivalents or substitutes (Pearson, 1990; OECD, 1989).

With these developments, trend for research network building was accelerated.[1] (Metcalfe, 1990; NBIA, 1998; Oh & Masser, 1995; Porter, 1998; Rullani & Zanfei, 1988). Behind the development lies an important undercurrent of increasing complexity of science and technology and its application—that is, a system integration characteristic, in which one researcher's effort is less likely to make a decisive contribution.

In Europe, the concept of a National Innovation System (NIS) in which various levels of research units are interconnected has emerged (Freeman, 1987; OECD, 1997; Lundvall, 1992; Nelson, 1993). Similar and as part of a National Innovation System, science park or research-oriented cities were constructed with linkages being built around regional research units such as universities, firms and business incubators (Masser, 1989, 1991; Grayson, 1993; Kawamoto, 1992; Koschatzky & Kulicke, 1994; Luger, 1994; Sung, 1997; Tatsuno, 1986; Segal, 1985; Monck et al., 1988).

The Korean Context

While Asian countries like Korea have been known as serious practitioners of industrial policy (Kim, 1999, 2000; Galbraith & Kim, 1998), with increasing globalization and its accompanying international rules, it is less likely that these countries can exercise identical policies in the future. This has caused a change of policy direction from big-business-oriented industrial policy to a new focus on SMEs in Korea. A dilemma in promoting SMEs was that they lacked technological competitiveness (Kim, 1999). Against this dilemma, networking the SMEs with research units is becoming a viable policy option, while more studies are required to test feasibility for a policy. This policy environment endows this research an opportunity to question feasibility of research network building in Korea.

Theoretical Reference

With the background for building research networks, this research intends to find empirical data to support the feasibility of research network building in the Korea context. In doing so, this research finds theoretical reference on research networks from existing literature.

Definition of Research Network. A research network can be defined as both formal and informal linkages among research and production units such as universities, research institutions and firms that are connected to generate synergy while fulfilling their individual objectives (Freeman, 1991; Debresson & Amesse, 1991).

Functioning of the Network. From existing studies, it is possible to find some necessary conditions that would generate research networks, which include implicit or tacit nature of knowledge and factors of trust that would enable the networks to be sustained. With these theoretical conditions, it becomes possible to infer some conditions that would inhibit the development of networks (Levy & Samuels, 1991; Lynch, 1990; NBIA, 1997; Powell, 1990):

- when participants to the network lack experience of trust
- when there is no convergence on the long-term goal among network participants
- when participants expect virtually no interactions in the future
- when there exists communicational difficulties in exchanging and learning about knowledge and intellectual property

With these conditions, factors of trust, frequency, experience, and interaction seem to be the important variables that would enable or inhibit the formation and maintenance of research networks (Koschatzky & Kulicke, 1994; Gibson & Rogers, 1994; Hamel et al., 1989; Hipple, 1987). Thus, in an empirical survey to find feasibility of more formal research networks, it would be crucial to investigate the above variables.

METHODOLOGY

Data and Sampling Frame

In order to deduce the feasibility of a formal research network as a policy option, this research utilized the survey method to acquire a data set. In doing so, it became crucial to define the sampling frame. Since this research aimed at a research network between SMEs and government-funded research institutions, thinking that this element would be weakest link in designing a hypothetical National Innovation System with currently available research resources, an important step was to narrow down the concepts and scope of SMEs and researchers at government-funded research institutions or labs.

Definition of SMEs. In defining the SMEs, this research selected SMEs that are technologically intensive and advanced. On this definition, the following argument can be suggested. The policy of building a research network can be described as technologically intensive in nature, especially in the Korean context in which previous policies were mainly related to offering financing measures (Galbraith & Kim, 1998). This characteristic offers a clue in the sense that demand for this type of policy would mainly come from advanced SMEs whose competitiveness relies on technology itself (Porter, 1998; Rothwell, 1984).

Together with the hurdle of refining the scope of SMEs, another important task of sampling SMEs came from their geographical and industrial distribution. This issue shows a seemingly biased sampling result which, in fact, can be justified, considering the context of a country where this research was undertaken. Since this research focused on advanced SMEs in Korea where certain industrial sectors have technological competitiveness in the world market such as the electronics industry (Kim, 2000), it becomes quite natural that advanced SMEs in that country would be likely to emulate the industrial terrain of the mainstream industry (Reid, 1993; Reid & Anderson, 1993). In the Korean context, it would be the electronics, communication and information-related sectors. In other words, advanced SMEs in Korea, as of 1999 when the survey for this research was conducted, were predominantly in electronics and its related sectors. Thus, while the survey tried to balance a diverse industrial portfolio, the natural reflection of the SMEs' world itself is dominated by electronics related firms (see Figure 13.1 for the distribution of industries).

Figure 13.1
Overall Scheme

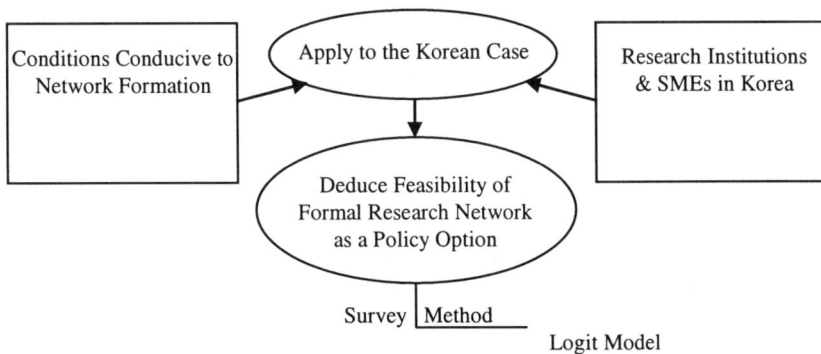

Similarly, those advanced SMEs in Korea tended to locate themselves (Reid, 1993; Reid and Anderson, 1993) in the metropolitan Seoul area where distance to downtown Seoul can be defined as about 1.5 hours. This is a justification for the seemingly biased sampling which in fact reflects the true distribution of the target industries under study. For the government-funded research institutes, this research selected those labs located in Seoul and Taejon area, since these two regions hold most of the advanced research labs in Korea.

Defining Researchers at SMEs and Government-Funded Labs. In selecting the people who actually were surveyed, this research sampled researchers at both SMEs and research labs. The fact that SMEs under the sampling frame are advanced ones denotes that there are in-house researchers in those firms. For the labs, this research selected those researchers with Ph.D. degrees and previous experience with SMEs regarding technical consultations, assistance or joint research to check whether their experiences of an informal network, which tends to be generated naturally, influenced their support for a more formal network. The condition that researchers at labs should have previous experience working with SMEs dramatically reduced the size of the population available in Korea. This decision, however, was thought to provide serious answers regarding the feasibility of network building.

With ideas for the sampling frame, this research surveyed 105 SMEs in which one questionnaire was asked per firm. For the labs, 102 Ph.D.-level experienced researchers from three major government labs were surveyed.[2]

Frame of Analysis and Research Methods

Overall Scheme. The overall scheme of this research starts with comparing the Korean case with the theoretical guidelines from the existing literature. This research finds crucial variables reported by existing literature such as trust, long-term nature of the relationship, frequency and satisfaction with the existing relationship. Prior hypothesis in designing the research is presented in Figure 13.2. First, we can be sure that the crucial variables mentioned above have a relationship with the development of an informal network, which develops naturally. The causal link is that, with the development of crucial variables, research performance would be affected. Prior expectation is that as the crucial variables develop in a way to activate informal networks, it will improve research performance, as indicated in Figure 13.2. The next step is the relationship between informal and formal networks. This research asked whether previous experience of informal networking would strengthen the respondents' preference for a more formal networking policy to improve research performance (Schrader, 1991; Perrin, 1988; Debresson & Amesse, 1991). In so doing, Logit regression was utilized to deduce the support for networking as a policy option. If the Logit model proves to be statistically significant, it would offer an opportunity to compare the Korean case with the theoretical hypothesis of whether the four crucial variables are important in the Korean context. If the four variables or a certain combination of the four turn out to be meaningful, it would imply that successful formal networking policy would depend on healthy informal networking. This would illuminate the direction in which networking policy would go forward.

Figure 13.2
Conceptual Scheme

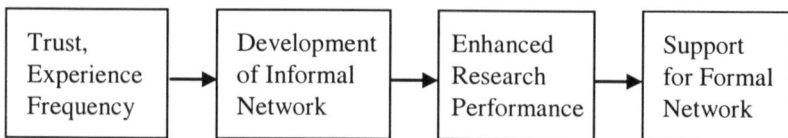

| Trust, Experience Frequency | → | Development of Informal Network | → | Enhanced Research Performance | → | Support for Formal Network |

Logic of Critical Case Study. In proposing the research design, it is also important to prepare for validity of research. In other words, it would be helpful if this research design can persuade the audience why the case study method in surveying is meaningful. For this, it is possible to refer to an influential forerunner in this tradition, namely the critical case study method by John Goldthorpe. (Goldthorpe, 1968). His contribution was an offspring of his research question: "would increased wealth change the attitude of blue-collar workers in the 1960s in Britain?" In actualizing his research, his research design was to find the most industrially advanced region of Britain, which was Lutton, to test his hypothesis. His logic was that if the hypothesis is rejected in Lutton, where workers' attitudes are most likely to be changed, the hypothesis is very likely to be rejected in all other industrial regions of the country. Thus, the case study method virtually apply the logic of statistical testing, especially hypothesis testing.

This idea can be broadly applied to this research. After defining the focus of the research on the network between advanced SMEs and government labs, the critical case study method offers this research a justification to defend the geographical and industrial concentration in Korea, due to the country's specific context illustrated in the previous section (Kim, 1999).

Logit Model

This chapter utilizes important variables that are crucial in forming networks. To fulfill the purpose of the research, the chapter has employed Logit modeling to deduce the feasibility for a formal networking policy between SMEs and government labs. It is possible to present the Logit models used for labs and SMEs. Due to minor differences in survey format (length and questions), specifications differ, although they eventually search for the same results.

Logit Model for Government-Funded Research Institutions. In designing the Logit model, the dependent variable is the necessity for a formal research network expressed in response to a direct question asking about the necessity, in which 0 denotes that it is unnecessary, while 1 denotes that it is necessary. For independent variables, this research employed years at work, age, experience of trust from the previous cooperative research or interaction and size of firms researchers worked with in the past.

With these variables, it is possible to present the specification as follows:

Model 1-1: Dependent Variable with a response from a direct question
Necessity for research network (0= not necessary/1= necessary) = constant + years at work (X1) + AGE (X2) + experience of trust from the previous cooperative research or interaction (X3) + firm size (X4)

An Indirect Measure for the Support of the Network. In Model 1-1, the dependent variable was the necessity for a formal research network as expressed in a direct question. A concern with this was the skewness of answer toward support for the network as a policy option. To verify the original specification that it is decent and to resolve the possible problem, this research had a built-in shadow question asking essentially the same content. The intention was to contrast whether the two specifications show similar results or not. The intention was to get a better fitting model considering the delicate nature of the question.

Model 1-2: Dependent Variable with a response from an indirect question
Necessity for research network (0= not necessary/1= necessary) = constant + years at work (X1) + AGE (X2) + experience of trust from the previous cooperative research or interaction (X3) + firm size (X4)

Logit for SMEs. In the SME case, the dependent variable is also the necessity for a formal research network, in which 0 denotes that it is unnecessary, while 1 denotes that it is necessary. For independent variables, years at work, age, frequency of cooperative research or interaction previously experienced with government-funded research institutions, satisfaction level for the current government policies and experience of communication efficiency from the previous cooperative research or interaction were employed. Due to differences in survey format, the trust variable was not utilized, but the communication variable works as a proxy in the sense that it is a determing variable to support the network.

With these variables, specification for SMEs' logit model can be presented.

Model 2: SMEs
Necessity for research network (0= not necessary/1= necessary) = constant + years at work (X1) + AGE (X2) + satisfaction level for the current government policies (X3) + experience of communication efficiency from the previous cooperative research or interaction (X4)

FINDINGS

Results from Logit Regression

Research Institutions. From Tables 13.1 and 13.2, both Models 1-1 and 1-2 showed their statistical significance. In both models, except for the variable, age (X2), all other variables' coefficients showed identical directions, which convinced the validity of the model. As for statistical significance of coefficients, Model 1-1 showed only one statistically meaningful variable, which was trust (at 0.01 level). In comparison, in Model 1-2, all variables except work years proved to be statistically meaningful. This difference can be attributable to the difference in the way the question was asked, whether it was a direct or an indirect one. Thus, an indirect question, in this case, was more effective in deducing a meaningful model.

The main implication from the result is the reconfirmation of the theoretical arguments on the importance of the pivotal variables in building and maintaining networks, namely trust. In addition, the model presented implications from other variables including work years, age and the type of firms researchers worked with.

Table 13.1
Logit Results from Model 1-1 (Research Institutions with Direct Question)

	Const.B0	WK YRS	AGE	TRUST	FIRM SIZE
Estimate	-1.874	0.0911	0.0245	1.452	-0.892
(t test)		(0.336)	(0.037)	(2.970)	(-1.347)
Antilog		1.0953	1.0248	4.271	-2.44
Chi-sq=11.777	P=0.0191	Loss Function: Maximum Likelihood	Final Loss: 33.0255	N of 0s: 13 N of 1s: 89	

To interpret the meaning of the coefficients from the Logit model, it is necessary to convert the figures into plain numbers by taking an antilog transformation. As Table 13.1 shows in the antilog row, as one unit of age increases, which is ten years in the survey, the probability that a researcher would support the formal research network building increases by 2.48%. Work years, in which three years is a unit, shows a 9.53% increase of support for the network with one unit (three years) increase of work experience by a researcher. The most influential variable, trust, presents a dramatic figure. As one level of trust in the 5-scale trust indicator increases, support for the network increases by 327%. A peculiar finding, which reflects the Korean context, comes from interpretation of the firm size variable. The result shows that co-research experience with big firms reduces the support for the network, which indirectly evidences that various factors in building a network, including communication, did not work well with the big firms. This, reflexively, suggests that networking policy may be started between SMEs and research institutions, since researchers at the government labs showed low preference for the big firms.

In Model 1-2, a similar interpretation can be presented, except age. In this specification, variables of age (at 0.05 significance level), trust (at 0.05 significance level) and firm size (at 0.10 significance level) all proved to be statistically meaningful, except the variable of work years (see Table 13.2). With one level increase in the 3-scale trust indicator, support for the network increased by about 102.4%, while with a unit increase of work years (three years), 5.8% was increased. Firm size variable also showed an identical direction, compared to its role in Model 1-1. In comparison, a highlight can be given to the age variable, which presented a coefficient in the opposite direction. With one unit of age (ten years) increases in Model 1-2, about 144% decrease for the support of network occurred. In interpreting this, it is possible to mention that the indirect question used in Model 1-2 may have revealed the hidden preference for the network related to age factor. Thus, it is reasonable to infer from the results in Model 1-2 that relatively young researchers tended to favor the idea of building research networks.

Table 13.2
Logit Results from Model 1-2 (Research Institutions with Indirect Question)

	Const.B0	WK YRS	AGE	TRUST	FIRM SIZE
Estimate	1.6217	0.056886	-0.89204	0.70537	-0.88041
(t-test)		(0.3125)	(-2.0156)	(2.19208)	(-1.8718)
Antilog		1.058	-2.44	2.024	-2.411
Chi-Sq=11.682	P=0.01989	Loss Function: Maximum Likelihood	Final Loss: 62.72937	N of 0s: 42 N of 1s: 59	

SMEs. For the SMEs, this research presents a single model, basically due to different survey environments. In other words, advanced SMEs tended to dislike long questionnaires, and this reduced the length of the questionnaire utilized in the survey of research institutions. Despite this, overall, it was possible to retrieve broadly similar results from the survey and the model (see Table 13.3).

Table 13.3
Logit Results from Model 2 (SMEs)

	Const.B0	WK YRS	AGE	SATF	COMM
Estimate	-2.51888	-0.33318	0.6352	0.5781	0.63685
(t-test)		(-1.456)	(1.550)	(1.9110)	(2.59335)
Antilog		-1.395	1.887	1.782	1.89
Chi-Sq=13.694	P=0.00835	Loss Function: Maximum Likelihood	Final Loss: 73.660	N of 0s: 32 N of 1s: 130	

In the SME model, as mentioned above, variables of satisfaction to existing policy (at 0.10 significance level) and communication (at 0.02 & 0.01 significance levels), which was designed to act as a proxy for the trust variable, was found to be statistically meaningful. The variable of trust was the most influential variable in the model by having the highest coefficient. With a unit increase of communication indicator (trust), support for network increased about 89%. Satisfaction for government policy measured in a 5-scale indicator showed that the support for network increased by 78.2% with a unit increase of the satisfaction level. In comparison to the model for government labs, the importance of work years and age variables expressed in the magnitude of coefficients is noteworthy. With a unit increase of age (ten years), support for network increased by 88.7%, while the work years variable showed a reduction in support for the network by 39.5% with a unit increase of work years (three years). The direction found in work years is opposite to that from the government lab models. People at advanced SMEs had a tendency to favor a network with increase of age, while showed disfavor for it as they accumulate work years.

Policy Implications

From the analysis in this chapter, it is possible to infer the following policy implications.

First, it is the rediscovery of the importance of informal network. Existing theories have told us the importance of those variables such as trust and frequency of interaction in the formation and maintenance of a network (Levy & Samuels, 1991; Lynch, 1990; Hipple, 1987; Hamel et al., 1989; Gibson & Rogers, 1994). This research confirmed the salience of these variables in the settings of research institutions and SMEs. What is more important regarding this finding is the possible link between basic elements of network building and feasibility of network as a policy option.

Figure 13.3
Relationship Between Conducive Factors and Support for Network

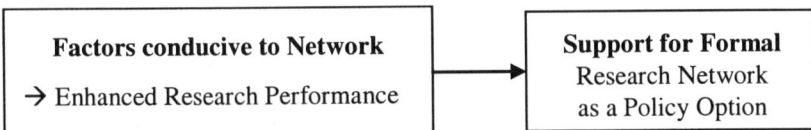

Factors conducive to Network → Enhanced Research Performance	⟶	**Support for Formal** Research Network as a Policy Option

As seen in Figure 13.3, when basic factors conducive to the development of an informal network are met, research performance was enhanced, which was reported in the survey of this research. When the link between conducive conditions for network and enhanced research performance is experienced, researchers at both government labs and advanced SMEs showed clear support for the research network as a feasible policy option.

Second, from the finding regarding the importance of an informal network, it is reasonable to argue that a policy option to promote a formal research network can be found meaningful only when informal networks can be maintained. In other words, maintenance of informal networks can be regarded as a sufficient condition that enables a formal network as a policy option.

Third, technology trading, for example, can be regarded as a type of policy available within the umbrella of formal research network that can foster the development of informal networks by increasing traffic volume of research needs at both government labs and SMEs. Therefore, it is reasonable to argue that there are mutually reinforcing relationships between informal and formal research networks (see Figure 13.4). And it would be the role of a government policy to make the first link between the two.

Figure 13.3
Mutual Reinforcing Nature of Informal and Formal Networks

CONCLUSION AND FUTURE TASKS: IMPLICATIONS FOR POLICY

Through the survey and research, this chapter has presented the dynamic of research networks. Despite the fact that this research setting was in Korea, it is quite reasonable to infer the generalization that can be applied to the relationship between firms and research labs in other regions and countries. The most important implication from this research is an insight on how to foster informal research networks, so that these can be functionally workable with more formal research networks that might be proposed by government policies. Results from this chapter showed that strengthening the basics of informal networks such as trust and frequency of interaction is the keystone (Hipple, 1987; Hamel et al., 1989; Gibson & Rogers, 1994; Levy & Samuels, 1991) in starting the research network as a policy option.

As a concluding remark, it is possible to mention the salience of this research. As we reviewed earlier in the chapter, networking has been widely accepted as the way to enhance research performance. The trend has been expressed in various forms and levels from macrolevel National Innovation Systems to science city and business incubators at the microlevel all over the globe. What is salient in this chapter is to approach a microfoundation in building the research networks at various levels. Also, as the final word, this research tradition would greatly benefit if an analysis could be carried out on an international basis.

NOTES

1. There exists a difference between Europe and the United States in adopting research networks. While the European tendency is to concentrate on the National Innovation System influenced by the Brain Drain phenomenon during the postwar period, the U.S. national labs have been isolated entities that were faithful to their original mission.

2. Survey questions for the SMEs and government labs overlap greatly, although the questions for SMEs are shorter, reflecting their work loads and prior contacts that restricted the length of survey.

BIBLIOGRAPHY

Berger, S. (1981). *Organizing Interests in Western Europe*. Cambridge: Cambridge University Press.

Boltho, A. (1982). *The European Economy: Growth and Crisis*. Oxford: Oxford University Press.

Boyer, R. (1988). *The Theory of Regulation* (translation in English). Manuscript. Paris.

Conceição, P., M. V. Heitor and P. M. Oliveira (1998). "Expectations for the University in the Knowledge Based Economy", *Technological Forecasting & Social Change*, 58(3): 203-214.

Debresson, C. and F. Amesse (1991). "Networks of Innovators: A Review and Introduction to the Issue". *Research Policy*. Special Edition.

Dosi, G. (1988). "Sources, Procedures, and Microeconomic Effects of Innovation". *Journal of Economic Literature*, 26, September.

Drucker, P. (1998). "The Discipline of Innovation". *Harvard Business Review*. November-December.

European Commission (1995). *Managing Science and Technology in the Regions*. Proceedings of the 5th STRIDE conference.

Freeman, C. (1987). "The Challenge of New Technologies". In *Interdependence and Cooperation in Tomorrow's World*. Paris: OECD.

Freeman, C. (1991). "The Nature of Innovation and the Evolution of the Production System". In OECD Report, *Technology and Productivity: Challenges for Economic Policy*, Paris.

Galbraith, J. K. and J. Kim (1998). "The Legacy of the Heavy and Chemical Industrialization (HCI) in Korea". *Journal of Economic Development*. Seoul: Joong-Ang University.

Gibson, D. V. and E. M. Rogers (1994). *R&D Collaboration on Trial*. Boston: Harvard Business School Press.

Goldthorpe, J. H. (1968). *The Affluent Worker: Industrial Attitudes and Behavior*. London: Cambridge University Press.

Grayson, L. (1993). *Science Parks: An Experiment in High Technology Transfer*. London: The British Library.

Hamel, G., Y. Doz and C. K. Prahalad (1989). "Collaborate with Your Competitors and Win". *Harvard Business Review*, 1.

Hipple, E. Von (1987). "Cooperation between Rivals: Informal Know-how Trading". *Research Policy*, 16(6).

Kawamoto, T. (1992). "An Assessment of Science Cities: Lesson from the Experience of Tsukuba Science City". *Proceedings of the International Symposium on the Development Strategies for Science Town*. Daejon.

Kim, J. (1999). "Firm Size and Inequality in Government Policy". *The Korea Public Administration Journal*. Seoul: The Korea Institute of Public Administration (in Korean).

Kim, J. (2000). "Empirical Approach to the Korean Industrial Policy". In P. Conceição, D. Gibson, M. V. Heitor and S. Shariq, eds., *Science, Technology, and Innovation Policy: Opportunities and Challenges for the Knowledge Economy*. Westport, CT: Greenwood.

Koschatzky, K. and M. Kulicke (1994). "Policies towards Technology Based Companies in a Regional Context". In K. Gonda, F. Sakauchi and T. Higgins, eds., *Regionalization of Science and Technology Resources in the context of Globalization*. Tokyo: Industrial Research Center of Japan.

Krugman, P. (1996). *Domestic Distortions and the Deindustrialization Hypothesis.* NBER Working Paper No. W5473 (March). Cambridge, MA: National Bureau of Economic Research.

Levy, J. D. and R. J. Samuels (1991). "Research Collaboration as Technology Strategy." In L. Mytelka, ed., *Strategic Partnerships.* London: Pinter Publishers.

Luger, M. I. (1994). "Critical Success Factors for High Tech Development Policy: Science Park/Innovation Centers in the U.S." *Proceedings of NISTEP Conference on Regionalization of Science and Technology Resources in the Context of Globalization.* Tokyo: National Institute for Science and Technology Policy.

Lundvall, B.-Å. (1992). *National Systems of Innovation.* London: Pinter Publishers.

Lynch, R. P. (1990). *The Practical Guide to Joint Ventures and Corporate Alliances: How to Form, How to Organize, How to Operate.* New York: John Wiley & Sons.

Masser, I. (1989). "Technology and Regional Development". *Regional Studies,* 24.

Masser, I. (1991). "By Accident or Design: Some Lessons from Technology Led Local Economic Development Initiatives." *Review of Urban and Regional Development Studies,* 3.

Metcalfe, S. (1990). "On Diffusion, Investment, and the Process of Technological Change." In Deiaco et al., eds., *Technology and Investment: Critical Issues for the 1990s.* London: Pinter Publishers.

Monck, C. S. P., R. B. Porter, P. Quintas, D. J. Storey and P. Wynarczyk (1988). *Science Parks and the Growth of High-Technology Firms.* London: Routledge.

National Business Incubation Association NBIA (1997). *Business Incubation Works: The Results of the Impact of Incubator Investments Study.* Athens, OH: National Business Incubation Association.

National Business Incubation Association (NBIA) (1998). *State of the Business Incubation Industry Report.* Athens, OH: National Business Incubation Association.

Nelson, R. and P. Romer (1996). "Science, Economic Growth and Public Policy". In B. L. R. Smith and C. E. Barfield, eds., *Technology, R&D, and the Economy.* Washington, DC: Brookings Institution.

Nelson, R. (1993). *National Innovation System: A Comparative Analysis.* Oxford: Oxford University Press.

OECD (1989). *Education in OECD Countries.* Paris.

OECD (1997). *National Innovation System.* Paris.

Oh, D. S. and I. Masser (1995). "High Tech Centers and Regional Innovation". *Habitat International,* 19(3).

Pearson, R. (1990). "Scientific Research Manpower: A Review of Supply and Demand Trends". Institute of Manpower, *Studies Report,* 169. University of Sussex.

Perez, C. and C. Freeman (1988). "Structural Crises of Adjustment, Business Cycles and Investment Behavior". In G. Dosi, C. Freeman, R. Nelson, G. Silverberg and L. Soete, eds., *Technical Change and Economic Theory.* London: Pinter Publishers.

Perrin, J. (1988). "New Technologies, Local Synergies and Regional Policies in Europe". In P. Aydalot and D. Keeble, eds., *High-Technology Industry and Innovative Environments: The European Experience.* London: Routledge.

Piore, M. and C. Sabel (1984). *The Second Industrial Divide.* New York: Basic Books.

Porter, M. E. (1998). "Clusters and the New Economics of Competition". *Harvard Business Review,* 76(6): 77-90.

Powell, W. W. (1990). "Neither Market nor Hierarchy: Network Forms of Organization", *Research in Organizational Behavior,* 12: 295-336.

Reid, G. C. (1993). *Small Business Enterprise: An Economic Analysis.* London: Routledge.

Reid, G. C., L. R. Jacobsen and M. E. Anderson (1993). *Profiles in Small Business.* London: Routledge.

Rosenberg, N. (1982). *Inside the Black Box: Technology and Economics.* Cambridge: Cambridge University Press.

Rosenberg, N. and R. Nelson (1996). "The Roles of Universities in the Advance of Industrial Technology". In R. S. Rosenbloom and W. J. Spencer, eds., *Engines of Innovation.* Cambridge, MA: Harvard Business School Press.

Rothwell, R. (1984). "Technology-based Small Firms and Regional Innovation Potential: The Role of Public Procurement", *Journal of Public Policy,* 4(4).

Rullani, E. and A. Zanfei (1988). "Networks between Manufacturing and Demand: Cases from Textile and Clothing Industries". In C. Antonelli, ed., *New Information Technology and Industrial Change: The Italian Case.* Dordrecht: Kluwer Academic Publishers.

Sandholtz, W. (1992). *High-Tech Europe: The Politics of International Cooperation.* Berkeley: University of California Press.

Schrader, S. (1991). "Information Technology Transfer between Firms: Co-operation through Information trading", *Research Policy,* 20(2).

Segal, Q. (1985). *The Cambridge Phenomenon: The Growth of High-Tech Industry in a University Town.* Cambridge: Segal Quince & Partners.

Shimada, H. (1991). "Humanware, Technology, and Industrial Relations". In *Technology and Productivity.* Paris: OECD.

Sung, T. K. (1997). "Comparative Study of the Development Strategy of Technopoleis in Korea". The First International Conference on Technology Policy and Innovation. July 1997, Macau.

Tatsuno, S. (1986). *The Technopolis Strategy: Japan, High Technology and the Control of the 21st Century.* Englewood Cliffs, NJ: Prentice Hall.

14

Innovation Clusters in Latin America

Isabel Bortagaray and Scott Tiffin

INTRODUCTION

Latin America is a major player in the world economy. It consists of Portuguese-speaking Brazil and 15 Spanish-speaking countries stretching from Mexico to Argentina[1]. Although there are numerous national distinctions, the countries can be thought of as a distinct region, sharing significant cultural, linguistic, political and economic characteristics. Brazil by itself is the eighth largest economy in the world, roughly equal in size to all the Spanish-speaking countries taken together. Mexico is one of the largest trading partners for the United States. There are large and modern resource, cultural, manufacturing, transportation and communications industries in Latin America.

However, this region is typified as "developing". In industrial terms, this translates into a landscape where large firms are usually dependent subsidiaries of multinationals, technology-based innovation in small firms is very infrequent and the productive sector tends to be isolated from research and knowledge-producing institutions. With significant exceptions, the region is characterized by deep-seated difficulties in creating locally owned, innovative industries based on scientific and technological knowledge.

Developed countries are racing to transform themselves into knowledge-based societies, where industry is constantly innovating new technology-based products that form the basis of new, entrepreneurial companies competing in global markets. As part of this race, in countries like the United States and Canada, it seems as though the director of local economic development of every medium-sized city is thinking of creating a business incubator, wondering how to link the community college or local university better to small firms, or dreaming of a science park[2]. Larger cities or governments with more resources at their disposal are attempting to create science cities such as Japan's Tsukuba, or local versions of the classic innovation clusters: Silicon Valley, Route 128, Research Triangle Park (Rogers and Larsen, 1984; Voyer, 1997).

Why all this recent attention to local innovation clusters? At the heart of innovation clusters are new, entrepreneurial enterprises based on commercializing new technology. Altenburg and Meyer-Stamer (1999) show clustering benefits as follows: "Clustering seems to enable firms, especially small and medium-sized enterprises (SMEs), to grow and upgrade more easily. SMEs may even become players in world markets if a high degree of interfirm specialization and their proximity to other firms performing complementary functions offset the disadvantages of being small. Clusters often create positive externalities which help managerial and technical learning. Empirical evidence shows that clustering is especially common among traditional small-scale and labor-intensive activities. Upgrading these activities contributes to a more balanced firm size structure and a more labor-intensive growth pattern" (p. 1693).

Innovation clusters create an image of success that instills a sense of dynamism, hope and future in their communities. They give jobs to young people in the community and attract new talent from outside. Clusters are strongly community-focussed, limited to distinct geographical areas, which means that the returns on public and private investment can be captured by the local investors. Marceau (2000) points out the critical importance of technology policy and management work that is oriented to be taken up by the policy system and politicians. Local innovation clusters are not the only way of promoting economic development, nor is the local area the only one that is important for policy, but it is a very effective way for municipalities to be involved (OECD, 1997a). In contrast to the difficulty many national governments face, especially in complex federal states, in developing useful science and technology policy, municipalities can be extremely effective in promoting local innovation clusters.

In a detailed analysis of the two paradigmatic innovation clusters, Silicon Valley and Boston's Route 128, Saxenian (1994) shows that both have shown a rapid and sustained growth in terms of numbers of jobs generated and numbers of new firms created. Of critical importance is that these jobs and firms are in the advanced technology sectors, with enormous positive impact on the rest of the economy (see Table 14.1).[3]

Table 14.1
High Technology Employment and Establishments

	High Technology Employment		Number of High Technology Establishments	
	Silicon Valley	*Route 128*	*Silicon Valley*	*Route 128*
1959	17.376	61.409	109	268
1975	116.671	98.952	831	840
1990	267.531	150.576	3.231	2.168

"Silicon Valley is now home to one-third of the 100 largest technology companies created in the United States since 1965. The market value of these firms increased by $25 billion between 1986 and 1990, dwarfing the $1 billion increase of the Route 128-based counterparts.... In 1990 Silicon Valley-based producers exported electronics products worth more than $11 billion, almost one-third of the nation's total, compared to Route 128's $4.6 billion" (Saxenian, 1994:2).

The positive benefits from industries locating close to each other has long been recognized in economic thought, beginning with Alfred Marshall, in his *Principles of Economics* (1890). His observations were based on a study of the uneven geographic concentration of firms involved in the English textile industry. For Marshall, there were several obvious positive externalities from this clustering:

• creation of a corps of workers highly specialized in the range of industry requirements
• provision of intermediate inputs to firms from local sources
• interchange of knowledge, information and ideas about improvements to production techniques and organizations

By now there is a general consensus in the literature that clustering benefits firms' economic performance (Cassiolato and Lastres, 1999; Echeverri-Caroll, 1997; Krugman, 1991; Malecki, 1997). The main reasons seem to be positive externalities (or knowledge spillovers, as Baptista and Swann state (1998, pp. 525-540), and the potential for the actors in the cluster to undertake joint action. There is another important factor, which Schmitz (1997, p. 3) stresses as joint action. The combination of both gives what he calls "collective efficiency". Quandt (1999) sees the same concept of collective efficiency as describing the nascent cluster around his city of Curitiba, in Brazil.

Preliminary investigation in 1997 by one of the authors (Tiffin) at the International Development Research Centre (IDRC, a Canadian public corporation which funds research in technology and industry in Latin America), suggested that there was little awareness in both the public and private sectors about the importance of technological innovation and its role in the new knowledge-based economy.[4] In addition, there seemed to be no mature innovation clusters

—groups of firms, universities and governments working together to create new technologies, new products and new enterprises. If innovation clusters are so important in developed countries, the obvious question is, how important are they in developing regions like Latin America? Can innovation clusters function in Latin America? Where are they? How do they work in this region? And most important, are they structures that will help to overcome some of the deep-seated constraints facing the conversion of Latin American industry to more knowledge-based forms?

These considerations led Tiffin to sponsor a feasibility study on the topic to better understand the situation.[5] Next came the hiring of a Research Intern (Bortagaray) to carry out a survey of innovation clusters in the region and run an electronic conference on the topic to locate researchers and bring the state of the art in Latin America to light. This chapter reports on Bortagaray's survey work, placed within the broader context of the IDRC effort to develop a major research program in the area.

OBJECTIVES

There are three major objectives to this chapter:

First, to create a definition of innovation clusters that makes sense for the Latin American context. Inputs for this definition come from the international literature and Latin American literature directly on this topic, and from literature on industrial innovation in Latin America. The definition must not only reflect the dynamics of what occurs in areas where innovation clusters work, but also highlight issues that are unique and critical for Latin America, at the same time.

Second, to use this definition to create a model of an innovation cluster that expands on the basic principles of the definition, allows the possibility of surveys and measurements and analysis on innovation clusters in Latin America to take place, using a concept that makes sense for the region.

Third, to use this model to examine available data in order to see where innovation clusters might be already in Latin America and draw any conclusions about how they are functioning. Key outputs would be a better understanding of how important innovation clusters might be to overcome the characteristic difficulties facing industrial innovation in the region, as well as how to support these structures by formal policy and management intervention. Given the very preliminary nature of the data we have at hand, we are likely to be able to give only a very preliminary assessment of the state of innovation clusters in Latin America, but future research will be able to go farther on this basis.

METHODOLOGY

This chapter draws on concepts from IDRC-sponsored feasibility studies, mentioned before, as well as an electronic conference run by IDRC from February to April 1999, involving 73 people (30 participated actively) from 7 countries in Latin America, in addition to Canada and the United States. The electronic conference was followed up by a workshop held in Montevideo where 27 people met to discuss key aspects around innovation clusters for three days (Bortagaray, 1999). Next, Bortagaray undertook a literature search using the resources of the IDRC library and the Perry-Castañeda Library (PCL) at the University of Texas at Austin. Field trips were made to visit cluster sites in Buenos Aires (Argentina), Porto Alegre, Recife and Curitiba (Brazil), Montevideo (Uruguay), Havana (Cuba) and San José (Costa Rica). The concepts in this chapter have been further refined in managing the development of a major research program on this topic which IDRC funded in 2000.

DEFINING INNOVATION CLUSTERS

What exactly is an innovation cluster? At present, there is no single definition of what the term is and only limited consensus on its meaning. In practice, it seems to be used interchangeably with technopoles, science cities, incubators, industrial clusters, science parks, networks and systems of innovation. In Latin America, the word "cluster" is usually used without translation, and is preferred to *aglomeración-aglomeração*, or *conglomeración-conglomeração*, which would be more exact translations. It is necessary to try to clarify the basic concepts, in order to move ahead to make a practical model and then test it on real data. Obviously present in the term are two basic concepts: industrial production in organized groups; and knowledge, learning and technological change. We examine what the literature has to say about these concepts.

Industrial Production in Organized Groups

There is a large literature dealing with *Industrial Districts*. The basic consideration is as expressed by Lerer (1979, p. 82) as: "un área de terreno delimitada y convenientemente localizada, subdividida por la planificación, y en el cual se encuentran dispoinibles medios y condiciones necesarios y favorables a la actividad industrial". Here the emphasis is on a sharply defined geographical space where firms are physically present. The concept has evolved considerably since, now emphasizing the "social environment of the ideal-type industrial-district" in terms of a common culture, frequent face-to-face relations, and "norms of reciprocity accompanied by relevant social sanctions" (Dei Ottati, 1994, p. 530 cited in Schmitz, 1997, p. 9). Altenburg and Meyer-Stamer (1999) define it as "those local business networks in which a dense social fabric based on shared cultural norms and values and an elaborate network of institutions facilitate the dissemination of knowledge and innovation".

A great deal of investment has been made by local governments to create *Industrial Parks*. These are physical spaces created to supply the basic infrastructure that firms will need, often with subsidized conditions, such as tax exemptions, to encourage them to locate in the district. A study of industrial parks in Latin America defined them as "aglomeraciones industriales conjuntamente planeadas y equipadas con una infraestructura complete [y] además una serie de servicios decentralizados" (Jonas, 1979, p. 16). In this sense, the Industrial Park is a re-creation of the original concept expressed by Lerer above, with an awareness of the importance of the need to promote links among the firms in the park. Jonas (1979) says that the proximity between firms "permite el desarrollo de relaciones interindustriales con costos mínimos de transporte. También permite la concentración de varias funciones empresariales y su manejo a través de compañiás especiales de tipo cooperativo" (p. 16).

Saxenian (1994) emphasizes the value of the relationships which form and characterize industrial districts, which she refers to as *Regional Network-based Industrial Systems*: "In these systems, which are organized around horizontal networks of firms, producers deepen their own capabilities by specializing, while engaging in close, but not exclusive, relations with other specialists. Network systems flourish in regional agglomerations where repeated interaction builds shared identities and mutual trust while at the same time intensifying competitive rivalries" (p. 4).

Schmitz (1997) focuses explicitly on *Industrial Clusters*, meaning "a sectoral and geographic concentration of firms". Ramos (1998, p. 5) emphasizes that industrial clusters generate significant advantages of "exernalidades, economías de aglomeración, 'spillovers' tecnológicos e innovaciones que surgen de la intensa y repetida interacción entre las empresas". Altenburg and Meyer-Stamer (1999) present a very useful summary of the various attributes of an industrial cluster:

- positive external effects emanating from the existence of a local pool of skilled labor and the attraction of buyers;
- forward and backward linkages between firms inside the clusters;
- intensive information exchange between firms, institutions and individuals in the cluster, which gives rise to a creative milieu;
- joint action geared to creating locational advantages;
- the existence of a diversified institutional infrastructure supporting the specific activities of the cluster;
- a sociocultural identity made up of common values and the embeddedness of local actors in a local milieu which facilitates trust.

They define a cluster as "a sizeable agglomeration of firms in a spatially delimited area which has a distinctive specialization profile and in which interfirm specialization and trade is substantial. This excludes agglomerations of the EPZ-type [Export Processing Zones], as these do not build upon intensive linkages" (p. 1695). Porter (1998) also defines a cluster as a "geographic concentration of interconnected companies and institutions in a particular field".

Knowledge, Learning and Technological Change

Baptista and Swann (1998) introduce a new concept to the industrial cluster, of the formal knowledge component "... a strong collection of related companies located in a small geographical area, sometimes centred on a strong part of a country's science base" (p. 525). This emphasis on scientific and technological knowledge is taken up by Magalhães Tavares (1998, p. 312) for a structure he calls a *Technopolis*: "Enquanto paradigma, a tecnópolis consiste em três zonas integradas, uma zona industrial compreendendo indústrias, locais de distribuição e setores administrativos; um núcleo de universidades, centros de pesquisa públicos e privados; e zonas residênciais para os pesquisadores e suas famílias".

Gómes (1999) uses the term *Science Park* to include technopoles and industrial clusters that are based on science and technology, and describes them as "arreglos institucionales mediadores, que se proponen ejercer un papel de articulación-gestión política-operacional en favor de los intereses-necesidades de empresas de base tecnológica localizadas en un espacio geográfico, normalmente dispersos en el ámbito de una ciudad" (p. 205).

There is also a large literature on *National Innovation Systems*, which puts the aspects of knowledge and learning in a distributed context of multiple actors as the core concern. An OECD report (1997a, p. 10) summarizes this research and the key contributors. This concept is like an industrial system except that it deals with the interrelated organizations producing, transferring and using knowledge:

1. "the networks of institutions in the public and private sectors whose activities and interactions initiate, import, modify and diffuse new technologies" (Freeman, 1987).
2. "the elements and relationships which interact in the production, diffusion and use of new, and economically useful, knowledge ... and are either located within or rooted inside the borders of a nation state" (Lundvall, 1992).
3. "a set of institutions whose interactions determine the innovative performance ... of national firms" (Nelson, 1993).
4. "the national institutions, their incentive structures and their competencies, that determine the rate and direction of technological learning (or the volume and composition of change generating activities) in a country" (Patel and Pavitt, 1994).
5. "that set of distinct institutions which jointly and individually contribute to the development and diffusion of new technologies and which provides the framework within which governments form and implement policies to influence innovation process. As such it is a system of interconnected institutions to create, store and transfer the knowledge, skills and artefacts which define new technologies" (Metcalfe, 1995).

In large countries with multiple layers of government, López and Lugones (1999, p. 2) point out that there are multiple and very different systems of innovation—national, local and sectoral. Obviously, there are also regional systems which act at the supranational level as well.

Use of the term *Innovation Cluster* is much more recent and, so far, more limited in scope. An innovation cluster, according to Voyer (1997), who has had the benefit of working with this topic for many years as an academic, a policy manager and a cluster promoter in the private sector, is "regional or urban concentrations of firms including manufacturers, suppliers and service providers, in one or more industrial sectors. These firms are supported by an infrastructure made up of universities and colleges, research institutes, financing institutions, incubators, business services and advanced communications/transportation systems" (p. 2). Voyer (1997) presents a list of eight factors which the consulting industry tends to use when contracting to local development authorities on the likelihood of creating an innovation cluster in their community[6]. These factors are important to highlight the main characteristics of an innovation cluster as seen by professional practitioners, and they stress:

1. the recognition of the potential of knowledge-based industries by regional/local leaders;
2. the identification and support of regional strengths and assets;
3. the catalytic influence of local champions;
4. the need to have an entrepreneurial drive and sound business practices;
5. the availability of various sources of investment capital;
6. the cohesion provided by both informal and formal information networks;
7. the need for educational and research institutions; and most importantly,
8. the need to have "staying power" over the long term (p. 4).

Lundvall (1994) stresses that innovation clusters feature processes of interactive learning and collective action. In this regard, Porter's definition of clusters emphasizes the importance of "willingness to cooperate" and "closely knit social-cultural links" (1990).

Gibson et al. (1999)[7] use the term *Learning & Innovation Pole (LIP)* defined as "an evolutionary improvement to the concept of technopolis [which interactivity links technology commercialization with the public and private sectors to spur economic development and promote technology diversification (p. 1)] that is more attuned to the needs of developing countries" (p. 2). They analyze the importance of the environment in the development of technology-based regions. For them, this "environment consists of regional networks of talent, technology, capital, and know-how that provide support essential to successfully commercializing innovations and new technology".

Definition

Given the above discussion, it seems most useful to conceptualize an innovation cluster as a special kind of industrial cluster where the core is technology-intensive or knowledge-intensive firms, and scientific and technological knowledge drive the development of new products and firms. An innovation cluster is similar to the concept of local innovation system as well, except the emphasis is on firms rather than on technology. Innovation clusters are local structures focussed on geographically identifiable communities, as opposed to sectors or networks, which may spread out over an entire country. Innovation clusters should be found within science parks, which are administrative structures that are supposed to promote their development (focussing on the most technology-intensive types of industries). Given the predominance of the literature on industrial clusters and the fact that the term cluster is already in popular use, we therefore choose to call these structures clusters (as opposed to local systems), but distinguish them from industrial clusters with the word "innovation".

We suggest the following definition:

> an innovation cluster is an organizational structure that creates new products and enterprises by means of collective industrial production within restricted geographical boundaries, based on high concentrations of knowledge exchange, interactive learning and shared social values.

It must be stressed that an innovation cluster is not something that can be easily seen or touched; the physical components work together with invisible information exchange networks, or communities of people with shared values. The tangible components of a cluster—a university, a new technology-based firm, an incubator—are elements of a cluster, but not the cluster itself.

Table 14.2
Innovation Cluster Typology

Cluster Type	Description
Dependent or truncated	Composed of branch plants, which are installed from another region or country and specialize in very limited activities, e.g., assembly (maquiladora) or resource extraction and processing. Technology is mature and arrives in fully packaged form of installed process equipment.
Industrial	A group of firms working together, focussing on producing mature goods and services. Very limited engagement of knowledge sources except for maintaining routine quality control and hiring skilled graduates. Limited entry of new firms.
Innovative industrial	An industrial cluster with strong product upgrading, quality improvements, creation of new enterprises and seeking of new markets. Routine engagement with local consultants, labs and universities to inject new knowledge into the cluster.
Proto innovation	An innovative industrial cluster which is aware of world markets, the need to be at international best practice levels, is focussed on rapid acquisition of cutting edge technology to create new products and supports a limited growth of new knowledge-intensive firms. Some key stakeholders typically missing and not clear if will continue to develop in medium term.
Mature innovation	A cluster which defines the social structure of the community it is in, creates a dynamic, expanding group of firms based on cutting-edge scientific knowledge, sucks in talent from around the world, generates venture capital and drives the pace and direction of scientific and technological research.

An innovation cluster is obviously a continuum of structures which range from those that are not innovative, or involved with science and technology, to those that are so closely bound up with R&D that they drive the scientific and technological frontiers forward. This suggests the typology showed in Table 14.2, which will be important to apply when considering the Latin American situation.

INDUSTRIAL INNOVATION IN LATIN AMERICA

A review of the Spanish-language literature on Latin America shows that there has been relatively little formal attention to innovation clusters. The literature in Portuguese about Brazil, however, has a good number of studies on structures very close to innovation clusters. Fortunately a good deal is known about industrial innovation in the region. These sources allow us to get a strong image of the constraints facing industrial innovation based on science and technology in Latin America. From this survey, we can see if it provides any guidance on what local characteristics innovation clusters must have in order to maximize their chance for success.

To start, it appears that a major feature of the innovation landscape in Latin America may be what Sutz (1998) calls innovation circuits, "entendiendo por tales a los ... procesos donde se generan innovaciones tendientes a resolver situaciones de importancia clave para problemáticas productivas específicas" (p. 33). Industrial innovation does occur but it is not self-sustaining, let alone growing. Clients request specific innovations from specific firms. This is the innovation circuit, which produces the innovation, but then disappears after the one-off task is complete. This "encapsulated" type of innovation plays an important role, but it does not generate the kind of broad industrial benefits that innovation systems create. Sutz's work is based on a detailed analysis of industrial innovation in Uruguay, but this may be a widespread phenomenon in Latin America.

Gómes (1995) has carried out a survey of science parks and technology poles in five cities in Brazil: São José dos Campos, São Carlos, Campinas, Campina Grande and Florianópolis. He notes a great deal of variation in terms of the regulation and management of these structures, as well as the support from state governments, but does find some strong commonalities in terms of the low degree of knowledge the park management has of the industrial and economic profiles of the technology firms involved; the scarcity of venture finance (with the exception of Campina Grande and the Fundo Constitucional de Desenvolvimento do Nordeste-FNE); the difficulty of involving universities and research institutions in a formal way; and the low capability of the park management for mobilizing and coordinating its members. In short, Gómes is remarkably sceptical as to whether these initiatives have any real substance in terms of what we are calling innovation clusters in this chapter. He qualifies these initiatives as fragile and vulnerable to changing political whims, often based more on abstract speeches rather than real links and outputs.

Gómes's view is not the last word, fortunately. With a shift in perspective from the science parks to the more organic form of bottom-up cluster, Quandt (1997) also studied Campinas and came to very different conclusions about this predominantly microelectronics-informatics cluster: "its specialization in technology-intensive industries combines endogenous efforts by the region's institutions and entrepreneurs with equally significant initiatives from extra-regional sources such as outside companies and the federal government.... There is a strong correlation between the regional specialization in high-technology industries and their interaction with top-level universities and other organizations that support high-technology industrialization, such as R&D centers, research institutes and industrial associations" (p. 13).

Now, we turn to the much larger literature on industrial innovation, and point out the highlights. A major difficulty in establishing innovation clusters is the generally low level of investment in research and development, almost an order of magnitude less than in the leading developed countries, as Table 14.3 shows.[8]

Table 14.3
Gross Expenditure on Research and Development (GERD)

Country	% total of R&D in relation to GDP in 1998
Argentina	.42
Bolivia	.33[a]
Brasil	.76[b]
Colombia	.41
Costa Rica	1.13[b]
Cuba	.86
Chile	.62
Ecuador	.08[c]
El Salvador	.08
Mexico	.34[b]
Nicaragua	.12[b]
Panama	.34
Perú	.06
Canada	1.61
United States	2.61

a. Data for 1996.
b. Data for 1997.
c. Data for 1995.

Not only is the public sector a minimal investor in R&D, the percentage of the overall investment carried out by firms is much lower than in developed countries, as Table 14.4 shows, also from RICYT (1999).

Table 14.4
Spending on Science and Technology by Financing Sector (1998)

Country	Government	Firms	Educación Superior	Non-profit Private Organizations	Extranjero
Argentina	42.8	27.4	24.5	2.2	3.0
Bolivia[a]	30.0	24.0	12.0	22.0	10.0
Brazil	64.0	31.8	4.1		
Colombia	65.0	14.0	17.0	4.0	
Costa Rica[a]	53.4	17.4	14.8	4.5	9.9
Cuba	55.3	44.8			
Chile[b]	70.7	15.2	7.6		6.5
Ecuador[c]	39.8	32.5		4.9	22.9
El Salvador	51.9	1.2	13.2	10.4	23.4
Mexico[b]	71.1	16.9	8.6	.9	2.5
Panama	40.2	0.0	2.5	1.3	56.1
Venezuela[b]	31.5	44.8	23.7		
Canada	13.1	62.0	23.7	1.2	
United States	7.7	75.1	14.1	3.1	

a. Data from 1996.
b. Data from 1997.
c. Data from 1995.

There are good reasons for this, of course. Much of Latin American industrial infrastructure is oriented to the production of natural resources. The typical dynamic for a natural-resource-based economy is to export raw and semi-processed materials, and import the sophisticated equipment to do the basic extraction and transportation operations. In developed countries like Canada, there is significant investment in new technology on the extraction and processing, but this tends to be done predominantly by the public sector. Innovation clusters in these areas are more constrained to working in a few areas (although important) like environment, surveying, prospecting and instrumentation. This truncated pattern can be broken, but only a few exceptional countries—like Finland and Sweden—have managed to do this (Tiffin, 1989). Agriculture is perhaps more amenable to local innovation cluster development, but even here, the control of global markets by a few chemical and biological multinationals constrains action.

The heritage of decades of forceful public policies promoting import substitution has also left its mark. The following comment by Perez (1989) about the inability of Brazilian firms to innovate, due to the import-substitution heritage, is very telling. "A maior parte das empresas não foi constituida para evoluir. A maioria o foi para operar tecnologias maduras, supostamente já otimizadas. Não se esperava que as empresas alcançassem competitividade por elas próprias. A lucratividade era determinada por fatores exógenos, como a proteção tarifária, subsídios á exportação e numerosas formas de auxílio governamental, em vez de capacidade de a própria empresa aumentar a produtividade ou qualidade. As empresas não são conectadas [tecnicamente] ... [e tem sindo] difícil a geração de sinergias nas redes e complexos industriais"[9]. Or as Sutz (1996a) puts it, "La producción moderna de bienes y servicios y la producción de conocimientos nació así divorciada" (p. 9). It is not importing substitution per se that is the problem, it is the conjunction with other forces such as low investment in R&D that causes this serious basic problem of "enterprises not being designed to evolve".

Industry, government and university tend to operate very separately throughout much of Latin America. This problem was pointed out by Jorge Sabato in the 1970s, and the "Sabato Triangle" model, with the three groups each occupying a corner, has not been significantly changed in the intervening three decades. At a recent Triple Helix conference in Rio de Janeiro (the Triple Helix being a rediscovered image of the Sabato Triangle), many authors noted the continuing mutual isolation of the three essential partners to innovation.[10]

Tiffin, Couto and Bas (2000) have pointed out the impact on creation and growth of new technology-based firms from the extreme lack of venture capital throughout most of Latin America. Not only professionally managed venture capital is missing, so is the early-stage, informal angel or seed capital. A study on technopoles, technoparks and incubators in Brazil (Medeiros et al., 1992, p. 231) said the following about venture capital: "De todas a pré-condições necessárias para florescimento dos pólos científico-tecnológicos, a menos presente tem sido o capital de risco, ... Ausente este instrumento, as empresas de base tecnológica usualmente ficam numa espécie de limbo financeiro: por serem do setor privado não têm acesso a finaciamentos a fundo perdido; por não possuírem um capital que sirva de garantia real, não conseguem empréstimos junto ao sistema financeiro".

Recent events related to the craze for investing in electronic commerce companies have sparked a boom in venture capital in Brazil, but looking at the early signs of this boom, the reality still shows some disturbing trends. According to one observer partly responsible for a national Brazilian program in venture capital (Baptista, personal communication, 2000), the capital is almost wholly of foreign origin and it is oriented only to Internet investments whose business plans show payoff in less than two years. This is very different than venture capital associated with local innovation clusters, and in a sense is a repeat of previous history in Brazil of foreign-managed resource booms and busts.

Enterprises in Latin America seem to have a culture that limits their ability to cooperate, according to some authors. Albuquerque (1995, cited in Borges Lemos and Campolina Diniz, 1999) describes this for Brazil, and Sutz (1996b) points out the same thing for Uruguay. "...la escasa configuración del sector empresarial como actor colectivo, frente a actividades que requieren, para ser eficientes, de la disposición de los destinatarios a sumar voluntades y capacidades". This makes innovation more difficult, as close relationships among suppliers, innovators and customers are critical for success.

In circumstances where enterprises have difficulty working together and the Sabato triangle (or Triple Helix) is not linked up, it is obvious that the role of the state in promoting this communication becomes greater[11]. Unfortunately, as a general rule in Latin America, the state tends not to send clear, long-term signals favouring the need for innovation and sponsoring the elements that make it happen. Gómes (1999) refers to this problem in his study of science parks in Brazil: "la inserción de estas entidades en la agenda política de los gobiernos es sin excepción dependiente de actores políticos individuales que en un momento dado deciden apoyar tales iniciativas.... No existen políticas públicas consistentes, con instrumentos de aplicación general, dirigidas a apoyar las entidades gestoras de los polos tecnológicos y sus empresas. Existen proyectos y obras en marcha, pero pueden sufrir paralizaciones en cualquier momento, en función de las condiciones político/partidarias vigentes"(p. 206).

Various authors signal a common problem about sharing information. Latin America tends to have a culture where information is hoarded, not shared. This results in a duplication of efforts, underutilization of available resources and a mismatch between supply and demand. A Brazilian federal government official states "Brasil invirtió mucho en tener centros de información tecnológica sectorial pero la información no es usada por los empresarios"[12]. Villaschi Filho (1999, p. 240) states "o principal fator inibidor/minimizador de potencialização da capacidade de inovar no SCI é a quase total desarticulação entre seus diversos componentes. No desenvolvimento deste trabalho foi marcante a falta de informação mínima que as diversas organizações têm sobre os trabalhos desenvolvidos em outros componentes del sistema".

Obviously, from the preceding discussion, industrial innovation in Latin America faces significant and deep-rooted difficulties. Our interest in the topic of local innovation clusters is not only because they are important to create a knowledge-based economy, but because they may be structures that offer a way around some of the deep-seated innovation difficulties. For example, if national governments have trouble articulating innovation policies, a focus on clusters may help avoid this being a problem, because clusters are an extremely local phenomenon. Municipal governments and community business associations can step in and play a much more effective role. The next section takes these characteristic limitations into consideration, in designing a model of innovation clusters that makes sense for Latin America.

A PRACTICAL MODEL OF AN INNOVATION CLUSTER

In this chapter, we elaborate a model of an innovation cluster. This model is based on the previous discussion of literature, the prior experience of Tiffin working in this field and IDRC's workshop on the topic. Like all models, it is a simplification of a complex reality. It highlights the elements that are problematic for Latin America, as well as the basic elements found in a mature cluster in a developed country. The model consists of a list of stakeholders and their interactions. It will be tested in the next section to see how well it helps us measure innovation clusters in practice[13]. Note that some other stakeholders are involved in clusters, like the stock markets and industrial research funders, but these do not need to be local. In most cases they are supplied quite effectively from a single national source, or even international, so they do not show up in our discussion below.

Our model stresses both tangible and intangible elements. The tangible elements are:

- knowledge-based firms
- knowledge inputs
- specialized consulting services
- specialized inputs
- markets
- cluster support
- financing

The intangible elements are:

- supportive social climate
- links and interactions among individuals and organizations
- quality of life for people working in the community where the cluster operates

Tangible Elements

Knowledge-based Firms. At the core of any innovation cluster is the group of new knowledge-based firms. The firm is the core element because it produces goods and services to sell, as well as innovating new goods and services. However, the firm is also a major actor helping the innovation cluster to evolve as a system in its own right. As Kozul-Wright (1995) points out, "the firm is in a position to fulfill a number of critical conditions for innovation: (i) by acting as an organization for storing knowledge (including tacit knowledge); (ii) as an enduring institution which can reproduce that knowledge and inculcate it in new entrants or share it with other firms, and (iii) as a social agent which can establish trust and cooperation".

There are several different kinds of firms involved at the core. They include the much-desired "anchor" firms, which are large sources of technology, markets and expertise (like Bell Northern Research in Ottawa or Dell Computers in Austin). There should be a swarm of small firms and a constant flow of spin-offs and start-ups coming out of the large firms and the technology centres (labs and universities). These firms must be located near their suppliers as they need to have close relationships in the innovation process, understanding and modifying the technical inputs. Large firms often act as miniature innovation systems in their own right, supplying incubation space to employees, financing their start-ups, providing technical expertise, product specifications and initial markets. Large firms also provide a steady flow of trained people which the small innovating firms can hire.

Knowledge Inputs. The knowledge on which firms base their new products comes from universities, public R&D labs and other sources of technology, either as publications or "on the hoof" from skilled individuals. These elements represent the main knowledge inputs on which innovating firms draw. The more they are available locally, the easier will be the lines of communication and transfer of knowledge from the sources to the innovating firms. In a more complex model, we could distinguish between the stock of knowledge, the institutions that produce it and the institutions that train and educate people who create, diffuse and apply the knowledge. Since Latin America, as a whole, has a mature educational and training system, this is not such an important point to make, and we prefer for current purposes to focus more on the groups that produce scientific and technological knowledge for this model.

Consulting Services. Spin-offs and start-ups tend to focus on the technology innovation and research aspects first, then incorporate incrementally other areas of expertise as they grow. Therefore, the roles of a whole range of specialized consulting services are critical to the functioning of a cluster. Even for large firms, many specialized functions are still outsourced to consultants. Some of the most important ones are:

- technology: transfer and commercialization
- legal services: patents, trademarks
- accounting for small firms
- industrial design for creation of new products
- industrial engineering: focussing on devising new production processes and scaling up current ones
- marketing: both local and international
- business associations which act as clubs providing moral support, pressure to public-sector agencies and a source of contacts

Most of these service providers operate independently of each other, coming together in business association meetings which aim to support the development of the cluster. They will also get to know each other through the contracting activities of the individual firm that hires them. Generally speaking, these services are available throughout the industrial areas of Latin America, with the possible exception of industrial design (and fashion, when dealing with garment and leather industries).

Specialized Inputs. Every innovating firm needs inputs of materials, instrumentation and equipment. For some industries, this is fairly universal (e.g., computers or the Internet). For others, there may be an outstanding requirement to locate next to a laboratory with a unique facility. Easy intercommunication of staff between the input companies and the innovating firm will make the innovation more successful. In some cases, the labs or specialized materials inputters will even participate in the innovation itself, as they need to modify specifications or improve their own products to match the specifications of the new innovation. Because the inputs required can be so specific to each type of firm, it is difficult to make any general remarks about the strength or weakness in Latin America.

Cluster Support. Many cities have created organizations to promote the creation and management of local innovation clusters. These organizations typically have a very small staff of one to three professionals, whose coordination and promotion roles are critical. They will orchestrate the connections with their peers in the business incubator, public-sector regulators at different governmental levels, technology transfer agents and business associations, to promote effective development of the cluster. In addition, they will promote linkage among all the other stakeholders.

Most developed countries have networks of technology transfer agents that also work to promote the functioning of local innovation clusters. In Canada, the Industrial Research Assistance Program (IRAP) network maintains hundreds of scientists and engineers to serve innovating firms throughout the country. In addition to the technology transfer services they offer linking to federal R&D labs, these agents also have at their disposal industrial research grants (essentially allowing them to act as early-stage angel capitalists), and they provide advice on product design, marketing and suppliers. A major part of their work is to liaise with other cluster support agents, including those who run local business clubs and innovation support networks in the private sector.

From time to time, these public and private innovation support agents will deal with regulators in the public sector, on specific issues that need resolution to assist groups of innovating firms to move ahead. For instance, in biotechnology, restrictions on importing live cultures can seriously impede product development. If the cluster promoters work actively with the government regulators early on, these legal problems can be more quickly resolved.

The other major element in this group of cluster support agents is the incubator. Attention is lavished on this element of specialized infrastructure, which in far too many cases is mistaken for the innovation cluster itself. It is typically attached to a university, where it serves a double role: to facilitate the incorporation of professors and senior students into entrepreneurial business; and to provide the physical facilities and specialized services (often at a deep discount) to help start-up firms learn how to stand on their own during the first critical years of their creation. "Incubators ... can act as a laboratory for commercialising the ideas of academics and provide a training ground for entrepreneurs" (OECD, 1997b, p.7).

Incubators serve as a meeting place between the university, the productive sector and the state, and because of this and the fact that they are obvious physical symbols of rather intangible, often esoteric activities, they have enormous symbolic power in the community. Incubators are symbols of modernity, progress, community pride and hope. In Latin America especially, they may provide a unique source of low-cost, low-risk, high-potential-value investment prospects for venture capitalists, thus acting as the focal point for a small cluster[14].

There is a debate, however, as to the direct importance of incubators in local innovation clusters. There are many horror stories about investments in incubators that have been unproductive.[15] In most cities of Latin America, there are already many choices for new firms to find physical premises. While there is no denying the need for minimum levels of infrastructure, and the need for superb infrastructure of specialized types (e.g., fiber-optic communications for software development), a concentration on the visible physical often leaves support for the links and social relationships unanswered. Ruffiex (1987) (cited in Gómes, 1999) states that "a mera criação de infra-estrutura é insuficiente para promover uma eficiente rede de comunicação e relacionamento entre as empresas, e de cada uma delas com a universidade" (p. 15). As with the incubators, we are in the midst of a critical reassessment of the utility and efficiency of investments in more complex infrastructures like technopols, science parks and science cities. Gómes (1999) presents an exhaustive discussion of the literature on this topic.

Markets. The innovation processes also involve the customers, as they define the performance requirements of the new products and test them out. Having lead customers immediately at hand is a critical support to a firm creating a new product. In a vibrant cluster, the firm's technical and managerial people may even move across the boundaries between the customer and supplier companies rather easily. Governments have a key role to play as purchasers of new products to provide large markets for risky new products. Many knowledge-intensive products aim for global markets from the outset or very early on, hence it is usually very important for the local cluster to have communications and transport facilities to world customers.

Financing. Seed capital, venture capital and knowledge-based banking are essential local elements in an innovation cluster. A paper by Tiffin, Couto and Bas (2000) surveying venture capital in Latin America gives a good summary of the current situation. Seed capital is typically very small sums of equity (i.e., for the purchase of shares in a company, not loans), supplied by private individuals, sometimes working together in informal groups. Seed investing is nearly always only within the community. Venture capital is required later on by the growing enterprises, in much larger amounts (typically from $500,000 to $2 million). It comes from professionally managed firms, which typically will only invest in enterprises up to an hour or two of travel time distant. Venture capital firms also usually invest in groups. It is well known that angel and venture capitalists not only supply money, but equally important, expert management services in the form of contractual relationships with the new firms and via informal mentorship. These management and related technical services from experienced investors are worth their weight in gold to the new firms. Information about deals often flows from one group to the other, and then to the bankers who are becoming more involved with early stage corporate loan instruments. This area is called Knowledge-based Banking, and is one of the growth areas for banks. They try to move upstream as far as they can with traditional loan instruments, and always want to be involved with information about growing firms, from early stages on. Knowledge-based banking is also a very local phenomenon.

Intangible Elements

Culture. The cultural values held by the local society that spawns the innovators, as well as the values of the innovators and entrepreneurs themselves, are a central example of the intangibles that allow innovation clusters to work. As well as an underlying favourable business climate, there must also be a broadly held social mindset that understands, values and rewards small business, competitive cooperation, risk-taking, research, innovation and entrepreneurship (Pérez, 1990). Underlying all this is the sociocultural environment in the surrounding community that values novelty, insists on quality and promotes education and learning. Popular magazine articles portray these powerful, eccentric and novel cultures well.[16] It is difficult, perhaps impossible, to create innovation clusters in societies that are not open to innovation. While the state cannot single-handedly create an innovative society, it can put in place a system of education and economic and symbolic rewards that encourage the other key actors to participate in the long-term transformation.

While not all the characteristics associated with successful innovation clusters may be positive to all people—there is a dark underbelly of compulsive consumption and display of wealth, destruction of established values and personal relationships that are important to society—no one can deny the immense social and symbolic power of these innovation clusters. Once sleepy, grey government towns, Austin and Ottawa have been transformed in a few decades into two of the most dynamic, creative and wealthy parts of the world by their innovation clusters.

Integration. Integrating the visible organizations mentioned in the previous section are the invisible community links. Flows of information, money, technology and people between firms are intense and ceaseless in mature innovation clusters. Most important, and hardest to measure, is that innovation and learning are occurring not just at the individual level or the firm level but, in some way, at the overall system level of the cluster as a whole. Face-to-face contacts and personal relationships facilitate this learning and impose limits of distance and size, which we do not yet fully comprehend, on the geographic nature of the cluster. A great deal of the knowledge that is interchanged among the stakeholders of an innovation cluster is tacit, as opposed to formal knowledge (Wilson, 1998). Integration and explicit community self-awareness are essential constituents supporting innovation. In the words of Castells, an innovative environment is a "sistema de estructuras sociales, institucionales, organizativas, económicas y territoriales que crean las condiciones para una generación continua de sinergias y su inversión en un proceso de producción que se origina a partir de esta capacidad sinérgica, tanto para las unidades de producción que son parte de este medio innovador como para el medio en su conjunto"[17]. In a knowledge-based economy, firms seek interactions with other firms to develop learning strategies which allow them to reduce costs and risk related to the innovation process, access new research results, acquire key technological components of a new product or process and share costs of production, distribution and marketing (OECD, 1996, p. 16).

Geographical proximity of the organizations favours this interchange of formal and tacit knowledge, by formal and informal means. It is important to note that dense interchange of information implies a set of social relationships where there is a community value promoting a balance of cooperation and competition. This is especially important for small firms. Trust in group processes, sharing of information and openness of communications are critical to form working clusters. Public and private agents specializing in this information exchange are equally important to ensure it happens.

Quality of Life. This is a somewhat controversial element of a cluster. The literature seems to show quite clearly that cities where the quality of life is perceived as high can more easily attract the mobile, highly educated people who are innovators and entrepreneurs, as well as the skilled professionals who support them in consulting roles and the researchers and educators who create the pools of trained staff and new technical discoveries. However, quality of life is something that is to a significant extent dependent on the individual's unique perceptions; one person may value a bustling city environment, but another may detest it. In the literature on Latin American clusters, this debate has not seemed to surface at all so far. It is significant, though, that business magazines are picking up on this issue, for normal business location decisions. A recent edition of *EXAME* magazine (2000) had a feature article on the best cities in which to do business in Brazil, in which various indices of quality of life featured prominently.

The indicators are rather obvious that should be considered in measuring quality of life. Housing costs are a key. This has been stated as one of the key reasons for the rapid growth of second-tier clusters like Austin, Texas—its housing prices are significantly below those in the Silicon Valley area.[18] Other factors to consider that are measurable include cultural amenities, recreational facilities, urban services, commuting time, clean environment, good schools for children, jobs for the spouse who is usually highly educated as well. This element may become much more important in Latin America.

Visual Representation of the Model

Figure 14.1 summarizes the previous discussion on a practical model. It is made up of two elements: the circles represent stakeholders, participants or critical factors; the lines represent flow of goods, ideas or money. A more detailed representation could show the lines in terms of thickness to represent the intensity or importance of a specific flow, along with the direction. Within each of the large circles lie a number of smaller ones, to represent more detailed elements. Some of these are also interlinked by flows: for example, we represent the Financing circle as being made of up seed capital, venture capital and knowledge-based banking, all of which need to be in constant communication with each other. Around the whole model is another circle indicating that the cluster is restricted to a small geographical area. In the form presented, this model is amenable to computer modelling, which might be a useful step, once more firmly elaborated, to display the "ecological" dynamics of a local innovation cluster.

Figure 14.1
Visual Representation of the Model

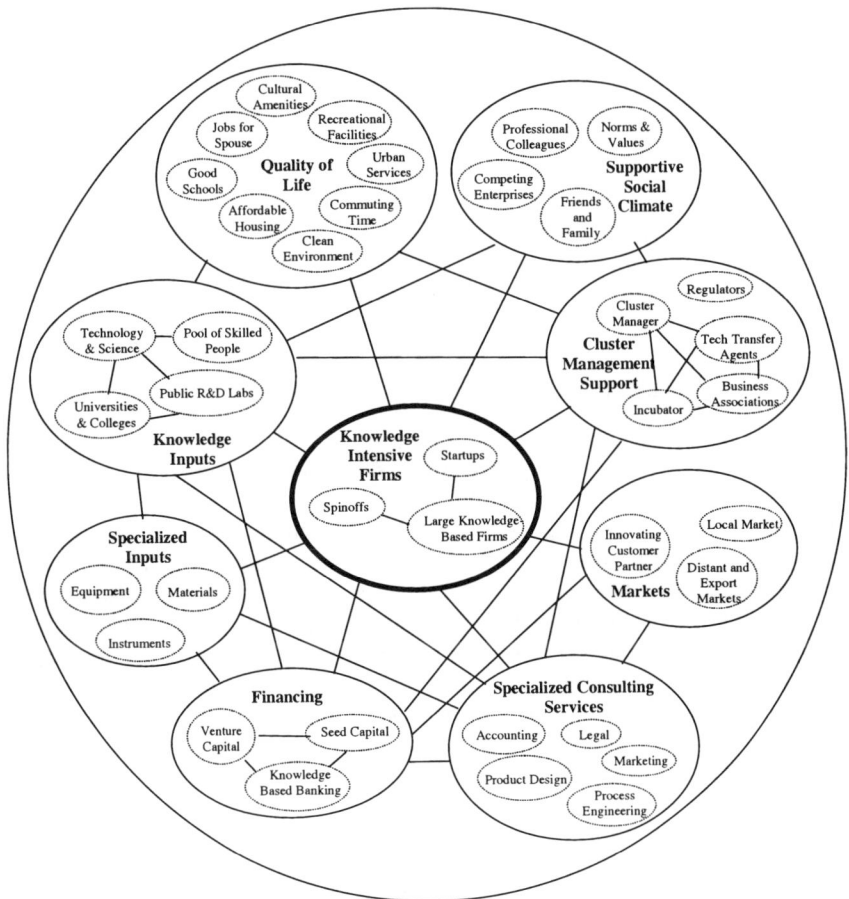

INNOVATION CLUSTERS IN LATIN AMERICA

The model we have generated arose from a combination of literature analysis and field trips in the region. Since the data on innovation clusters in Latin America are very fragmentary, and come from different and noncomparable sources, it is not possible to claim the results presented in this chapter are anything more than very preliminary. It is highly likely they are biased according to where there are researchers interested in the topic. We are likely, for example, to be overemphasizing Brazil, and within Brazil, skewing the data to cities like Rio and Campinas, which have strong university teams with doctoral students involved with clusters and innovation policy. Still, there are no better data at a Latin American level that have been brought into a single document, and a bias towards Brazil has good reasons to exist. For example, Brazil has made a huge investment to promote innovation clusters, by means of creating incubators. There are now more than 100 incubators (52 technology-oriented, according to the public information, and maybe 20 or so really doing this, according to one incubator manager) in that country.[19] In addition, one of the biggest universities in Brazil, the Federal University of Rio de Janeiro, has land which it is proposing to develop as a science city[20].

Field Studies

Bortagaray visited Porto Alegre (Brazil), Buenos Aires (Argentina) and San José (Costa Rica), and studied Uruguay in some detail, to gather preliminary impressions about the potential of clusters in these cities. In the course of routine IDRC work, Tiffin gathered opportunistic data on clusters during visits to Havana, Monterrey (Mexico), Recife (Brazil) and Curitiba (Brazil). Innovation clusters in these cities are discussed in varying degrees of detail as follows.

Porto Alegre. This city, capital of the southernmost state in Brazil, is well known for the excellence of its urban planning and management, with a strong emphasis on democratic participation of local communities and improving the quality of life of the urban and regional environments. In 1995, the Projecto Porto Alegre Tecnópole was initiated with a mission to transform the metropolitan region to a knowledge-based economy. It is led by the Universidade Federal de Rio Grande do Sul and the Prefeitura Municipal de Porto Alegre. There appears to be strong and frequent dialogue among the promoters and the local stakeholders, many of whom are working together on other projects. The personal relationships are said to cut through bureaucratic obstacles.

The project has several distinct and clearly articulated components:

- the *Tecnópole a domicilio* program which is basically a technology extension service for small enterprises
- a linkage among the technology-based business incubators in the region

- a series of specific investment projects (electronics-informatics, health, a science-technology campus and the linkage of the two main universities with the city industrial park)
- a *teleporto descentralizado* project which is extending the fibre optic cable network throughout the city

There are three incubators focussing on new technology-based firms:

- Centro de Empreendimentos de Informática, created in 1996 as part of the Instituto de Informática of the Universidade Federal de Rio Grande do Sul (UFRGS)
- Centro de Biotecnologia do Estado do Rio Grande do Sul, at the UFRGS
- Incubadora Empresarial Tecnológica de Porto Alegre, created in 1991, which is managed by the city government and linked to a variety of universities and organizations (UFRGS, PUC/RS, ULBRA, PROCEMPA, IEL/FIERGS, FUNDATEC, CIENTEC, BANRISUL, PETROBRAS, REFAP, CNPq).

Porto Alegre has the distinction of being one of two cities in Brazil with a private—and successful—venture capital firm, Companhia Riograndense de Participações (CRP). However, it does not appear that the CRP is closely linked to the formally planned elements of this cluster, preferring to work independently with prospective enterprises.

We looked more closely at the health industry cluster being promoted. Here, the situation seems identical to the broad outlines: very strong and clear communication links among the main players; a clear sense of mission and shared goals; a feeling that the goals will be accomplished with hard work and commitment for the long term; an impressive science base which is oriented to practical outputs in the private sector; a close linkage between the researchers, the specialized inputs community and the markets, mediated by formal cluster managers and strong incubators.

Costa Rica. Costa Rica has worked consistently over recent decades to develop, based on its democratic traditions, social harmony and high levels of primary and secondary education. Since the 1970s, Costa Rica has not only been synonymous with environmental tourism, it has managed to set the standards for this new and rapidly growing industry. As a study financed by CEPAL (Comisión Económica para América Latina y el Caribe) clearly shows (Acuña, Villalobos and Ruiz, 2000), the environmental tourism industry has attained the complexity of a robust tourism industry cluster. Field work by Tiffin found important elements relating to the development of a strong set of management and research services oriented to tourism, principally located in INCAE, the university in San José with strong links to Harvard, and the numerous nongovernmental organizations (NGOs) working there. In addition, a number of local and regional venture capital companies have been set up (such as Empresas Ambientales de Centroamerica S.A.), as well as other regional banks, focussing on the environmental business and technology areas (Tiffin, Couto and Bas, 2000). With these related elements, we might be seeing the development of a tourism innovation cluster and that of an environmental services and management innovation cluster as well. These two clusters, being focussed on services instead of products, may require a different way of viewing the model elaborated before.

However, the country has managed to lay the basis for yet another innovation cluster in the last decade, based on microelectronics and software. The anchor for this cluster is, of course, capturing the Intel Corporation plant investment in San José, which has acted as a spark and catalyst for much more cluster development, not just in terms of infrastructure and companies, but in terms of institutional and social transformation. In the words of one senior university administrator (as recorded by Bortagaray in April 1999): "Entonces trabajamos muy duro, ellos [Comisión de Intel] vinieron en las vacaciones y nosotros sacamos a todos los profesores de las vacaciones y los pusimos a trabajar con Intel y comenzamos a generar un compromiso muy fuerte. Y así actuaron muchas de las instituciones de alrededor que tenían que atender la electricidad, agua, etc. Todo el mundo trabajó muy orquestadamente para poder lograr que Intel se viniera, porque sabíamos que Intel es Intel, e Intel provoca un montón de empresas más y una serie de condiciones que nos ayuda a mejorar una serie de cosas internamente".

For several decades, Costa Rica had been capturing small but important investments principally as a maquiladora in the textiles area. But the national government, industry and university leaders are engaged in formal and informal discussions about the future of the country and how to develop new industry based on the human resource potential of the nation. The installation of the Intel plan is seen as a watershed, drastically changing the self-confidence and the image of the city and surrounding region, which has acted to intensify the cooperation and imagination of the communities involved, to continue pushing for further development of the knowledge components of the informatics industry, with Intel as the anchor point. The following interview carried out by Bortagaray in April 1999 illustrates this point clearly: "Y hubo una época en donde una inversión, hace unos doce, diez años, de algunas empresas en el campo electrónico, como Motorola por ejemplo que se instalaron en Costa Rica, en ese momento encontraron atractivo en el asunto sin que hubiera toda una facilidad de diferentes formas. De eso al día de hoy ha pasado mucha agua bajo el puente, y ahora se acaba de modificar la ley de Zonas Francas, se está tratando de regular la economía y de regular los trámites para hacer más sencillo al inversionista, se han instalado ventanillas únicas tanto para exportación como para importación, de manera que el trámite esté en un solo lugar, en vez de que le hagan ir a veinte lugares diferentes. Se han ido buscando esos mecanismos de agilidad, se ha ido buscando integrar también la industria local con la industria internacional o que viene a estos regímenes de Zonas Francas para ir creando también integración industrial vertical y horizontal. Y se empezó a trabajar en un concepto, bueno estudiemos qué sectores podrían ser atractivos para que Costa Rica pueda promover la inversión tomando en cuenta también qué ofrecemos, qué recurso humano hay, qué infraestructura, qué capacidad y se hizo un primer análisis como estrategia del gobierno donde salió un grupo de áreas que podrían ser interesantes para el país ... qué podía Costa Rica ofrecer en el campo, por qué promover la Industria Electrónica y que ventajas y qué cosas estaban pasando, qué compañías, qué países estaban llevándose la atracción, por qué, entonces se hizo un estudio muy interesante que nos permitió tener muy claro y empezar a buscar industrias en ese sentido. Y ahí se inició un programa muy fuerte de atracción de inversiones cuya corona de oro para ese programa fue el poder capturar a Intel. ¿Por qué? Porque era la primera vez que uno de los dominadores de un área tecnológica como es los microprocesadores, estaba buscando un sitio que no fuera Asia o los Estados Unidos o Israel en donde normalmente tienen plantas, o Irlanda del Norte. Y resulta que entonces en ese concepto logramos demostrarle a Intel que nosotros éramos el país apropiado y tomó la decisión".[21]

In 1994 the Instituto Tecnológico de Costa Rica set up an incubator, under the management of the Escuela de Administración y Negocios. The project took shape with the support of the Ministerio de Ciencia y Tecnología and the Italian group Zeta, which was managing the industrial park on the outskirts of San José. The incubator's tenants were all working either in the informatics or biotechnology sectors at the time of Bortagaray's interviews. Cooperation among established firms based on R&D is promoted by the Cámara de Empresas de Base Tecnológica de Costa Rica, which was set up in 1992. Growth of some local firms in some instances has been spectacular: Bortagary interviewed one software firm set up in 1993 with four staff which now employs 90. With this kind of development Costa Rican software exports jumped from $10–20 million to $70–80 million over the last three years, making Costa Rica the largest exporter per capita of software in Latin America.[22]

Despite the striking achievements in recent years, we do not see any of these areas as yet representing a mature innovation cluster. Costa Rican industrial and technological development has taken off from a very limited base, in a very small country quite a distance from markets. The venture capital recently located in San José tends to be more regional in scope, focussing on traditional investments, as opposed to local high-tech start-ups. For these firms, our field work showed both a shortage and an absence of techology-oriented venture and angel funding. The incubators find it a hard task to fill their space, and need to spend a great deal of effort selling their facilities. Entrepreneurs need a lot of specialized training (as everywhere), but the only place to do it is in the incubator itself. There still is a shortage of many specialized inputs and consulting services.

Recife. Recife is a mid-sized Brazilian city in the extreme Northeast coast. Historically basing its industry on sugarcane, the region is managing recent growth based on rather traditional industries and services, such as a modernized port and transportation facilities from the interior, for raw materials and agricultural products. One of the raw materials available for technology-intensive transformation is a very diverse supply of plants that can provide alternative inputs to the local pharmaceutical industry. The state of Pernambuco has 33 firms producing pharmacochemical products, corresponding to 47% of the total

Northeast region of Brazil[23]. Most of this output goes to the Northeast region market.

In recent years, there has been a considerable change in production technology and products of this industry, spearheaded by interaction with research teams at the Federal University of Pernambuco. A government research institute called LAFEPE is working with the university and a private pharmaceutical firm HEBRON to develop new products based on local biological inputs. The initiative now under way is to see how this kind of work can be increased and formalized by working with other state agencies centred on Recife, and local industries to link the pharmacochemical industry backwards into local biological resource production. This is an incipient innovation cluster, with strong support from a variety of local stakeholders in university, industry and government.

Curitiba. Curitiba is the capital city of the southern Brazilian state of Paraná. Long overshadowed by massive São Paulo just to the north, Paraná has a tradition of an agricultural base and a small population. Dynamized by a charismatic mayor in the early 1990s and a population highly receptive to organized change, Curitiba has surged to a leadership position in Brazil in terms of urban planning, environmental awareness and university development. Researchers, politicians and public managers from all over the world now come to the city to see how it has developed and expanded its urban management systems, principally transportation. The lead in creating an innovation cluster centred around environmental management and software-informatics, however, seems to have been taken by the state government.

One of the key leaders is the top civil servant in the science, technology and higher education ministry, Ramiro Wahrhaftig. He has a technical background and maintains active links with one of the best "think-do tanks" in the United States where cluster research and promotion is done, the IC^2 Institute at the University of Texas at Austin. In addition, there are numerous other industrial leaders who move back and forth between the private sector and the public programs like Softex that support the growth of technology-based industry. They work closely together in the state of Paraná, promoting this growth in a highly effective and efficient manner.

More recently, the main universities have begun to follow the lead of government and industry to support the quickening of the links between knowledge and enterprise. In Curitiba, the main private university, the Pontifícia Universidade Católica do Paraná (PUC-PR), has begun to implement a strategy to transform itself into a research-based university instead of a teaching enterprise. One of its steps is the creation of a graduate specialty in Knowledge Management.

There is a well-established incubator, but it does not seem to be a central actor in the emerging technopol, as it is focussed so strongly on its immediate role of supporting technology start-ups. There is no venture capital, but negotiations have begun recently to set up a branch of an established venture capital firm in the city.

Curitiba is recognized as the city in Brazil with probably the highest quality of life and is extremely active in promoting this status to attract environmentally friendly industries and knowledge workers. Partly as a result, Curitiba is the fastest-growing city in the country. The influx of new, high-technology skills can be imagined to be rapidly filling the previous gaps in specialized consulting services.

Buenos Aires. Located in Buenos Aires is the Polo Tecnólogico Constituyentes (PTC), made up of several large and powerful institutions: Comisión Nacional de Energía Atómica, the Instituto de Investigaciones Científicas y Técnicas de las Fuerzas Armadas, the Instituto Nacional de Tecnología Industrial—INTI, the Servicio Geológico Minero Argentino and the Universidad Nacional de General San Martín. The pole is intended to generate and transfer scientific and technological knowledge among its members, within Mercosur and around the world, as well as promoting linkage with this knowledge to the private sector. It emphasizes the signing of formal protocols with other technology poles around the world to project and develop the national scientific and technological systems of Argentina.[24]

Its principle lines of work are said to cover materials, environment, energy, transport, support to regulatory bodies and public services, quality control, instrumentation and monitoring industrial establishments, developing human resources and technical information, as well as the development of basic and applied research projects in biotechnology. There are plans to create a program for stimulating new enterprises based on technology, which include an incubator, courses in entrepreneurship, entrepreneurship competitions and the offer of technical assistance.

In the brief time available for our field work, we were not able to get much information on how this group is actually working, but it appears that it is more a "technopol" than a cluster in the sense we use the term in this chapter. The emphasis is strongly on technology transfer out of large government research labs, and the seeking of relationships with the private sector. The group proposes to develop more elements of an innovation cluster, but the lack of technical focus, and the continued need for a public-sector push from the research side, leaves us doubtful of its long-term economic potential without other, major, intervening transformative factors.

Uruguay. Snoeck (1998) has carried out a detailed study of the wine industry in Uruguay. While this is a small cluster, and specialized around a single agricultural product, it seems to show some characteristics that could metamorphose into an innovation cluster. The firms at the centre of the cluster are wineries. As agricultural enterprises, they show few of the characteristics of knowledge-based spin-offs, but in Uruguay a significant segment of the industry has converted itself into specialized, export-oriented companies producing small quantities of high-quality wines. Faced with financial ruin after several decades of stagnation, a group of firms in the industry decided to work together to implement radically new and different strategies, based on continuous quality and knowledge upgrading, inputs of best-practice technology and close links with customers for developing specialty products. Overall, this industry has achieved significant success since the early 1990s.

In terms of our model, Snoeck points out the following:

- strong integration and a highly supportive social climate, including the creation of limited time and function consortia
- strong support from regulators, technology transfer agents and business associations
- transfer of skilled people, technology and science
- purchase and application of best-practice process technology and genetic stock
- utilization of specialized bank credits.

On the other hand, it should be pointed out that there was virtually no support from the national school set up to train enologists, and that the main university only very recently has begun to link with the producers in terms of research and development. Most of the specialized consulting services and the technology and equipment inputs seem to come from foreign sources. In terms of our categorization scheme set earlier, this would be an innovative industry cluster.

Earlier in the chapter we mentioned Sutz's (1998) concept of innovative circuits. She studied four in Uruguay:

- Animal health, bioengineering, wool-textiles and informatics. The first two of these are worthy of mention in slightly more detail. The animal health circuit revolved around a national lab (EUBSA) which was producing a vaccine against aftosa fever. There were both local and international clients in the public and private sectors participating in the innovation. However, work came to an end when a legal injunction stopped EUBSA from producing the drug. The loss of this drug was also a serious blow to the Uruguayan meta-exporting industry which was struggling to be labled "aftosa-free".
- The second circuit, bioengineering, related to the production of electronic pace-makers by the firm Centro de Construcción de Cardioestimuladores del Uruguay (CCCU). Sutz says this circuit "tiene varios de los componentes que cualquiera puede imaginar como importantes: tradición de investigación de calidad en los dos polos cognoscitivos extremos del circuito—medicina clínica e ingeniería; tradición de enfoques mancomunados de ambos polos para la resolución de problemas; construcción concreta de dispositivos en estadio de experimentación desde el punto de vista de la aplicación y de prototipo a nivel de fabricación; empresa exitosa nacional e internacionalmente, de fabricación de dispositivos muy sofisticados de bioingeniería" (p. 39). However, it was never able to generate other links, markets or the critical mass to get past experimentation with prototypes.

Neither the government of Uruguay nor the city of Montevideo has an effective policy recognizing or promoting innovation clusters. While there has been sporadic talk over the past decade about science parks and incubators, it is only now that one incubator may finally be getting off the ground (*El Pais*, 2000). The discussion still focusses on the need to create a physical infrastructure of the science park type. The national technology lab system (Laboratorio Tecnológico del Uruguay, LATU) is leading a consortium to establish a technology-oriented venture capital fund, with support from the InterAmerican Development Bank (IADB), but it is not clear if this long-standing project is moving ahead and if it will be led from the private sector, which is essential. The most important strength in Uruguay is a highly educated populace, but social norms and values do not favour entrepreneurship and innovation; the society as a whole is extremely conservative and has not moved significantly from its old vision of living off the rent of agricultural produce, which brought such prosperity early in the 20th century. From this we could conclude that it will be some time before an innovation cluster takes shape in Uruguay.

Havana. On the outskirts of the capital city of Cuba, the national government has made enormous investments to create biotechnology research and product development capability. This investment was originally made as part of a commitment towards excellent, universal health care, with an emphasis on valorizing local pharmaceutical raw materials and herbal traditions (Tancer, 1995). Estimates are that in the period from 1959 to 1991, the government invested about US$300 million in this pharmaceutical-medical-biotechnology system. There are now about seven major research centres, employing some 1,131 research scientists and technicians.[25] It does not appear, however, that this set of laboratories is able to convert itself into an innovation cluster because of the extreme difficulties in commercializing products in a communist regime. There is no ability to create start-up firms, few specialized business services and many restrictions on marketing and sales activities. In addition, the severe financial and regulatory restrictions on access to the Internet make it difficult for researchers to participate in cutting-edge bioscience.

However, several laboratories are acting as partners in an international innovation system with Canadian venture capitalists and basic research labs. A venture capital company in Toronto, York Medical, has created a venture capital fund which pays for teaming up the Cubans with Canadian researchers to commercialize new drugs. The Canadians excel in basic research, the Cubans in applied research, and this work seems to be generating a fruitful partnership. The drugs will be commercialized in Canada and the Cubans paid royalties under this scheme. The difficulty may be that the easily commercialized drugs are rapidly exploited, while there is insufficient continuing investment maintaining the upstream supplies new products from research.

The Cubans are making strenuous efforts to extract commercial benefits from this prodigious investment, within the limitations of the Castrist system, including the export of many highly trained physicians to Latin America and the licencing of drugs and related medical techniques to biotechnology companies in the region. Overall, this export is reported (Nash, 1996) to be bringing in over US$100 million per year.

Monterrey. This booming industrial city in the northeast of Mexico has great potential for creating innovation clusters in a number of areas. Conditions are very favourable: it is close to one of the most successful innovation clusters in the United States, Austin, Texas, and enjoys close industrial, cultural and educational ties. Foreign high-tech investment is pouring into this maquiladora centre. One of the biggest and best technical universities in Mexico—and Latin America—the Instituto Tecnológico de Monterrey is located in the city. It is a central point in training, research, consulting and testing, with ambitious plans to expand its reach and depth in the community, and as well its influence internationally.

However, as a 30-year debate on branch plants and innovation in Canada attests, it is not at all clear that a maquiladora strategy leads to an innovation future. There is much literature to show that it can be a dead end, trapping the community in a dependent, truncated position. There seems to be no venture capital available in the urban region for new technology-based ventures. In addition, the social climate is still promoting employment in large firms, not entrepreneurship in start-ups. The research base seems to be currently oriented to testing and trouble-shooting with the local industrial community. This is an excellent first step, but only time will tell if the technological and business structures come together in a deeper partnership to form an innovation cluster.

Summary. Table 14.5 summarizes our interpretation of the innovation cluster status of these cities. We use a scale of asterisks with 5 being the rating of a mature innovation cluster and 1 having a few basic elements. This is a highly personal and preliminary rating, but does give a more complete picture at a single glance of all the clusters.

The subtotal of points awarded to each location give some preliminary and very rough measure of how we think the regions may compare to each other and to a fully mature cluster. Note how the individual elements show much variation and that there is little consistency for each city. For example, Havana gets a very high rating for its investment in knowledge infrastructure, but very low for business-related potential. On the other hand, Buenos Aires has a very strong service infrastructure, but extremely weak support for an innovation cluster. We rank the cities in terms of relative strength. The cities that appear strongest score about two-thirds of what we might consider the score for a mature innovation cluster, indicating a respectable potential to emerge. Community will to succeed can make a huge difference in these scores, even if the industrial base is small, as in the case of San José, Costa Rica. Despite the imprecision of these preliminary data, it is interesting to note that the numbers for Curitiba and Porto Alegre generally coincide with what some business investors agree in terms of the highest growth areas in Brazil and the best places to do business (*EXAME*, 2000).

Table 14.5
Summarized Features of Innovation Clusters

Location	Quality of Life	Knowledge Inputs	Supportive Social Climate	Cluster Management Support	Specialized Inputs	Markets	Knowledge Intensive Firms	Specialized Consulting Services	Integration	Finance	Relative Strength
Porto Alegre	****	****	****	****	***	**	***	**	***	****	1
San José	*****	**	*****	*****	**	**	***	**	***	****	1
Curitiba	******	***	****	****	***	**	***	**	***	*	2
Monterrey	***	****	**	*	***	*****	***	****	***	**	3
Buenos Aires	***	****	**	*	****	****	***	****	*	*	4
Recife	***	**	**	**	***	***	**	**	**	**	5
Havana	***	*****	***	*****	**	*	*	*	***	*	6
Uruguay	******	**	*	*	**	**	**	***	*	*	7
Relative Strength	1	2	3	3	4	5	6	6	7	8	

In terms of the different cluster elements, it is interesting to note that finance comes up as the weakest element overall, followed by weak integration and then a probable lack of support from specialized consulting services and lack of market support, both local and access to global. On the positive side, some cities have a respectable quality of life. This of course is measured in terms of the amenities available to those with jobs and education, as many of these cities suffer from extreme income disparities in different neighbourhoods. As well, knowledge inputs score high, reflecting the mature and significant investments the public sector has made in education and research over the past decades.

Our tentative conclusions from this exploratory field work? No mature clusters; significant potential in a variety of countries and cities. The model should be useful if expanded further, for more detailed questionnaires and research.

Additional Data from the Montevideo Workshop and Literature

Several innovation clusters were named during the electronic conference and discussed during the Montevideo meeting. The participants in the workshop made additional suggestions in a session on the topic, based on criteria elaborated in the first day. These criteria stressed geographical focus, industrial innovation as the underlying goal, scientific and technical knowledge being the basis for innovation, invisible social relationships as critical to define the feeling of a community with shared values, the dense interchange of tacit knowledge.

The list in Table 14.6 was not discussed afterwards by the group, so it is presented here to show very preliminary and tentative suggestions only. In this list, we add some references to clusters mentioned in the workshop where there has been an academic study, and, as well, a few more clusters which the literature discusses, which were not mentioned in the workshop. (We do not repeat references to studies already discussed earlier, for example, Uruguay.)

In the workshop, the Argentinian participants indicated that there were no organized clusters, although there have been isolated attempts to create and grow them. In retrospect, we feel they were being far too modest relative to the Brazilian participants, and using criteria much stricter about what should be counted as a cluster. In discussions with the Argentinian participants afterwards, it became obvious that they were referring to fully mature clusters. Therefore, we have added the two for which we found literature analyses, Rafaela and Mar del Plata. It is clear these data are skewed to countries where the participants came from: with no participants from Columbia, Peru, Chile and Venezuela, these countries simply do not appear, unfortunately, in this preliminary list. In addition, this list obviously includes entries that are too small, too weakly innovative or too weakly focussed in a geographical sense to be considered real clusters. For example, the previous section discusses in some detail the three suggestions for clusters in Uruguay, and concludes that both biotechnology and software innovation exist, but they are characterized more by Sutz's model of transitory innovative circuits. The wine industry definitely has some elements of innovation cluster, but it is very new and very small, as well as missing some key elements. For several entries, such as Oceanography in Ensenada, Mexico, we have no additional data to give us a better understanding, but we accept this as possible, given the close proximity to the world's largest such cluster, Scripps at La Jolla, California.

Table 14.6
Summary of Data

Country	City/Region	Focus
ARGENTINA	Mar del Plata[26]	Chemistry Metal-mechanical
	Rafaela[27]	Food (agroindustrial) Metal-mechanical
BRAZIL	Bahia[28]	Petrochemicals
	Belo Horizonte	Biotechnology
	Cachoeiro de Itapemirim, NovaVenécia	Metal-mechanical, marble and granite
	Campinas[29]	Microelectronics Computer science Telecommunications
	Curitiba[30]	Telecommunications Environmental industries Software
	Espírito Santo[31]	Software Automation engineering
	Linhares	Furniture
	Londrina	Agrobusiness
	Maranhão	Cultivation and processing of soya
	Novo Friburgo	Garments
	Paraíba and Ceará	Textiles and garments
	Pernambuco State	Tropical fruits Tourism
	Porto Alegre and region	Leather and shoes Auto parts Farm machinery and implements Health Software
	Porto Real	Software
	Resende/Porto Real	Automotive parts

Table 14.6
(continued)

	Rio de Janeiro[32]	Deepwater drilling for offshore petroleum around Petrobras Electricity technologies around CEPEL/Eletrobrás Audiovisual entertainment around Rede Globo de Televisão Software
	Santa Catarina[33] (Florianópolis and region)	Electro-metal-mechanical Frozen foods Textiles and garments Ceramics and tiles
	Santa Rita do Sapucaí[34]	Electric-electronic Fiat network of automotive parts suppliers and the associated metal mechanical complex
	São Carlos	Advanced materials
	São Francisco	Tropical fruits
	São José dos Campos[35]	Aeronautics
	São Leopoldo	Computer science
COSTA RICA	San José	Biotechnology
		Software
		Ecotourism, environmental technologies and management
CUBA	Havana	Biotechnology
MEXICO	Cuernavaca	Electronics
		Environment
	Ensenada	Oceanography
	Guadalajara	Electronics (known locally as the Mexican Silicon Valley)
	Mexico City	Pharmaceuticals
	Monterrey	Informatics
	Tijuana	Communications
URUGUAY		Biotechnology Wine Software

Analysis

Here we consider all together the results of the field studies, the Web conference and the literature. Since these data have been presented in a way that errs on the side of being inclusive rather than exclusive, future research should investigate the cases signalled before as having potential for being or becoming innovation clusters. Now we should look at the same set from a more limited, exclusive viewpoint and see what a tighter application of our model will portray. The categorization scheme used is the one developed earlier in the chapter.

We categorize all natural resource industrial clusters as dependent or truncated. While they may be functioning industrial clusters, it is our estimation that they have relatively few of the fundamental characteristics that enable them to transform easily into innovation clusters. In earlier research by Tiffin (1989) comparing innovation strategies in the Canadian mining industry with Finnish strategies, it was seen that the latter was clearly able to create a dynamic innovation cluster, but the Canadian industry was dominated by a strategy promoting import of machinery and equipment from abroad. To our knowledge, this is the strategy overwhelmingly followed by the Latin American resource industries of mining, forestry, oceans and, to a lesser extent, agriculture. In both the Canadian and the Latin American cases, the industries are dominated by multinationals using Latin America to produce relatively simple raw materials. More recently, the Canadian industry has been able to create clusters around management, control and instrumentation aspects of resource exploitation, as well as pollution mitigation systems, but it seems doubtful to us that this is happening in Latin America at present, although we stand to be corrected by future research.[36]

In agriculture, there may be noteworthy exceptions that we are ignoring. We therefore considered wine clusters. However, innovation clusters that focus on lower quality wines for mass markets are in our opinion less likely to be the first breeding grounds for transformation to innovation clusters than those that focus on fine quality, specialty products, where knowledge of markets, design, quality and constant innovation are fundamentals instead of low-cost, large-scale production. Hence our interest in reviewing the Uruguyan case, where this is the niche strategy followed. In the absence of any studies on this topic in Chile and Argentina, we are left not including wine clusters in either country, unfortunately. However, it is very likely that there are candidates for agroindustry innovation clusters, for example in Londrina, southern Brazil, where in addition to the strong agro-industry cluster, there is a committed community will to create a technopol[37]. It would not be realistic to assume that a single crop, however important like soya, could be the basis for an agroindustry cluster, so we exclude this category mentioned earlier. Brazil's specialty tropical fruit production is so new that, although there is great promise here, we reluctantly choose not to include these entries at the present time. Small scale and newness also caused us to remove the Recife initiative in pharmaceuticals, despite its promise for the future.

Automotive parts could be either industrial clusters or dependent-truncated clusters. In our understanding, these sectors, although highly technology-intensive, tend to be largely branch plants controlled by foreign firms, using technology implanted by the head office, much of it significantly off the technological frontier. There will be many small firms involved, under local ownership and management. However, recent research in this sector by Katz (2000) claims that much of the potentially innovative capacity of small, locally owned firms has disappeared with the opening of national markets to international competition. In addition, much of this production is not generated within a distinct and limited urban environment, but often involving many different cities and even countries. We include this as a marginal case of an innovative industrial cluster.

There is frequent mention of metal-mechanical innovation clusters. We feel these may be very dynamic industrial clusters of considerable economic importance, with a strong local content, so label them innovative industrial clusters.

Similarly with textiles and garments, leather and footware. Although working with limited knowledge, we do not see the fundamental role of knowledge, science and technology, venture capital, the creation of new products and the integration of new production technology with new products. Mature innovation clusters are definitely possible in these areas, as many studies of the industry in Italy have shown, but to us, these are currently, overwhelmingly, innovative industrial clusters.

Tourism is mentioned several times. Two detailed studies sponsored by CEPAL (Acuña, Villalobos and Ruiz, 2000; Barbosa and Zamboni, 2000) on tourism clusters show they are both industrial clusters in our terminology, and even the mature Costa Rican cluster lacks many elements of an innovation cluster. Therefore, we exclude all the tourism clusters mentioned except for the Costa Rican case, which seems to have very strong stakeholder will and resources to continue to advance, giving it the status in our eyes of a proto-innovation cluster.

We do not consider the Polo Tecnologica Constituyentes in Buenos Aires an innovation cluster, from the limited data at hand. It is based too much on public research institutions trying to find industrial applications and customers, and is not focussed enough. New, technology-based firms do not seem to have a strong presence yet. Obviously, this organization has potential for creating innovation clusters in a variety of fields over the long term (for example, two decades of concerted, consistent effort).

The most difficult cases are those where we have insufficient data to make any judgment. Here we have relied on the quality of the source of the data. For instance, in the Mexican cases, the data were supplied at the Montevideo workshop by several Mexicans who were expert in the field and were working on very similar definitions of innovation clusters to what this chapter proposes. Therefore, we include them, except for Tijuana, which we feel is more like Monterrey; dynamic, but currently more industrial innovative in character. We label the clusters as shown in Table 14.7 as either proto- or innovative industrial (Inn. Ind.).

Note in the table that none of the entries classify as mature innovation clusters. There are a good number of protoclusters, which have some of the most significant elements of an innovation cluster and they have potential to develop farther. Their principal limitations seem to be:

- extremely limited to nonexistent risk equity funding,
- weak and intermittent social interaction and integration,
- lack of specialized consulting services.

Table 14.7
Proto- and Innovative Industrial Clusters

Country	City/Region	Focus	Category
Argentina	Mar del Plata	Chemistry, metal-mechanical	Inn. Ind.
	Rafaela	Food (agroindustrial), metal-mechanical	Inn. Ind.
Brazil	Bahia	Petrochemical	Inn. Ind.
	Belo Horizonte	Biotechnology	Proto
	Cachoeiro de Itapemirim, NovaVenécia	Metal-mechanical, marble and granite	Inn. Ind.
	Campinas	Microelectronics Computer science Telecommunications	Proto
	Curitiba	Telecommunications Environmental industries Software	Proto
	Espírito Santo	Software Automation engineering	Proto
	Linhares	Furniture	Inn. Ind.
	Londrina	Agrobusiness	Proto
	Novo Friburgo	Garments	Inn. Ind.
	Paraíba and Ceará	Textiles and garments	Inn. Ind.
	Porto Alegre	Health Software	Proto
	Porto Alegre region	Leather and shoes Auto parts Farm machinery and implements	Inn. Ind.
	Porto Real	Software	Proto

Table 14.7
(continued)

	Resende/Porto Real	Automotive parts	Inn. Ind.
	Rio de Janeiro	Deepwater drilling for offshore petroleum around Petrobras Electricity technologies around CEPEL/Eletrobrás Audiovisual entertainment around Rede Globo de Televisão Software	Proto
	Santa Catarina (Florianópolis and region)	Electro-metal-mechanical Frozen foods Textiles and garments Ceramics and tiles	Inn. Ind.
	Santa Rita do Sapucaí	Electric-electronic Fiat network of automotive parts suppliers and the associated metal mechanical complex	Inn. Ind.
	São Carlos	Advanced materials	Proto
	São Francisco	Tropical fruits	Inn. Ind.
	São José dos Campos	Aeronautics	Proto
	São Leopoldo	Computer science	Proto
Costa Rica	San José	Software Ecotourism, environmental technologies and management	Proto
Cuba	Havana	Biotechnology	Proto
Mexico	Cuernavaca	Electronics Environment	Proto
	Ensenada	Oceanography	Proto
	Guadalajara	Electronics	Proto
	Mexico City	Pharmaceuticals	Proto
	Monterrey	Informatics	Inn. Ind.
	Tijuana	Communications	Inn. Ind.
Uruguay		Wine	Inn. Ind.

To give a visual impression of where the protoclusters are located in Latin America, we present them on a map (Figure 14.2), each with a star. Note the close correspondence with existing industrial centres in Mexico and Brazil.

Figure 14.2
Distribution of Proto Innovation Clusters in Latin America

CONCLUSIONS

1. It is important for Latin America to consider innovation clusters. They seem to grow more firms, grow them faster and make them more profitable than can be done elsewhere. Innovation clusters are seed beds out of which the new knowledge-based economy takes root. Local governments can make successful policy interventions to promote these structures.
2. Although the data are very preliminary and incomplete, it seems likely that there are no mature innovation clusters in Latin America.
3. There are innovative industrial clusters and protoclusters that exhibit a significant number of characteristics that a mature innovation clusters would

have. The concept of an innovation cluster is not black and white, on or off; industrial communities evolve and integrate knowledge in many different forms. The leading communities discussed in our report are what we call protoclusters, illustrating not only some of the important characteristics of mature clusters, but most important, potential to evolve into mature clusters.

4. With the previous conclusion in mind, we stress that investing in physical infrastructure is not as important as investing in the mechanisms that promote the invisible parts of community and integration. Large-scale projects to create science parks and technopols should be approached with great caution, as the past experience is so ambiguous about their success and the potential for using up large amounts of scarce resources in an unproductive manner is so high.

5. Innovation cluster development can be stimulated by community will and the will of individual champions. In this sense, we see the potential for growth of innovation clusters in some areas of Latin America as high in the short- and medium-term future.

6. One of the most obvious strategies to create local innovation clusters is to work with existing industrial clusters. Increasing the availability of venture finance, community integration and linkage with the local science and technology institutions could be a quick and low-cost way of making these structures be more knowledge- and innovation-intensive. However, overcoming social barriers limiting cooperation among businesses and linking to universities may prove difficult for some countries.

7. Conversely, it may be difficult to create an innovation cluster in a city without the existence of a strong industrial cluster—except, of course, in areas where the science and technology involved are sufficiently revolutionary that they create remarkably new industrial possibilities, and the markets tend to be global.

A DIRECTION FOR FUTURE RESEARCH

It would be appealing if we could conclude with the remarks above, pointing out simply that with more research along the lines already undertaken, we would be well on our way to improving the functioning of public and private investments for innovation clusters. However, there might be a significant technological event intruding on the current situation which will rapidly alter the local nature of innovation clusters as we know them today—the diffusion of the World Wide Web and the Internet. Thus we are obliged to conclude with the following caveat, opening up still more questions.

Ernst (1999) has begun to pose some important questions about how national systems of innovation are beginning to reshape themselves under the simultaneous influence of forces promoting globalization, regionalization and localization. As the communications infrastructure rapidly becomes cheaper, better and more widespread, the balance of tacit and codified knowledge exchange and development may shift significantly. Some of the requirements for proximity may weaken dramatically as virtual exchanges on the Web grow.

In a report on this topic to IDRC, Gibson et al. (1999) ask: "A key question for the proposal research project for the 21st century, therefore, is how necessary and sufficient is the regional development of 'smart' infrastructure in all its aspects (i.e., talent, technology, capital and know how) or physical infrastructure (i.e., science parks, incubators, and high tech corridors) in the emerging Internet-based economy where the movement of knowledge is increasingly through ICT? And it may be asked, which sectors or components of this infrastructure must be physically co-located or digitally networked at different stages of a firms becoming globally competitive."

Lalkaka (1998) illustrates the potential of the World Wide Web to strengthen the links between different components with the "Netcelerate Web Site" at the Georgia Institute of Technology (in Atlanta). "This is a virtual community of companies inside and outside the incubator, mentors and accredited investors. It offers selected participants entry to discussion groups, information library, directories of qualified accountants, lawyers and consultants, and access to potential angel-investors" (p. 7). The University of Texas at Austin's IC^2 Institute has recently set up a Global Business Accelerator, which is working to provide services from the innovation cluster at Austin to international start-up firms; a partial concretization of the concepts elaborated by Gibson et al. (1999) about virtual innovation clusters.

In addition, clusters may be moving into a more complex global environment, as they begin to interact more directly with other clusters and take on complementary or specialized roles. Both Ernst and Saxenian suggest this[38].

NOTES

1. We include Cuba and the Dominican Republic from the Caribbean but not the French, English- and Dutch-speaking countries of Belize, Guyana, Guyane Francaise and Surinam. This choice is made only to emphasize the linguistic and cultural commonalities of the Iberoamerican group for the purposes of this chapter.

2. Tiffin (1987) has seen this field grow from the early 1980s in Canada and the United States. A typical early approach for building innovation clusters in small communities can be seen in his consulting study for the diversification of a nuclear-research-based town in Canada.

3. Both employment and number of establishments are refered to the following sectors: (i) computing and office equipments, (ii) communications equipment, (iii) electronic components, (iv) guided missiles, space vehicles, (v) instruments, and (vi) software and data processing.

4. Recently, articles on the topic have begun to appear in the popular press. El Mercurio (1999) describes Chilean preoccupations, and a discussion in El Pais (2000) with the Dean of the Chemistry Faculty, Dr. Alberto Nieto, presents some important Uruguayan actions.

5. Gibson, Conceição, Nordskog, Burtner, Tankha and Quandt (1999); and Quandt, (1999).

6. Researchers sometimes criticize the creation of simple lists which describe critical factors for building local innovation clusters, alluding to the complex and so-far poorly understood processes involved. However, these lists are very popular with policy pro-

fessionals and cluster managers, because they give them some anchor and direction which is better than none.

7. Gibson et al. (1999).

8. Data taken from Red Iberoamericana de Ciencia y Tecnología (RICYT, 1999). Note that data come from several recent years.

9. Pérez (1989) cited in Cassiolato and Lastres (1999), p. 14: "The present wave of technical change: implications for competitive restructuring and for institutional reform in developing countries", text prepared for Strategic Planning Department of the World Bank (Washington, DC: The World Bank, 1999), p. 32.

10. The Endless Transition: Third Triple Helix International Conference. Rio de Janeiro, 26-29 April 2000.

11. We are not stating, however, that the state has the only responsibility for this, or can do it very effectively; all stakeholders are responsible to different degrees. Tiffin visited a large technology transfer exhibition in Mexico City in 1998 where there were dozens of excellent booths with eager, competent staff, all public-sector organizations trying to promote transfer and linkage, but virtually no visitors were present—a telling metaphor of the difficulty of state promotion.

12. João Bosco from Brazilian Science and Technology Ministry, at the Innovation Clusters in Latin America Workshop, IDRC, Montevideo, May 1999.

13. The model as presented here is static and pictorial only. However, it should be able to be greatly extended through mathematical techniques in ecology and systems analysis.

14. Jose Pimenta-Bueno, closely involved with the PUC-Rio incubator in Brazil, has commented that this seems to be a key service to venture capitalists, and is thinking of packaging and marketing its firms explicitly to further lower costs and risks to investors.

15. As one example, the University of Calgary, in Canada, created a beautiful building with a wide range of services to incubate firms, just a few years before an economic crash in the city slashed the cost of office rental space. As costs of accommodation fell in the city, the incubator rapidly found its tenants leaving to cheaper office space downtown. David Gibson, associated with the innovation cluster at Austin Texas, through the university's IC^2 Institute, tells a story of a neighbouring city which built an incubator, equipped it and then found no "tenants" showing up.

16. "Silicon Valley: Droles d'indigenes!" *GEO*, No. 247, Sept. 1999, pp. 36-50.

17. Castells (1984); Andersson (1985a); Aydalot (1986a); Hall (1990), cited in Castells and Hall (1994).

18. Personal communication from David Gibson, IC^2.

19. ANPROTEC "Panorama 99. As Incubadoras de Empresas no Brasil", 1999, taken from http://www.anprotec.org.br. The number of incubators is continually rising.

20. Personal communication from Maurizio Geddes, Director of the UFRJ Incubator, 1998.

21. Interview carried out by Bortagaray at the Centro Nacional de Alta Tecnología (CENAT), Costa Rica, April 1999.

22. Personal communication from Ricardo Aguilar, Vice Rector, Research and Extension, Instituto Tecnológico de Costa Rica.

23. Information for this industry comes from ADM & TEC (2000), a research proposal prepared for IDRC on innovation clusters.

24. http://www.unsam.edu.ar/polo/espa.htm

25. Data compiled from York Medical information circulars. Tiffin explored the possibility of IDRC investing in Cuban biotech commercialization with York Medical and the Canadian Medical Research Council in 1998.

26. Gennero de Rearte et al. (1999).

27. Bosherini et al. (1999).
28. Rodas Veras Filho (1999).
29. Gómes (1999); Voyer (1997).
30. Krüger Passos (1999).
31. Villaschi Filho (1999).
32. Martins de Melo (1999).
33. Ramos Campos et al. (1999).
34. Borges Lemos and Campolina Diniz (1999).
35. Gómes (1995); de Souza and Garcia (1999).
36. IDRC is planning to sponsor further research in this area. Personal communication from Dr. Andres Rius of IDRC, Montevideo.
37. There is a Londrina Techapolis project, created in 1998, and the city is now into its 7th Jornada Tecnológica Internacional de Londrina conference and exhibition. (www. adetec.org.br is the address of the managing group, the Associação do Desenvolvimento Tecnológico de Londrina.)
38. "The concept of global production network (GPN) allows us to analyze the globalization strategies of a particular firm with regard to the following four questions: (1) Where does a firm locate which stages of the value chain? (2) To what degree does a firm rely on outsourcing? What is the importance of inter-firm production networks relative to the firm's internal production network? (3) To what degree is the control over these transactions exercised in a centralized or descentralized manner? And (4) how do these different elements of the IPN hang together?" (Ernst, 1999, p. 13).

Saxenian states the following: "the creation of regional clusters and the globalization of production go hand in hand, as firms reinforce the dynamism of their own localities by linking them to similar regional clusters elsewhere" (Saxenian, 1994, p. 4).

REFERENCES

Acuña, M., Villalobos, D. and Ruiz, K. (2000). *El Cluster Ecoturístico de Monteverde/ Costa Rica*. Santiago de Chile: CEPAL.

ADM & TEC (2000). *Knowledge Based Industries, Local Development and the Role of Innovation*. Recife: Instituto de Administração e Tecnologia.

Altenburg, T. and Meyer-Stamer, J. (1999). "How to Promote Clusters: Policy Experiences from Latin America", *World Development*, 27(9): 1693-1713.

Baptista, R. and Swann, P. (1998). "Do firms in clusters innovate more?", *Research Policy*, 27: 525-540.

Barbosa, M. A. and Zamboni, R. (2000). *Formação de um "Cluster" em torno de natureza sustentável em Bonito—MS*. Brasília: IPEA.

Borges Lemos, M. and Campolina Diniz, C. (1999). "Sistemas locais de inovação: o caso de Minas Gerais". In Cassiolato, J. E. and Lastres, H. M. M. (eds.), *Globalização e Inovação Localizada. Experiências de Sistemas Locais no Mercosul*. Brasília: IBICT.

Bortagaray, I. (1999). *Innovation Clusters in Latin America and the Caribbean*. Paper prepared for the IDRC/LACRO Workshop on Innovation Clusters, Montevideo, 24-26 May.

Boscherini, F., López, M. and Yoguel, G. (1999). "El desarrollo de capacidades innovativas de las firmas en un medio de escaso desarrollo del sistema local de innovación". In Cassiolato, J. E. and Lastres, H. M. M. (eds.), *Globalização e Inovação Localizada. Experiências de Sistemas Locais no Mercosul*. Brasília: IBICT.

Cassiolato, J. E. and Lastres, H. M. M. (eds.) (1999). *Globalização e Inovação Localizada. Experiências de Sistemas Locais no Mercosul.* Brasília: IBICT.

Castells, M. and Hall, P. (1994). *Las Tecnópolis del Mundo: La Formación de los Complejos Industriales del Siglo XXI.* Madrid, Spain: Alianza Editorial.

Dei Ottati, G. (1994). "Trust interlinking transactions and credit in the industrial district", *Cambridge Journal of Economics*, 18(6): 529-546.

Echeverri-Caroll, E. (1997). "Japanese style networks and innovation in high-technology firms in Texas", Bureau of Business Research, Graduate School of Business, University of Texas at Austin, mimeo.

El Mercurio (1999). "Formación de conglomerados productivos: la unión hace la fuerza". Saturday, 25 December.

El Pais (2000). "Una incubadora de empresas", 2 April, p. 4, section 3 (Montevideo).

Ernst, D. (1999). "How Globalization Reshapes the Geography of Innovation Systems: Reflections n Global Production Networks in Information Industries" (First Draft). Prepared for DRUID 1999 Summer Conference on Innovation Systems, Aalborg.

EXAME (2000). "As melhores cidades para fazer negócios", 3 May, 713(9): 77-110.

Freeman, C. (1987). *Technology and Economic Performance: Lessons from Japan.* London: Pinter.

Gennero de Rearte, A., Lanari, E. and Alegre, P. (1999). "La capacidad innovativa de nucleos impulsores de firmas en entornos territoriales dinámicos: el caso de Mar del Plata, Argentina". In Cassiolato, J. E. and Lastres, H. M. M. (eds.), *Globalização e Inovação Localizada. Experiências de Sistemas Locais no Mercosul.* Brasília: IBICT.

Gibson, D., Conceição, P., Nordskog, J., Burtner, J., Tankha, S. and Quandt, C. (1999). *Incubating and Sustaining Learning and Innovation Poles in Latin America and the Caribbean.* Montevideo: IDRC.

Gómes, E. J. (1995). "A experiência brasilera de pólos tecnológicos: uma abordagem político-institucional," M.Sc. Dissertation, Instituto de Geociências, UNICAMP, São Paulo (mimeo).

Gómes, E. J. (1999). "Polos tecnológicos y promoción del desarrollo: hecho o artefacto?" *REDES*, November, 14(7): 177-216.

Jonas, R. (1979). "Conceptos, funciones y aspectos sociales de parques industriales". In *Parques Industriales en América Latina: Bolivia, Brasil, Colombia, Chile, Ecuador.* Bogotá: ILDIS-CENDES-Soc. de Ediciones Internacionales Ltda.

Katz, J. (2000). "Reformas estructurales, regimes sectoriales y desempeño industrial en America Latina en los años noventa". Santiago de Chile: CEPAL.

Kozul-Wright, Z. (1995). *The Role of the Firm in the Innovation Process.* Geneva: UNCTAD.

Krüger Passos, C. A. (1999). "Sistemas locais de inovação: o caso do Paraná". In Cassiolato, J. E. and Lastres, H. M. M. (eds.), *Globalização e Inovação Localizada. Experiências de Sistemas Locais no Mercosul.* Brasília: IBICT.

Krugman, P. (1991). *Geography and Trade.* Cambridge, MA: MIT Press.

Lalkaka, R. (1998). "Support systems for small enterprises and their clusters". Presented at the United Nations Conference on Trade and Development Promoting & Sustaining SME Clusters & Networks, Expert Meeting, Geneva, 2-4 September.

Lerer, J. I. (1979). "La importancia de los parques industriales en el desarrollo regional. El caso de Brasil". In *Parques Industriales en América Latina: Bolivia, Brasil, Colombia, Chile, Ecuador.* Bogotá: ILDIS-CENDES-Soc. de Ediciones Internacionales Ltda.

López, A. and Lugones, G. (1999). "Los sistemas locales en el escenario de la globalización". In Cassiolato, J. E. and Lastres, H. M. M. (eds.), *Globalização e Inovação Localizada: Experiências de Sistemas Locais no Mercosul*. Brasília: IBICT.

Lundvall, B.-Å. (ed.) (1992). *National Systems of Innovation: Towards a Theory of Innovation and Interactive Learning*. London: Pinter.

Lundvall, B.-Å. (1994). "Innovation policy in the learning economy". Paper presented at the International Seminar on Policies for Technological Development. CIDE, Mexico City.

Magalhães Tavares, H. (1998). "Pólos Tecnológicos, Meio Urbano e Planejamento do Território". VIII Seminário Nacional de Parques Tecnológicos e Incubadoras de Empresas, Belo Horizonte.

Malecki, E. J. (1997). *Technology and Economic Development: The Dynamics of Local, Regional and National Competitiveness*. Essex: Longman.

Marceau, J. (2000). "Innovation and industry development: a policy-relevant analytical framework". Third Triple Helix Conference, Rio de Janeiro, 26-29 April.

Marshall, A. (1920). *Principles of Economics*. London: Macmillan (first edition published 1890).

Martins de Melo, L. (1999). "Sistemas locais de inovação: o caso do Rio de Janeiro". In Cassiolato, J. E. and Lastres, H. M. M. (eds.), *Globalização e Inovação Localizada: Experiências de Sistemas Locais no Mercosul*. Brasília: IBICT.

Medeiros, J. A., Medeiros, L. A., et al. (1992). *Pólos, Parques e Incubadoras: A Busca da Modernizaçao e Competitividade*. Brasília: CNPq.

Metcalfe, S. (1995). "The Economic Foundations of Technology Policy: Equilibrium and Evolutionary Perspectives". In Stoneman, P. (ed.), *Handbook of the Economics of Innovation and Technological Change*. Oxford: Blackwell Publishers.

Nash, M. (1996). "Fidel Castro's Most Idiosyncratic Venture Pays Off", *TIME Magazine*, May 13, vol. 147.

Nelson, R. R. (ed.) (1993). *National innovation systems: A comparative analysis*. New York: Oxford University Press.

OECD (1996). *The Knowledge Based Economy*. Paris: OECD.

OECD (1997a). *National Innovation Systems*. Paris: OECD.

OECD (1997b). *Technology Incubators: Nurturing Small Firms*. Paris: OECD.

Patel, P. and Pavitt, K. (1994). "The Nature and Economic Importance of National Innovation Systems", *STI Review*, 14: 9-32.

Pérez, C. (1989). "The present wave of technical change: implications for competitive restructuring and for institutional reform in developing countries". Paper prepared for the Strategic Planning Department of the World Bank, Washington DC: The World Bank.

Pérez, C. (1990). *Tecnología, Desarrollo y Sistema Nacional de Innovación*. Ponencia presentada en el Seminario Internacional sobre el Nuevo Contexto de la Política de Desarrollo Científico y Tecnológico, Montevideo, mimeo.

Porter, M. (1990). *The Competitive Advantage of Nations*. New York: Free Press.

Porter, M. (1998). "Clusters and the new economics of competition", *Harvard Business Review*, November-December, 76(6): 77-90.

Quandt, C. (1997). *The Emerging High-technology Cluster of Campinas, Brazil*. Paper prepared for IDRC-International Development Research Centre, Technopolis 97, Ottawa, September.

Quandt, C. (1999). *The concept of virtual technopoles and the feasibility of incubating technology-intensive clusters in Latin America and the Caribbean*. Montevideo: IDRC.

Ramos, J. (1998). *Una Estrategia de Desarrollo a partir de los Complejos Productivos (clusters) en torno a los Recursos Naturales.* Santiago de Chile: CEPAL.

Ramos Campos, R., Nicolau, J. A. and Ferraz Cário, S. A. (1999). "Sistemas locais de inovação: casos seleccionados em Santa Catarina". In Cassiolato, J. E. and Lastres, H. M. M. (eds.), *Globalização e Inovação Localizada. Experiências de Sistemas Locais no Mercosul.* Brasília: IBICT.

RICYT (Red Iberoamericana de Indicadores de Ciencia y Tecnología) (1999). *Principales Indicadores de Ciencia y Tecnología 1990-1997.* Buenos Aires: Organización de Estados Americanos (OEA)—Programa Iberocamericano de Ciencia y Tecnología para el Desarrollo (CYTED).

Rodas Veras Filho, R. (1999). "A Rodovia Salvador-Camacari: Possivel Versão Brasileira da Route 128". Research Proposal, Department of Geography, Universidade Federal da Bahia.

Rogers, E. M. and Larsen, J. K. (1984). *Silicon Valley Fever: Growth of High-Tech Culture.* New York: Basic Books.

Saxenian, A. L. (1994). *Regional Advantage: Culture and Competition in Silicon Valley and Route 128.* Cambridge, MA: Harvard University Press.

Schmitz, H. (1997). *Collective Efficiency and Increasing Returns.* Brighton: Institute of Development Studies, University of Sussex.

Snoeck, M. (1998). "Transición, aprendizaje e innovación en la industria vinícola uruguaya". Comisión Sectorial de Investigación Científica de la Universidad de la República Oriental del Uruguay.

de Souza, M. C. and Garcia, R. (1999). "Sistemas locais de inovação em São Paulo". In Cassiolato, J. E. and Lastres, H. M. M. (eds.), *Globalização e Inovação Localizada. Experiências de Sistemas Locais no Mercosul.* Brasília: IBICT.

Sutz, J. (1996a). *Universidad, Producción, Gobierno: Encuentros y Desencuentros.* Montevideo: CIESU-Trilce.

Sutz, J. (1996b). *Una Aproximación Primaria al Sistema Nacional de Innovación de Uruguay.* Montevideo: CIESU-Trilce.

Sutz, J. (1997). "(Innovación y Desarrollo): condiciones de siembra y cosecha" In Sutz, J. (ed.), *Innovación y Desarrollo en América Latina.* Caracas: Nueva Sociedad.

Sutz, J. (1998). "La caracterización del Sistema Nacional de Innovación en el Uruguay: enfoques constructivos", Instituto de Economia da Universidade Federal do Rio de Janeiro (mimeo).

Tancer, R. (1995). "The pharmaceutical industry in Cuba", *Clinical Therapeutics,* 17(4): 791-798.

Tiffin, S. (1987). *Capturing Local Economic Growth from Science in the Town of Deep River.* Deep River, Ontario, Canada.

Tiffin, S. (1989). *A Comparison of Canadian and Finnish Technology Strategies in the Mining and Forestry Industries.* Québec: Conseil des Sciences du Québec.

Tiffin, S., Couto, G. and Bas, T. (2000). "Venture capital in Latin America". Third Triple Helix International Conference, Rio de Janeiro, 16-29 April.

Villaschi Filho, A. (1999). "Alguns elementos dinâmicos do sistema capixaba de inovação". In Cassiolato, J. E. and Lastres, H. M. M. (eds.), *Globalização e Inovação Localizada. Experiências de Sistemas Locais no Mercosul.* Brasília: IBICT.

Voyer, R. (1997). "Emerging high-technology industrial clusters in Brazil, India, Malaysia and South Africa". Paper prepared for IDRC.

Wilson, J. (1998). *The Creation and Management of Tacit Knowledge.* Montevideo: IDRC.

15

Globalization and Industrial Restructuring in Mexico: The Electronics and Automobile Industries

Cristina Casanueva Reguart

INTRODUCTION

This study explores the effects of globalization and industrial restructuring in Mexico on the automotive and electronic industries, two of Mexico's most dynamic industrial sectors over the last decade.[1] The focus of the study is on computer and telecommunication equipment and the auto parts industries.[2] The analysis took place in the regions of Guadalajara, for the electronics industry, and Saltillo-Ramos Arizpe-Monterrey, for the automotive industry. These locations were selected based on their important contribution to production, in relation to other regions in the country, since the opening of Mexico's economy in the mid-1980s.

The first issue examined in the study relates to the restructuring of the organization of production, analyzing the extent to which companies have modified their organization and management practices according to lean production systems. This reorganization of production demands a complex and rigorous system of cooperation from participating companies based on the use of just-in-time systems, total quality control strategies and the organization of productive activities around cells or work teams.

The second issue examined is the formation of the supply chain and the interaction between companies in the regions, and the effect of the new production systems in the formation of these chains.

Finally, the study poses the question of to what extent the emergence and growth of productive activities in these regions have led to the formation of "clusters" in the Guadalajara metropolitan area for the electronics equipment industry, and the Saltillo-Ramos Arizpe-Monterrey region for the auto parts industries.

This study attempts to contribute to the discussion of the spatial organization that results from the integration of regions within the global economy; whether these regions relate to the model proposed by literature as clusters or industrial districts or these regions remain as maquiladoras or export processing zones.

According to the objectives, the study briefly describes the policy framework in Mexico, for the opening of the markets, leading to industrial restructuring. Then it presents the methodology used in the study and criteria in the selection of regions under study.

In the following section a conceptual framework is presented. It deals with the primary aspects of lean production systems, and the impact of globalization on the acquisition of technological capabilities by the firms and their participation in innovation activities, and the theoretical discussion on the spatial organization that results from the integration of industrial regions to the global economy. This theoretical framework constitutes the basis for the main research questions guiding the empirical analysis, first on the reorganization and flexible production systems and, second on the formation of supply chains.

POLICY FRAMEWORK: LIBERALIZATION AND INDUSTRIAL RESTRUCTURING IN MEXICO

In the 1980s the Mexican government began the implementation of a package of policies that resulted in an increase of the export orientation of the Mexican economy and in a greater degree in the integration with global flows of investment, production and trade. Some of these policies had a significant impact on industrial restructuring by trade liberalization, by promoting an intense program of deregulation and a broader opening to foreign investment.[3]

The liberalization policy began in 1983 and was reinforced when Mexico joined the General Agreement on Tariffs and Trade (GATT) in 1986. Liberalization signified a substantial reduction in customs tariffs and the elimination of all types of qualitative barriers to trade (customs tariffs were decreased from 100% in 1982 to an average of 11.1% in 1993).

With the objective of consolidating the economic liberalization, Mexico began a program of negotiating free trade agreements with other countries and economic blocks. The most important of these have been with North America and the European Union, which culminated with the North American Free Trade Agreement (NAFTA) and the Mexico-European Union FTA, the former effective since 1994 and the latter since July 2000.

Foreign investment regulations were modified to reduce restrictions, allowing levels of up to 100% in most sectors of the economy. In addition, the procedures for authorizing foreign investment were simplified and made more expeditious.[4]

Within this policy frame, investment was viewed as a factor of international competitiveness that would facilitate access to advanced technology, leading to an expected impact on productivity. Deregulation and liberalization of foreign investment represented a central aspect of measures for promoting international integration. Parallel to these changes, technological transfer was also deregulated.

The changes in Mexico's economic policies permitted the economy to evolve, to a large extent, from an import-substitution industrialization model to an export-oriented one. This new orientation of the country's economy had an impact on the formation of different patterns of competition, specializing Mexico's exports on certain branches of the manufacturing industry, like textiles and clothing, automotive and auto parts, and the electric and electronic industries (Carrillo, Mortimore and Estrada, 1998). The greater specialization in productive processes and their integration into export markets are the result of both global strategic planning of companies and Mexico's competitive advantages.

The structural changes in the economy have led to the gradual emergence of a new economic geography, with the strengthening of manufacturing production in regions other than the greater metropolitan area of Mexico City and nearby zones, where during the import-substitution period, growth in manufacturing production was concentrated.

GLOBALIZATION, INDUSTRIAL RESTRUCTURING AND REGIONAL SPECIALIZATION

This section analyzes the growth in the electronic equipment industry in the Guadalajara region and the automotive industry in the Saltillo-Ramos Arizpe-Monterrey region, presenting the main factors that influenced the decisions to locate companies in these regions.

During the past decade, the production of computer and telecommunications equipment in the Guadalajara region rose at an impressive rate. This was also the case for the automotive industry in the northern region of Saltillo-Ramos Arizpe-Monterrey.

In 1993 the Guadalajara-Jalisco region's share of the national production of computers and telecommunications equipment was 39.4% and by 1997 it had more than doubled to 78.5%. In comparison, the share of the Mexico City metropolitan area (including Mexico City and the surrounding areas of the State of Mexico) in the production of electronics decreased from 54.9% in 1993 to 19.09% in 1997.[5]

Between 1993 and 1997, the electronic equipment industry in the Guadalajara-Jalisco region grew yearly by 71.4% on average. In contrast the same industry in the Mexico City metropolitan area experienced a negative growth (–4.7%) with respect to national production during this period.[6]

Examining the share of national production of the automotive industry, the Saltillo-Ramos Arizpe-Monterrey region (Coahuila and Nuevo León), this increased from 18.5% in 1993 to 23.5% in 1997. During the same period, the Mexico City metropolitan area's share of national production in auto parts decreased from 36.4% to 34.5%.[7]

While the northern region production activity experienced an annual average growth rate of 18.97 % (almost twice the average national growth rate), auto parts production the Mexico City metropolitan area production grew only by 5.56% between 1993 and 1997.[8]

These statistics clearly show the emergence of a different economic geography, in the distribution of manufacturing activities increasing in certain regions. Specifically, in the metropolitan area of Guadalajara in the case of the computer equipment industry, and in the northern region of Saltillo-Ramos Arizpe-Monterrey for the automotive industry.

Factors Influencing the Location of Electronics Companies in the Region of Guadalajara

Labor costs and skill levels have been the main factors influencing location decisions in computer electronics companies.[9] Another relevant factor has been companies' interest in complying with the local content requirements of the North American Free Trade Agreement. This factor became critical since they import a significant number of components from Asia. NAFTA creates incentives for the Asian-owned companies to establish themselves in Guadalajara in order to comply with the rules of origin, supplying parts and components to terminal assembly companies in the region.[10]

In the view of some company directors interviewed, the decision to locate their plants in the Guadalajara region is to a large extent based on its proximity to one of the world's largest markets, the United States, with costs much lower than in this country. They agree on the fact that Guadalajara is endowed with a satisfactory infrastructure for transportation of commodities. In this context an announced strategy of creating a center of manufacturing and design was made for this region, because of the cost advantages offered by its proximity to the United States. This strategy has the objective of successfully confronting the competition that companies from the Asian Pacific region present for U.S. companies.[11]

In contrast to labor costs and NAFTA Rules of Origin, government promotion and incentives appear to be minor contributing factors in location decisions by electronic manufacturing plants in the metropolitan Guadalajara region.[12] This finding should not come as a surprise, since nine of the ten companies surveyed were established in Jalisco between 1974 and 1998, earlier than the Investment Promotion Law, enacted in 1998.[13]

Nevertheless, information provided by representatives of the main electronic companies settled in the Guadalajara region in the 1970s (final assembly plants mainly) mentioned the impact of the decentralization policy set forth by the government in 1970 aimed at alleviating the excessive concentration and demographic congestion experienced by Mexico City. The Decentralization Plan provided fiscal incentives in federal taxes between 50% to 100% to companies that would move to areas where economic activities and employment were in high demand.[14]

Factors in the Location of Automotive Companies in the Saltillo-Ramos Arizpe-Monterrey Region

Different from the case of the computer electronics industry, in the automotive companies the most relevant factor is the presence of final assembly companies or original equipment manufacturers (OEMs). This finding reveals the importance of the location of suppliers close to the OEMs, with the objective of gradually consolidating a synchronized system of production and a just-in-time (JIT) practice, perhaps as part of their process of becoming JIT complexes.

The second most important factor for the location of automotive firms in the region is the skill level of labor (there is more emphasis on the level of skills, than cost of labor, different from the electronics industry). Finally, the third most important factor consists of labor costs.[15]

METHODOLOGY

According to the objectives of the study, it was necessary to identify the regions to be examined for each of the two industries based on the regions' relative performance since the early 1990s. Once the regions were selected, an in-depth examination was conducted of a total of 21 companies, 10 in the electronics computer industry in the Guadalajara region, and 11 in the auto parts industry in the Saltillo-Ramos Arizpe-Monterrey region. Due to the exploratory nature of the study, it did not attempt to generalize its results to the universe of firms in each industry, in the two regions selected.[16]

The selection of regions was based on secondary sources, such as Industrial Census Reports (1988 and 1994) and Annual Industrial Surveys (1993-1997)[17], supplemented by information available from industrial chambers. The two regions were selected on the basis of their gross production value growth, and their share in the production value in relation to other regions in the country.

Surveys and interviews with executives and other industry representatives which were conducted during in-person visits, provided information on the companies. The visits also allowed for observations regarding the implementation of organizational practices. Interviews with company executives and high-level personnel were complemented with interviews with government representatives in charge of promoting investment and representatives of industrial chambers in the auto and electronic equipment industries.

The quantitative information supplied by the surveys was examined through factor analysis and the results gathered were complemented by the qualitative information collected by the unstructured interviews.

The next section presents the primary aspects of lean production systems and a discussion on the effect of globalization on the innovation behavior of firms and on their effort aimed at developing their own technological capabilities. This section also has a theoretical discussion on the spatial organization that results from the integration of industrial regions to the global economy.

GLOBALIZATION, LEAN PRODUCTION, TECHNOLOGICAL CAPABILITIES AND THE NATURE OF REGIONS

Lean Production and Restructuring Manufacturing Activities

The objective of the lean production system is to achieve flexibility as the ability to adapt to changes in the demand for products within a competitive environment, while reducing costs and increasing product diversity.

The lean production system incorporates changes in product and process design and new technologies, and generates a trend of subcontracting production services from other companies. Within the organizational restructuring perspective, lean production implies three major transformations: the reorganization of production based on a strict system of inventory control, known as the just-in-time system; total quality control; and new type of relationships with suppliers. These interrelated changes are held to create a different production system, based on principles which contrast to the "Fordist mass-production" system (Humphrey, 1995).

Mass production has traditionally been organized in assembly lines and requires large inventories, stock reserves, temporary workers and extra space for its operation. In contrast, lean production eliminates these excesses by applying inventory control and the "kanban," JIT, total quality and continuing improvement or "kaizen" practices.

"Kanban" is a practice of internal control over the company's production, implemented also by suppliers, that leads to the production of components as they are needed, eliminating unnecessary inventory (Rao, 1999). The JIT practice refers to a complex organization of buyer-seller relations between companies. Under JIT, suppliers provide parts within the required time limit and with zero defects. For these reasons, JIT is considered to be the center of production reorganization in various regions (Schoenberger, 1987; Rubenstein, 1992; Kenney and Florida, 1993; Reid, Solocha and Uallachain, 1995; Uallachain and Reid, 1997).

"Kaizen" or continuing improvement practice is based on the principle that it is always possible to improve any process within production. Kaizen requires a structure of disciplined workers working in what are referred to as "cells" or work teams which, among other things, facilitates the establishment of total quality practice (Rao, 1999). Mass production has been characterized by the highly specialized assets and economies of scale maintained by final assemblers or OEMs and tends to establish short-term contracts with suppliers, based on costs and not quality.

In contrast with the mass production system, in lean production the most important factor in the selection of companies by OEMs is quality and their capability to deliver on time. The instrumentation of these practices leads to the establishment of more stable relationships between OEMs and their suppliers. These modifications have had a fundamental impact on the spatial configuration of regions, because of the need for final assembly and supply companies to be located in the same region.

On the subject of labor organization, a mass production system requires workers to be specialized in a single activity and to perform their jobs rigidly organized in assembly lines. In contrast, under lean production, the training of personnel is designed for workers to acquire various skills that can be applied in a variety of tasks. This is what is known as worker flexibility, or multitask training practice. With this focus in labor organization, there is no longer a need for worker reserves, or dependence on temporary workers, in order to confront absenteeism or modifications in production demand that results from the dynamics of product design and markets.

In terms of organization, there is a tendency to eliminate departments specialized in inspection and maintenance tasks. Instead, the workers actively participate in these tasks, sharing responsibility for quality control and the operation and control of the manufacturing process. In this context there are mechanisms for receiving suggestions and proposals from workers on how to process and product design. The successful implementation of this form of production requires ongoing efforts in the area of training (Andersen Consulting, 1994; Micheli, 1994).

Subcontracting of Production Services

The final assembly companies increasingly subcontract production to third parties as part of these companies' strategies to succeed in an environment characterized by fierce competition and market volatility (Sturgeon, 1998). Subcontracting manufacturing facilitates the increase or decrease of demand products with little prior notice. Final assembly companies are capable of easily adjusting to fluctuations in the market, since they are a supplier of a number of different companies, and have an adequate mix of products. By subcontracting production, the OEMs remain competitive without having to use their facilities below their capacity or absorb the high costs of maintaining an inventory of products. In addition, subcontractor manufacturers, as buyers of parts and components, create new backward and forward linkages in the production chain in the same geographical region.

By offering subcontracted production services, subcontractors induce efficiencies, based on their economies of scale, allowing cost reductions. Manufacturing subcontractors are gradually acquiring a new role in the industry, by incorporating tasks such as the design of subassemblies or "modules", the coordination and logistics for acquiring parts and the inspection and quality control of parts. In the past, final assembly companies exclusively performed these tasks.

In recent years, the trend among subcontractors has been to install manufacturing and design centers in different locations around the world. The two factors that appear to have an influence on decisions regarding location are (1) the possibility of being near final assembly companies or OEMs, and (2) the relative costs of personnel, including both workers and specialized staff who work in the area of design.

Formation of Supply Chains

The new spatial configuration of parts supplying based on synchronized production systems or JIT affects the way in which suppliers and final assembly companies are geographically located, in what has been defined in the literature as flexible or JIT complexes (Ramírez, 1997). This spatial configuration also affects the formation of chains in which suppliers from the various levels of the production chain tend to establish themselves, in the same region as the final assembly companies or OEMs.

In the first level of the chain are found companies that supply directly to final assembly companies, including the producers of modules and complete systems. Their capacity for R&D activities in their areas of expertise allows them to take on the tasks of assembling, logistics and coordination necessary for operating in coordination with second- and third-tier suppliers. It also means they can deliver where the assembling takes place, in the quantity required, with consistent punctuality and reliability (Chappel, 1994; Thorndike, 1991).

Decisions on the location of final assembly companies are based on cost factors and important considerations on the proximity of these companies to the major markets. International suppliers establish subsidiaries in the countries where final assembly companies are located, including developing countries, and transfer the learning process to any place in the world (D'Cruz and Rugman, 1993; Humphrey, Mukherjee, Zilbovicius and Arbix, 1998; Stevens, 1995; Sturgeon and Florida, 1997).

Globalization, Industrial Restructuring and the Nature of Regions

One of the objectives of this study is to contribute to the discussion of the spatial modifications that result from the restructuring of production and the integration of regional economies to the global economy. The literature on globalization and industrial restructuring poses the question on the extent to which the integration to global markets is conducive to the formation of clusters or just the expansion of maquiladoras or export-processing regions.

In the literature "cluster" is defined as a critical mass of companies in a specific field, located in a geographic locality, where intense horizontal cooperation and exchange occur in regard to investments in infrastructure, training, sharing large orders and bringing together productive capacity. Clusters include companies dedicated to distribution and commercialization, producers of complementary goods and other institutions that provide specialized training, vocational training and higher education, research and technical support. This cooperation takes place in some cases through business associations, combined investments, in centers of technology and training, in the establishment of quality standards and the corresponding consulting services, access to credit, marketing and even legal, accounting and consulting services.

In the definition of cluster an emphasis is placed on the presence of strong intercompany relations, vertical or horizontal, which are necessary between production and distribution and among the different economic agents. An essential characteristic of clusters is the relations established among companies located in the same area. In particular, the division of labor, the specialization among producers, the supply of products or specialized services, the presence of suppliers that provide raw materials, components and equipment, and the presence of business associations, all promote collective efficiency (Porter, 1998).

In contrast to the emergence of clusters, the maquiladora regions or export processing zones represent a different type of spatial organization (Gereffi, 1996). These forms of productive organization in certain regions respond to the logic of international production sharing, and although the companies are concentrated in the same region, economic relations by the integration of forward or backward linkages among these companies rarely occur.

Within a spatial dimension, examples of maquiladora regions or export processing zones which consist of industrial agglomerations located in the northern border area of Mexico have arisen in respond to the Border Industrialization Program. This program began in 1965 allowing foreign companies to be established in a 20-kilometer area from the Mexico-U.S. border (in the 1960s, extensions to these areas were authorized). The Program also contemplated companies' rights to import capital goods and components tax free. The so-called export processing zones were also a response to new shared production programs, which consist of shipping goods to be assembled to regions characterized by their low manpower costs.[18]

In their origin, production processing zones were based on the intensive use of labor, as well as their proximity to the United States. In their beginnings in the 1960s, these companies' decisions on where to locate their operations, in the case of Mexico, did not include the development of a network of local suppliers. In fact, an analysis of the maquiladora industry indicated that the contribution of domestic inputs has thus far not exceeded 3% (Carrillo and Hualde, 1997; Galhardi, 1998).

The discussion on the status of maquiladoras is particularly relevant, because the nature of Border Industrialization is being modified as a result of NAFTA,[19] including those firms whose origin is not from the countries of the Agreement (Canada and Mexico).[20]

In sum, the relevance of the previous discussion on spatial organization, production restructuring and global integration raises the question on the extent to which the regions of Guadalajara and Saltillo-Ramos Arizpe-Monterrey correspond to the model proposed by the literature as clusters or are these maquiladoras or export processing regions?

RESEARCH QUESTIONS

This section presents a series of research questions that have emerged from the above discussion. These questions are posed with the purpose of guiding the empirical research on the effects of industrial restructuring in the Guadalajara region for the electronics industry and the Saltillo-Ramos Arizpe-Monterrey region for the automotive industry.

- Are the companies in the electronics and automotive industries in the Guadalajara and Saltillo-Ramos Arizpe-Monterrey regions engaged in a process of restructuring production, management and labor organization, according to the schemes of lean and flexible production?
- Have forward and backward linkages been formed, between suppliers of components and final assembly in the same region where the companies of these two industries are concentrated?
- To what extent is the formation of supply chains in these regions (if any) a result of the implementation of synchronized programming systems of production (or approximations to JIT)?
- Do the supplier companies exclusively supply what the final assembly companies demand? Or do companies export some part of their production?
- Which agents or institutions have played a critical role in the formation of supplier chains and their integration with local final assembly companies?
- How important is the presence of national supplying companies (Mexican capital)? Which agents or institutions have played a critical role in the emergence of national suppliers?
- To what extent have the emergence and growth of productive activities in these regions led to the formation of clusters? Or have these regions remained as maquiladoras or export processing zones?

The following section presents the empirical evidence gathered in regard to the theoretical discussion and the questions that were raised above.

COMPANIES' RESTRUCTURING AND
FLEXIBLE PRODUCTION SYSTEMS

This section deals with the question of the extent to which the electronics firms in the Guadalajara region and automotive companies in the Saltillo-Ramos Arizpe-Monterrey region were involved in a restructuring of production processes, according to the schemes of lean and flexible production.

Electronics Industry

Synchronized Reception-Delivery Practice (JIT). Among the various companies in the computer and telecommunications industry located in the Guadalajara region, the implementation of a synchronized reception and delivery practice as JIT is not fully applied. This is true in the coordination with suppliers and in the terms of the delivery of products to clients. Nevertheless, most of the companies are implementing approximated versions of JIT, consisting of programs or logistics aimed at reducing the cycles that begin with the delivery of parts and components and end with the assembly of the final product. Most companies reported that it is difficult to implement rigorous systems for synchronizing deliveries, since some of the suppliers are located in distant countries in Asia.[21]

A certain level of production excellence has been achieved in some countries and regions, particularly in Asia, where there are original contributions in technical and design improvements, resulting in a "geographic division of labor". Therefore, a variety of patterns of specialization and parts-supplying is found among these regions.

However, in recent years major first-tier suppliers and manufacturing subcontractors have established themselves in the region, attracted primarily by the presence of final assembly companies (or invited by them) and by labor costs; some of them are the world's largest contractor manufacturers companies.[22]

While strict forms of JIT have not been fully implemented, as mentioned earlier, there are systems of extended programming that have facilitated coordination between suppliers of parts and subassemblies, on the one hand, and the final assembly companies, on the other. The result is greater efficiency in production and more rigorous quality control, with reduced downtime and waste rates. These efforts by companies have led to greater discipline in terms of costs, making it possible to increase competitiveness within the industry.[23]

Total Quality Systems. All the companies surveyed at the different levels of the production chain (final assembly companies, parts suppliers and manufacturing subcontractors) have implemented total quality processes, with the corresponding documentation, and have received certification of quality in accordance with international standards.

Flexible Organization of Personnel. The processes of continuing improvement (kaizen) and preventive maintenance are not generalized practices in the industry. Workers do not directly assume responsibility for kaizen or continuing improvement, nor for preventive maintenance; however, they participate more actively in taking responsibility for quality control.[24]

Worker teams have been adopted by nearly all the companies, illustrating the efforts to modify labor organization according to lean production schemes.

In half of the companies, direct participation by workers in taking responsibility for production control and rotation of tasks among personnel can be observed. [25]

Subcontracting. Only the OEMs or final assembly companies are subcontracting production services. Contractor manufacturers subcontract other services such as personnel hiring through human resource companies, as well as maintenance, transportation, cafeteria, security, among others, thus concentrating their efforts in their fundamental activity of assembling large volumes of products. In sum, cases of subcontracting are found, but in terms of production services, only to a significant degree within final assembly companies or OEMs. It is worth mentioning that subcontracting of other types of services has a positive spillover effect for other companies in the Guadalajara region.

Automotive Industry

Synchronized Production and Delivery System. A process of moving toward establishing JIT has begun in the Saltillo-Ramos Arizpe-Monterrey region. Especially noteworthy are the introduction of the kanban system or on-line programming, and the use of the JIT system for deliveries to final assembly companies. This is not the case, however, in receiving inputs from suppliers. [26] That the JIT system has not been completely implemented in the region can be explained by the insufficient number of suppliers established in the region and by the continued importing of components by final assembly companies. The data suggest that the establishment of auto parts suppliers is in process and final assembly companies continue to import a significant portion of their components.

Quality systems stand out from the dissemination of new production systems in the region. The use of these systems has extended to suppliers at different levels in the production chain. Next in importance, in terms of impacts analyzed, is production efficiency, with less downtime and lower waste rates, as well as a reduction in inventories. This suggests that some significant aspects of the lean production system have spread to a significant part of production in the region. [27]

Total Quality Systems. Mexico's vehicle parts companies have adapted their systems according to total quality and programming of deliveries. Mexico's supply companies—primarily the first-tier suppliers—have applied the ISO-9000 and QS quality standards and guarantee systems in their labor processes and manufacturing operations. [28] Parallel to these internationally recognized certifications of quality granted by independent certifying agencies, these companies have also received acknowledgments and awards that final assembly automotive companies grant to their most outstanding suppliers. This has not been an easy process and companies continue to confront difficulties in maintaining these certifications.

Flexible Organization of Personnel. As we also find in developed countries, the companies surveyed have modified labor organization and workers' responsibilities. In 10 of the 11 companies surveyed, workers are those directly responsible for quality control, continuing improvement and preventive maintenance.[29]

These companies have also managed to substantially modify their ways of working with flexible organization of personnel. They have created work teams, have incorporated workers into responsibility for production tasks and in half of the companies they have implemented rotation of tasks among workers. In most companies, workers participate in inspection and maintenance tasks. Nevertheless, companies do not have formal mechanisms for collecting suggestions or proposals from workers on how to improve the processes.[30] In sum, there is a tendency in the region toward labor organization in accordance with lean production systems and this is transforming workers' tasks and responsibilities.

Subcontracting. Contractor manufacturers are not found in this sector; subcontracting is not a common practice in car production in the region (as it is in the electronics industry in Guadalajara). In this region relationships between final assembly companies and their suppliers are based on purchase-sale contracts—in the case of first-tier suppliers. In the contracts the final assembly companies not only specify prices, but also penalties for noncompliance with quality standards or late deliveries made by producers.

The information presented on the organization of production practices suggests that the surveyed companies have adopted organizational practices, along the lines of lean and flexible production.

FORMATION OF SUPPLY CHAINS

Electronics Industry

Exporting and Location of Clients. The period 1995-1999 was a very dynamic one in terms of new supply companies and manufacturing subcontractors being established in the Guadalajara region. During this same period, these companies began to direct their supplying to the regional market, whereas previously all of their production was for exporting.[31] This has resulted in the formation of supply networks in the Guadalajara region that extend from final assembly to the second and third tier of the productive chain.

Among second-tier suppliers are companies with a high level of technological sophistication, which includes the production of mother boards, printed circuit cards, relays, electric surge protectors, cables and harnesses—these latter designed according to the client's specific technical requirements.

During the last part of the 1990s, final assembly companies worked hard to bring their suppliers to the Guadalajara region. As of this writing, seven of the world's largest contractor manufacturing companies (SCMs) are in the region. The recent arrival of manufacturing subcontractors has served as the main agent in the process of consolidating the supply network in the region, and in generating a constant flow of demand for production from second- and third-tier suppliers.[32]

Most of these firms are subsidiaries of components supply companies and contractor manufacturers that have supplied final assembly companies in other countries. The final assemblers in Guadalajara have stressed attracting proven suppliers, and to a lesser extent the development of national suppliers in the region.

Efforts to bring international companies to the region have been supported by the Jalisco state government, which passed an Investment Promotion Law in 1998 (mentioned before in the location factor section). Additional support has been brought by Cadena Productiva de la Electrónica, CADELEC (Electronic Productive Chain). This organization has been assisting the establishment of suppliers in the region, facilitating the integration of local companies, nationals and internationals with the local production networks. Among the services CADELEC offers is information through a database (constantly updated) on the more pressing requirements of production inputs by the local electronics industry, assisting companies that would consider establishing themselves in Jalisco. CADELEC has also worked as a linkage between investors and government authorities, assisting investors by providing information, and assisting investors' representatives in identifying locations and serving as a link with the local government.

The study found forward linkages providers (first to third tier) with the terminals in the region, but also with the international chains of production. The manufacturing subcontractors and final assembly companies in the region do not represent the main demand for the supply companies in the region. The primary demand for the supply companies remains international. The companies under analysis export 84.6% of their production, according to their scale, mainly to the United States, where final assembly companies are located, and represents the largest market in the hemisphere (Angel and Engstrom, 1995).[33] Nevertheless, of the total production of supplier companies, 15.4% remain in the region and are sold to the OEMs or final assembly located in the Guadalajara metropolitan region.[34]

It is also worth mentioning that the emergence of regional chains has not been accompanied by increased growth of national suppliers (companies with 100% Mexican investment). Parts and components manufacturing by national companies is only beginning, and is mostly focused in the area of packing materials, labels and printed matter (documentation); also indirect inputs such as energy, banking services, cleaning and maintenance services and other inputs of a relatively low added value.

The limited number of national suppliers may be partly explained by the financing problems faced by Mexican companies. These companies have been required to invest large sums of money in order to introduce quality control systems and synchronized programming of production—necessary to satisfy requirements for quality, volume and timely deliveries.

The formation of supply networks up to the second tier in the productive chain is an important factor that distinguishes electronics companies in the Guadalajara region from a maquiladora complex. Although this is a necessary factor, it is not sufficient to distinguish this conglomerate of electronics companies from a maquiladoras complex. This is especially true if we take into account that both first-tier (including manufacturing subcontractors) and second-tier final assembly companies import most of their inputs and components from the United States and Asia, and also export most of their production.

Second-Tier Suppliers. The major suppliers for the electronics industry in the Guadalajara region are located in the United States. This country not only represents the hemisphere's major market, but it is also the primary producer of components as well as finished equipment.

Asia is the geographic area second to the United States in terms of supplying components. These data confirm the geographic distribution of production areas that we identified in the discussion on the industry's geography. As indicated earlier, when sources of the needed components are located at a distance, companies have difficulties in strictly implementing the JIT system.

In sum, the factors explaining the new configuration of supply networks in the Guadalajara region are the result of a process of profound change, which has made it possible to evolve from an international maquiladora-type assembly enclave, to a cluster composed of a supply network. This supply network extends from the final assembly to the second tier and third tier—which are directly or indirectly moving toward becoming part of global trade.

The factor that explains the emergence of a supplier network around the final assembly companies in Guadalajara is the implementation of lean production systems, in particular delivery programming and quality control processes. This has led to first-tier suppliers (including contractor manufacturers) establishing themselves in the same region as the final assembly companies.

If the tendency for international suppliers to locate in the Guadalajara region continues, supply chains will become consolidated, largely as a result of efforts by final assembly companies to encourage their suppliers to establish themselves in the region where they are located. It will also be a result of the strategy on the part of final assembly companies to convert this region into a manufacturing and design center for the hemisphere's electronics industry—with the objective of confronting the competition from the Asian Pacific region, as established earlier.

Automotive Industry

Exports and Location of Clients. The spatial configuration of parts-supplying based on the JIT system that is emerging in the Saltillo-Ramos Arizpe-Monterrey region is leading to the formation of supply chains—as in other regions of the world. In this region, firms directly supplying final assembly companies make up the first level of the chain. These firms may be subsidiaries of international supply companies, or companies with national capital. Of the companies surveyed, four are new plants owned by international suppliers that have contracts with GM and Chrysler, and were invited by these two automotive companies to establish plants near their installations—for the purpose of complying more efficiently with their goals for exporting automobiles.

Five other companies—with primarily national capital—are also first-tier suppliers in the region. These companies were established during the import-substitution era, and have managed to maintain their positions as first-tier suppliers due to an intense process of production restructuring, supported by strategic alliances with international partners. They have been able to adapt to changing conditions because of these alliances. As it will be shown, this group of companies has come to be consolidated as first-line suppliers for the final assembly companies established in Mexico. In addition, companies reported that second-tier parts-supplying is not well developed in the region yet. Suppliers' networks are beginning to expand to the second tier—although this is still in the early stages. To have access to components from second-tier suppliers, companies must frequently import these products.

The links in the supply chains are not limited to this region alone; there are links with international suppliers located in other parts of the world. All the companies surveyed have clients in the region and in the country, with the exception of one that exports all of its production. Together, these companies directly export an average of 43% of their production, primarily to the United States and, to a lesser degree, to Europe and Asia. It is worth mentioning here that an even greater percentage of their production is exported indirectly through the final assembly companies that export 80% of assembled vehicles.[35]

Second-Tier Suppliers. The process of vertical integration in final assembly companies affects both direct suppliers as well as suppliers at the second tier in the chain. With the objective of satisfying their clients' requirements, direct suppliers also demand more reliable, high-quality products from second-tier suppliers—who are also expected to be located in the same region.

For this reason we can begin to see the traces of change in the group of second-tier suppliers of parts and components. Most of the companies surveyed (72%) have suppliers in the same region or in nearby areas, but they continue to depend on receiving parts from companies located in the United States (80%) and, for a smaller percentage (20%), from companies in Europe.[36]

The requirements of first-tier suppliers, including those that manufacture systems or modules, have obliged their suppliers to make modifications in their productive processes in order to satisfy their own demands for proximity, price and quality.

In sum, the lean production system has begun to be established in the region and, as a consequence, the terminal assembly companies have modified their demands to the first-tier suppliers, and these latter, in turn, to their own suppliers. While considerable segments of suppliers and clients are located in other parts of the world, we can speak of a tendency for an increasing number of suppliers to locate in the same region.

In other words, there is a trend in the region toward creating supply networks and also toward a future creation of a spatial configuration similar to that found in developed countries. In this spatial configuration there are first-tier suppliers capable of satisfying the demands of the final assembly companies, and in which there are important interrelationships among suppliers of different levels.

CONCLUSIONS

Information gathered on the organization of production practices reveals the surveyed companies' adoption of organizational best practices, along the lines of lean and flexible production and specifically including the implementation of production synchronization or JIT systems, total quality or kaizen systems, organization of labor, and the subcontracting of production services.

Most of the companies surveyed, in both the electronics and automotive industries, have established systems for programming production and deliveries. The implementation of the JIT system is not, however, completely generalized in these industries. The difficulties they experience in this regard are based on their dependence on importing key parts and components. This dependence has hindered the strict application of JIT in the receiving of production inputs.

There are, however, differences between the electronics and the automotive industries. While companies in the electronics industry have not managed to introduce JIT in the receiving of inputs or in deliveries to clients, most of those in the automotive industry use JIT in their deliveries to clients. It is worth mentioning that the companies confronting difficulties use a system of programming deliveries that is increasingly rigorous and seeks to reduce the cycle that begins with the delivery of parts and components, and ends with the assembly of the final product.

The companies' implementation of total quality systems in both the electronics and auto parts industries has enabled them to obtain internationally recognized certification. This certification has become a general requirement for terminal assemblers or original equipment manufacturers and for at least first-tier suppliers. This quality requirement is gradually being applied to all the levels of the supply chain. In terms of programming and coordination of deliveries, all the companies in the sample implement some form of production programming.

Productive organization applied in the two industries has led to the use of work teams, as well as the implementation of task rotation among workers, which is more generalized in the auto parts industry. Also observed in the latter industry is the participation of workers in quality control, in continuing improvement and preventive maintenance.

During the last part of the 1990s, final assembly companies in electronics worked diligently to bring their suppliers to the region. Most of these companies are subsidiaries of components suppliers and manufacturing subcontractors that supply terminal assemblers in other parts of the world.

In Guadalajara, manufacturing subcontractors have been the agents that have encouraged the formation of chains, integrating parts and components of second- and third-tier suppliers in their production.

The role of manufacturing subcontractors is not seen so clearly in the automotive industry, where this role is rather played by first-tier suppliers that directly supply final assembly companies. The network of suppliers in the region is beginning to expand, although very gradually, to including the second tier.

Final assembly companies in the electronics industry have emphasized bringing proven suppliers to the region and, to a lesser degree, the development of national suppliers. In contrast, in the auto parts industry there are Mexican business groups that have managed to technologically and organizationally restructure their companies by establishing strategic alliances—in terms of capital and technology—with international suppliers. This is what has enabled them to successfully confront industrial restructuring and has taken them to the level of top-of-the-line suppliers.

Parts-supplying at the different levels of the chain is not limited to the Guadalajara and northern Mexico regions, with supplying focused on final assembly companies located there. Rather, suppliers export most of their production, linking up with international supply chains.

In the case of the electronics industry, making connections with international chains follows the logic of a geographic division of production between zones of specialization, which can be explained to some degree by the differences in the distribution of technological capacities in the manufacture of particular parts and components. And therefore is difficult to strictly implement the JIT system. In the future, there will be a trend for the formation of regional chains in a more consolidated manner, as international suppliers establish subsidiaries in the regions where final assembly companies are located.

To date, the level of parts-supplying by Mexican companies is of lower added value. The scarcity of national suppliers can be explained, to some degree, by the lack of financial resources required by these companies in order to carry out technological and organizational changes. This restructuring for competitiveness by the Mexican firms is a requirement for their incorporation into both national supply and international chains. It is worth taking note that this restructuring has been possible in the auto parts industry, in which a group of companies (with a majority of Mexican capital) managed to consolidate themselves as world class suppliers, by establishing strategic alliances with foreign technological partners.

Unlike the opinion that these regions are export processing zones or maquiladora regions, the study found that there are conditions for the formation of clusters. Supplier relationships exist in the same regions, between final assembly companies or OEMs and first-tier suppliers, and/or subcontractor manufacturing companies; in the case of the electronics industry, also with the second-tier suppliers.

However, evidence suggests that parts and components suppliers do not work exclusively for the final assembly companies in their regions, but are also part of the international supply chains. This means that the regions' main activity is not limited to the exportation of final products, but also includes parts and components at various levels.

In sum, if it is not possible to identify these regions with the so-called clusters in a strict sense, it can be stated that there are industrial concentrations that are beginning to form supply chains at the local level. These companies have acquired technological capabilities in order to satisfy the needs of their customers and have also initiated the process towards the formation of more complex, integrated industrial structures.

NOTES

1. The author would like to acknowledge the invaluable assistance of Flor Brown, Oscar Fuentes and Juan Antonio Laguna.

2. In 1998 the automobile industry represented 11% of the manufacturing GNP, 20% of total exports, 22% of the manufacturing industry exports and 18% of total manufacturing employment. For the same year the electronics industry represented 9% of the manufacturing industry GNP, 18% of manufacturing exports and 16% of total exports (INEGI, 1993 and 1997).

3. These policies were also accompanied by an intense program of privatization that reduced the number of state-owned companies from a total of 744 in 1982 to 106 in 1992 (Carrillo, Mortimore and Estrada, 1998).

4. Previous to the new law, most sectors had been closed to foreign investment, or open only to minority participation, subordinated to national investment.

5. See Appendix I, Table 15.I.1, Regional Share in Mexico's Production of Electronic Equipment, 1993-1997.

6. See Appendix I, Table 15.I.2, Growth of Electronic Equipment, 1993-1997. Mexico City metropolitan area participation may even be overestimated, since this figure accounts for the Federal District (D.F.) and the State of Mexico, when, strictly speaking, Mexico City only accounts for the D.F. and its surrounding area, which is part of the State of Mexico.

7. See Appendix II, Table 15.II.1, Regional Share in Mexico's Production in Parts of Vehicles, 1993-1997.

8. This growth has been slower in comparison to the one observed in the electronics industry. Some of the auto assemblers and parts producers have remained in the State of Mexico area, in the cities of Toluca, Cuatitlán and Santiago Tianguistengo; although this is Central Mexico, it is not part of the Mexico City Metropolitan Region. BMW, Mercedes, Volvo and Noebus run assembly operations there. Chrysler, GM, Nissan and Ford have moved their main assembly operations to the northern region, but they run smaller operations of car assembly in Toluca. See Appendix II, Table 15.II.2, Growth of Parts of Vehicles, 1993-1997.

9. The factor analysis shows that manpower cost and level of skills explain 71.9% of the total variance, 54.7% for labor cost and 17.2% for level of skills. See Appendix I, Table 15.I.3, Location Factors: Electronics Industry Companies.

10. The factor analysis reflects that labor (both the cost and level of skills) and NAFTA's local content requirements explain 83.4% of the total variance. See Appendix I, Table 15.I.3, Location Factors: Electronics Industry Companies.

11. In the words of the director of Hewlett Packard, Mr. Jaime Reyes: "Our competition is not in the United States, because we are complementary (manufacturing) economies. The competition is with the Asian Pacific region, considered the traditional center of the electronics industry, and including countries such as Taiwan, Malaysia, Singapore and Hong Kong. These countries have the foundations for design and manufacture." López Villegas (1998).

12. Results of the factor analysis show that "Government Incentives" as a location factor explain 5.3% of the total variance. See Appendix I, Table 15.I.3, Location Factors: Electronics Industry Companies.

13. The law embodies incentives for investment, as contributions in infrastructure (electricity, water supply and road construction), in training at technical and professional levels, donation or partial contribution in the lease or acquisition of property. The law also includes fiscal incentives according to the extent that investment meets objectives of employment creation, level of wages and amount of investment and exports.

14. Tax breaks did not include companies located within Guadalajara City itself, but in the *municipios* (political division similar to counties) surrounding Guadalajara that complied with the Decentralization Decree of 1970.

15. The presence of final assembly companies or OEMs explains 45.4% of the total variance. The quality of the workforce, as determined by skill, explains 19.7% and labor costs explain 17.3%; the cumulative of these three factors explains 82.4% of the variance. See Appendix II, Table 15.II.3, Location Factors: Automotive Industry Companies.

16. See Appendix I, Table 15.I.4, Companies in the Electronics Equipment Industry: Guadalajara-Jalisco Region, and Appendix II, Table 15.II.4, Companies in the Automotive Industry: Saltillo-Ramos Arizpe-Monterrey Region. Given the exploratory nature of the study, these two groups of companies are not representative samples of the type of firms of the industries analyzed.

17. The Industrial Census and the Annual Industrial Surveys were the most recent statistics published.

18. The U.S. maquiladora firms located in the export processing regions frequently combined their fiscal status with the mechanisms of shared production by the U.S. International Trade Commission, which occurred when parts and components made in the United States were shipped for reassembly purposes to a region in countries with relatively lower labor costs (Harmonized Tariff Schedule Subheading 9802.00.80).

19. The companies are able to sell their products (as a local firm) to the domestic market or anywhere within the North American region, in accordance with the following calendar: in 1999, up to 85% of the total value of their exports in the past 12 months; in the year 2000, up to 85% of the total value of their exports in the past 12 months. As of the year 2001, such sales by maquiladora companies will not be subject to any limits.

20. The change of policy affects also those companies whose capital investment is not from the NAFTA region, as far the Rules of Origin are concerned. These companies, in order to obtain the tariff advantages of NAFTA countries, must meet a minimum level of use of parts and components from NAFTA member countries, according to chapter 3 of the Rules of Origin. This is the case of the Asian and European companies established in Mexico. This involves a modification of tariff classification in accordance with the stipulations of Annex 401 or that the good must meet the percentage requirements of national

integration corresponding to that annex, when a change in tariff classification is not required. The percentages of national integration can be 50, 60, or 70% according to the above-mentioned Annex or the product must be entirely produced in the territory of one or more NAFTA countries (Chapter 3, NAFTA).

21.   According to the factor analysis, deliveries programming explains 45.7% and kanban or inventory control practices explains 29.1% of the total variance of the factors related to the synchronized delivery practices in the companies. See Appendix I, Table 15.I.5, Synchronized Production and Delivery Practices (JIT).

22.   Seven of the computer and telecommunications main contractor manufacturers companies are in the Guadalajara region.

23.   Quality control explains 50% of the variance in the different impacts studied; greater efficiency in production that reduces downtime explains 23%; and reduction in the waste rate explains 14% of the total variance sample. See Appendix I, Table 15.I.6, Impact of the Synchronized Production and Delivery Practices.

24.   There is an active participation in the activities related to total quality control in six of ten of the electronics firms analyzed. See Appendix I, Table 15.I.7, Workers' Responsibilities in Quality Control.

25.   These three factors explain 90% of the factors considered in the area of labor organization. See Appendix I, Table 15.I.8, Flexible Organization of Work.

26.   The kanban system explains 67% of the variance in the factors of extended production, and JIT delivery to clients explains 25%. See Appendix II, Table 15.II.5, Synchronized Production and Delivery Practices (JIT).

27.   Quality systems explain 36% of the total variance in impacts analyzed; reductions in downtime explain 22%; and reductions in inventories explain 19%. See Appendix II, Table 15.II.6, Impact of the Synchronized Production and Delivery Practices.

28.   QS-9000 is a quality standard created by the Chrysler, Ford and General Motors automotive companies that is based on the ISO-9000 certification, to which they have added requirements for their suppliers, such as continuing improvement in their processes and reduction in the variability of processes and costs.

29.   See Appendix II, Table 15.II.7, Workers' Responsibilities in Quality Control.

30.   The formation of work teams explains 61% of the total variance in elements of flexible organization of production; responsibility in production explains 24%; and rotation of tasks explains 11%. See Appendix II, Table 15.II.8, Flexible Organization of Work.

31.   Between 1995 and 1997, the number of computer and telecommunications equipment companies increased from a total of 39 to 61, creating 28,000 new jobs and increasing exports to a level of US$3.5 billion (according to statistics from the Department of Economic Promotion in the Jalisco state government, 1998). The maquiladoras established since the mid-1970s exported their production directly or indirectly, by supplying other maquiladoras in the northern part of the country.

32.   SCMs such as Solectron, Dovatron, Flextronics, Jabil, SCI and Nasteel, among others.

33.   See Appendix I, Table 15.I.9, Exports and Location of Clients.

34.   See Appendix I, Table 15.I.10, Origin of Technology.

35.   See Appendix II, Table 15.II.9, Exports and Location of Clients.

36.   See Appendix II, Table 15.II.10, Location of Main Suppliers.

REFERENCES

Andersen Consulting (1994), *Worldwide Manufacturing Competitiveness Study, The Second Lean Enterprise Report*. Arthur Andersen & CO.

Angel, D. and Engstrom, J. (1995), "Manufacturing Systems and Technological Change: The U.S. Personal Computer Industry", *Economic Geography*, 71(1): 79-102.

Carrillo, J. and Hualde, A. (1997), "Maquiladoras de tercera generación. El caso de Delphi-General Motors", *Comercio Exterior*, September, 47(9): 747-758.

Carrillo, J., Mortimore, M. and Estrada, A. (1998), "El Impacto de las Empresas Transnacionales en la Reestructuración Industrial de México, El Caso de las Industrias de Autopartes para Vehículos y de Televisores". Santiago de Chile: United Nations, División de Desarrollo Productivo y Empresarial.

Casanueva, C. (2001), "The Acquisition of Firm Technological Capabilities in Mexico's Open Economy, The Case of Vitro", *Technological Forecasting and Social Change*, 66(1): 75-85.

Chappel, L. (1994), "It's survival of the biggest for suppliers in the '90s", *Automotive News*, March 7.

D'Cruz, J. and Rugman, A. (1993), "Developing International Competitiveness: The Five Partners Model", *Business Quarterly*, 58(2): 60-72.

Galhardi, R. (1998), "Maquiladoras Prospects of Regional Integration and Globalization". Geneva: International Labor Organization (ILO), Employment and Labour Market Policies Group.

Gereffi, G. (1996), "Mexico's 'Old' and 'New' Maquiladora Industries: Contrasting Approaches to North American Integration". In G. Otero (ed.), *Neoliberalism Revisited: Economic Restructuring and Mexico's Political Future*. Boulder, CO: Westview Press.

Humphrey, J. (1995), "Industrial Reorganization in Developing Countries: From Models To Trajectories", *World Development*, 23(1): 149-162.

Humphrey, J., Mukherjee, A., Zilbovicius, M. and Arbix, G. (1998), "Globalization, FDI, and the Restructuring of Supplier Networks: The Motor Industry in Brazil and India." In M. Kagami, J. Humphrey and M. Piore (eds), *Learning, Liberalization and Economic Adjustment*. Tokyo: Institute of Developing Economies.

INEGI (Instituto Nacional de Estadística, Geografía e Informática) (1993), *Censos Económicos 1993*. Mexico: INEGI.

INEGI (Instituto Nacional de Estadística, Geografía e Informática) (1997), *Censos Económicos 1997*. Mexico: INEGI.

Kenney, M. and Florida, R. (1993), *Beyond Mass Production: The Japanese System and Its Transfer to the U.S.* New York: Oxford University Press.

López Villegas, G. (1998), "Es Jalisco un valle del silicio", *Periódico Reforma*, November 30, Section A.

Micheli, J. (1994), *Nueva Manufactura, Globalización y Producción de Automóviles en México*. Mexico City: UNAM.

Piore, M. and Sabel, C. (1984), *The Second Industrial Divide*. New York: Basic Books.

Porter, M. (1998), "Clusters and the New Economics of Competition", *Harvard Business Review*, November/December, 76(6): 77-90.

Ramírez, J. C. (1997), "Los Modelos de Organización de las Industrias de Exportación en México", *Comercio Exterior*, January, 47(1): 1121-1131.

Rao, K. A. (1999), "Lean manufacturing", *Monthly Labor Review*, 122(1): 50-51.

Reid, N., Solocha, A. and Uallachain, B. (1995), "Japanese Corporate Groups and the Locational Strategy of Japanese Auto and Component Parts Makers in the United

States". In M. Green and R. McNaughton (eds.), *The Location of Foreign Direct Investment: Geographic and Business Perspectives*. Aldershot, UK: Avebury Press.

Roos, D., Jones, D. and Womack, T. (1991), *The Machine that Changed the World*. New York: Rawson.

Rubenstein, J. M. (1992), *The Changing US Auto Industry: A Geographical Analysis*. London: Routledge.

Sadler, D. (1994), "The Geographies of Just-in-Time: Japanese Investment and the Automotive Components Industry in Western Europe", *Economic Geography*, 70(1): 41-59.

Schoenberger, E. (1987), "Technological and Organisational Change in Automobile Production: Spatial Implications", *Regional Studies*, 21: 199-214.

Shaiken, H. and Herzenberg, S. (1987), *Automation and Global Production: Automobile Engine Production in Mexico, the United States, and Canada*. La Jolla, CA: Center for U.S.-Mexican Studies, University of California, San Diego.

Stevens, T. (1995), "Managing Across Boundaries", *Industry Week*, 244(5): 24.

Storper, M. and Scott, A. (1988), "The Geographical Foundations and Social Regulation of Flexible Production Complexes". In J. Wolch and M. Dear (eds.), *The Power of Geography: How Territory Shapes Social Life*. Boston: Allen and Unwin.

Sturgeon, T. (1998), "Technological Change and the Rise of Turnkey Production Networks for Electronics Manufacturing: Implication for Developing Places", Cambridge, MA: MIT mimeo.

Sturgeon, T. and Florida R. (1997), "The Globalization of the Automobile Production", International Motor Vehicle Program Policy. Massachusetts Institute of Technology (mimeo).

Thorndike, K. E. (1991), "Can companies manufacture stockless production?", *Harvard Business Review*, July/August, 69(4): 172-175.

Uallachain, B. and Reid, N. (1997), "Acquisition versus Greenfield Investment: The Location and Growth of Japanese Manufacturers in the United States", *Regional Studies*, 31(4): 403-416.

APPENDIX I: ELECTRONICS INDUSTRY

Table 15.I.1
Regional Share in Mexico's Production of Electronic Equipment, 1993-1997

(Thousands of pesos 1993=100)
Gross Production Value

Region	1993	Participation	1997	Participation
North				
Baja California	100,611	1.21%	161,814	0.80%
Coahuila	25,983	0.31%	17,739	0.09%
Nuevo León	117,645	1.41%	124,719	0.62%
Sinaloa	39,608	0.47%	33,791	0.17%
Tamaulipas	150,234	1.80%	118,261	0.58%
Subtotal	434,081	5.20%	456,324	2.26%
Center				
Aguascalientes	561,509	6.73%	1,071,452	5.30%
Colima	14,872	0.18%	4,937	0.02%
Jalisco	2,660,884	31.90%	14,441,923	71.41%
Puebla	45,846	0.55%	361,439	1.79%
Subtotal	3,283,111	39.36%	15,879,751	78.52%
Mexico City Metropolitan Region				
Distrito Federal	1,543,570	18.50%	1,669,226	8.25%
State of Mexico	3,035,648	36.39%	2,192,489	10.84%
Subtotal	4,579,218	54.89%	3,861,715	19.09%
South East				
Campeche	45,561	0.55%	25,029	0.13%
Subtotal	45,561	0.55%	25,029	0.13%
Total	8,341,971	100.00%	20,222,821	100.00%

Source: INEGI, 1993 and 1997.

Table 15.I.2
Growth of Electronic Equipment, 1993-1997

(Thousands of pesos 1993=100) Gross Production Value			
Region	1993	1997	Growth
North			
Baja California	10,061	161,814	12.61%
Coahuila	2,598	17,739	-9.10%
Nuevo León	11,764	124,719	1.47%
Sinaloa	3,960	33,791	-3.89%
Tamaulipas	15,023	118,261	-5.81%
Subtotal	43,408	456,324	1.26%
Center			
Aguascalientes	56,150	1,071,452	17.53%
Colima	1,487	4,937	-24.09%
Jalisco	266,088	14,441,923	52.63%
Puebla	4,584	361,439	67.57%
Subtotal	328,311	15,879,751	48.30%
Mexico City Metropolitan Region			
Distrito Federal	154,357	1,669,226	1.98%
State of Mexico	303,560	2,192,489	-7.81%
Subtotal	455,921	3,861,715	-4.17%
South East			
Campeche	4,556	25,029	-0.14%
Subtotal	4,556	25,029	-0.14%
Total	834,197	20,222,821	0.25%

Source: INEGI, 1993 and 1997.

Table 15.I.3
Location Factors: Electronics Industry Companies

FACTOR ANALYSIS		
Component	% Variance	Cumulative %
Labor cost	54.773	54.773
Level of skills	17.226	71.999
Rules of origin compliance	11.395	83.395
Transportation cost	7.31	90.705
Government Incentives	5.363	96.071
Proximity to the OEM	3.119	99.190
Proximity university and R&D centers	0.638	99.828
Proximity to local suppliers	0.17	99.945
Other NAFTA factors	5.453E-02	100.000

Source: Estimations based on surveys.

Table 15.I.4
Companies in the Electronics Equipment Industry: Guadalajara-Jalisco Region

Company	Year of Foundation	Foreign Investment Participation	Position in Supplier Chain	Main Product
Firm 1	1995	100%	Terminal assembly	Phones and answering systems
Firm 2	1975	100%	Terminal assembly	Laptops, desktops, hard disk's magnetic subassemblies operative system and network operative system software
Firm 3	1974	100%	Terminal assembly	Mobile phones and pagers
Firm 4	1982	100%	Terminal assembly	Laser printers
Firm 5	1996	100%	Contractor Manufacturer	Subassemblies, metal box assembly
Firm 6	1995	50%	Third Tier	PCB Motherboards
Firm 7	1995	100%	Third Tier	Cables and harnesses
Firm 8	1998	100%	Third Tier	Cables and harnesses
Firm 9	1997	100%	Contractor Manufacturer	Cellular phones, laser printers, network cards, laptop computers
Firm 10	1974	100%	Third Tier	Telecomm components and subassemblies

Source: Data from surveys.

Table 15.I.5
Synchronized Production and Delivery Practices (JIT)

FACTOR ANALYSIS		
Component	% Variance	Cumulative %
Programming	45.700	45.700
Kanban	29.167	74.867
JIT delivery	16.845	91.712
JIT clients	8.288	100.000

Source: Data from surveys.

Table 15.1.6
Impact of the Synchronized Production and Delivery Practices

FACTOR ANALYSIS		
Component	% Variance	Cumulative %
Quality	50.451	50.451
Dead time	22.936	73.387
Waste	13.590	86.977
Inventories	7.508	94.485
Raw material	3.621	98.106
Work	1.152	99.258
Safety	.742	100.000
Environment	3.268E-15	100.000
Production systems	1.379E-15	100.000
Energy	-2.151E-15	100.000

Source: Data from surveys.

Table 15.I.7
Workers' Responsibilities in Quality Control

Firm	Quality	Kaizen	Preventive Maintenance
Firm 1	–	–	–
Firm 2	✓	–	–
Firm 3	✓	–	–
Firm 4	✓	–	–
Firm 5	✓	–	–
Firm 6	–	–	–
Firm 7	–	–	–
Firm 8	✓	–	–
Firm 9	–	–	–
Firm 10	✓	–	–
Total	6	0	0

Source: Data from surveys.

Table 15.I.8
Flexible Organization of Work

FACTOR ANALYSIS		
Factors	% Variance	Cumulative %
Work Teams	50.551	50.551
Production responsibilities	39.511	90.062
Job shifting	9.938	99.99
Quality c. responsibilities	5.551E-15	100

Source: Data from surveys.

Table 15.I.9
Exports and Location of Clients

Firm	Exports %	Guadalajara Region	United States	Canada	Latin America	Europe	Asia
Firm 1	95	-	✓	-	-	-	-
Firm 2	90	-	✓	✓	✓	✓	✓
Firm 3	100	-	✓	-	-	-	-
Firm 4	90	-	-	-	-	-	-
Firm 5	98	✓	✓	-	-	-	-
Firm 6	60	✓	✓	-	-	-	-
Firm 7	90	✓	✓	-	-	-	-
Firm 8	90	-	✓	-	-	-	-
Firm 9	80	✓	✓	-	-	-	-
Firm 10	90	✓	✓	-	-	✓	-
Average	88	5	9	1	1	2	1

Source: Data from surveys.

Table 15.I.10
Origin of Technology

FACTOR ANALYSIS		
Components	% Variance	Cumulative %
Company headquarters	58.023	58.023
Client or final assembly	18.308	76.331
Suppliers	10.982	87.312
Machinery suppliers	7.716	95.028
Technological partner	2.808	97.837
Universities	1.536	99.372
Firms in the same group	.532	99.904
Local clients	9.579E-02	100.00

Source: Data from surveys.

APPENDIX II: AUTOMOTIVE INDUSTRY

Table 15.II.1
Regional Share in Mexico's Production in Parts of Vehicles, 1993-1997

(Thousands of pesos 1993=100)				
			Gross Production Value	
	1993	%	1997	%
North				
Baja California	84,131	0.36%	104,033	0.34%
Coahuila	3,023,941	13.05%	5,798,332	19.10%
Chihuahua	19,801	0.09%	22,140	0.07%
Durango	846,326	3.64%	232,766	0.77%
Nuevo León	1,272,282	5.48%	1,339,918	4.41%
San Luis Potosí	652,582	2.81%	985,649	3.25%
Sonora	187,562	0.81%	224,937	0.74%
Tamaulipas	14,438	0.06%	15,497	0.05%
Subtotal	1,804,840	7.77%	1,585,022	5.22%
Study's Automotive Region				
Coahuila	3,023,941	13.05%	5,798,332	19.10%
Nuevo León	1,272,282	5.48%	1,339,918	4.41%
Subtotal	4,296,223	18.53%	7,138,250	23.51%
Center				
Aguascalientes	2,621,287	11.29%	2,688,195	8.85%
Colima	337,508	1.45%	818,404	2.70%
Guanajuato	21,660	0.09%	17,246	0.06%
Hidalgo	127,574	0.55%	56,359	0.19%
Jalisco	439,742	1.89%	462,251	1.52%
Puebla	2,343,297	10.09%	3,263,981	10.75%
Querétaro	1,379,683	5.94%	1,889,658	6.22%
Tlaxcala	422,046	1.82%	365,745	1.20%
Subtotal	7,692,797	33.12%	9,561,839	31.49%
Mexico City Metropolitan Region				
Mexico City	2,605,575	11.22%	4,183,424	13.78%
State of Mexico	5,847,074	25.18%	6,311,650	20.79%
Subtotal	8,452,649	36.40%	10,495,074	34.57%
South East				
Campeche	6,913	0.03%	6,265	0.02%
Yucatán	963,062	4.15%	1,575,187	5.19%
Subtotal	969,975	4.18%	1,581,452	5.21%
Total	23,216,484	100.00%	30,361,638	100.00%

Source: INEGI, 1993 and 1997.

Table 15.II.2
Growth of Parts of Vehicles, 1993-1997

Region	1993	1997	1993-1997
Gross Production Value (Thousands of pesos 1993=100)			
North			
Baja California	84,131	104,033	5.45%
Chihuahua	19,801	22,140	2.83%
Durango	846,326	232,766	-27.58%
San Luis Potosí	652,582	985,649	10.86%
Sonora	187,562	224,937	4.65%
Tamaulipas	14,438	15,497	1.79%
Subtotal	1,804,840	1,585,022	-2.00%
Study's Automotive Region			
Coahuila	3,023,941	5,798,332	17.67%
Nuevo León	1,272,282	1,339,918	1.30%
Subtotal	4,296,223	7,138,250	18.97%
Center			
Aguascalientes	2,621,287	2,688,195	0.63%
Colima	337,508	818,404	24.79%
Guanajuato	21,660	17,246	-5.54%
Hidalgo	127,574	56,359	-18.47%
Jalisco	439,742	462,251	1.26%
Puebla	2,343,297	3,263,981	8.64%
Querétaro	1,379,683	1,889,658	8.18%
Tlaxcala	422,046	365,745	-3.52%
Subtotal	7,692,797	9,561,839	5.59%
Mexico City Metropolitan Region			
Mexico City	2,605,575	4,183,424	12.57%
State of Mexico	5,847,074	6,311,650	1.93%
Subtotal	8,452,649	10,495,074	5.56%
South East			
Campeche	6,913	6,265	-2.43%
Yucatán	963,062	1,575,187	13.09%
Subtotal	969,975	1,581,452	13.00%
Total	23,216,484	30,361,638	6.94%

Source: INEGI, 1993 and 1997.

Table 15.II.3
Location Factors: Automotive Industry Companies

FACTOR ANALYSIS		
Component	% Variance	Cumulative %
Proximity to the OEM	45.439	45.439
Workers' skill level	19.718	65.157
Labor cost	17.309	82.466
Transportation cost	8.264	90.731
Government Incentives	5.192	95.922
Proximity to university and R&D centers	2.879	98.801
Proximity to local suppliers	1.350	99.991
Rules of origin compliance	4.404E-02	99.995
Other NAFTA factors	5.056E-03	100.000

Source: Estimations based on surveys.

Table 15.II.4
Companies in the Automotive Industry: Saltillo-Ramos Arizpe-Monterrey Region

Company	Year of Foundation	Date of Restructuring	Foreign Investment Participation	Position in Supplier Chain	Main product
Firm 11	1979	1997	50%	First tier	Engine heads
Firm 12	1981	1995	20%	First tier	Engine heads aluminium
Firm 13	1966	1998	None	First tier	Brake, metal stamping and finishing
Firm 14	1962	1985	38%	First tier	Automotive glass
Firm 15	1995	_	100%	First tier	Pistons
Firm 16	1991	1999	100%	First tier	Metal stamping
Firm 17	1999	_	100%	Second tier	Assembled components for transmissions, engines, steering and braking systems
Firm 18	1968	1991	100%	Second tier	Filter
Firm 19	1996	_	None	Second tier	Assembled components
Firm 20	1956	1990	40%	Second tier	Combustion systems, suspensions
Firm 21	1956	1995	40%	First Tier	Metal stamping chassis

Source: Data from surveys.

Table 15.II.5
Synchronized Production and Delivery Practices (JIT)

FACTOR ANALYSIS		
Component	% Variance	Cumulative %
Kanban	7.264	67.264
JIT Clients	25.212	92.475
JIT Delivery	7.525	100.000
Programming	6.858E-16	100.000

Source: Data from surveys.

Table 15.II.6
Impact of the Synchronized Production and Delivery Practices

FACTOR ANALYSIS		
Component	Variance %	Cumulative %
Quality	36.299	36.299
Dead time	22.080	58.379
Inventories	18.736	77.115
Waste	9.049	86.164
Work force	7.815	93.979
Safety	3.222	97.201
Environment	2.149	99.350
Production systems	.634	99.984
Raw materials	1.617E-02	100.000

Source: Data from surveys.

Table 15.II.7
Workers Responsibilities in Quality Control

Firm	Quality	Kaizen	Preventive Maintenance
Firm 11	-	✓	✓
Firm 12	✓	✓	✓
Firm 13	✓	✓	✓
Firm 14	✓	✓	✓
Firm 15	✓	✓	✓
Firm 16	✓	✓	✓
Firm 17	✓	✓	✓
Firm 18	✓	✓	✓
Firm 19	✓	✓	✓
Firm 20	✓	✓	✓
Firm 21	✓	✓	✓
Total	10	11	11

Source: Data from surveys.

Table 15.II.8
Flexible Organization of Work

FACTOR ANALYSIS		
Component	% Variance	Cumulative %
Work teams	60.640	60.640
Quality c. Responsibilities	24.246	84.886
Job shifting	10.222	95.109
Suggestion schemes	4.891	100.000
Maintenance responsibilities	3.454E-15	100.000
Production responsibilities	3.551E-16	100.000

Source: Data from surveys.

Table 15.II.9
Exports and Location of Clients

Firm	Exports %	Mexico	United States	Canada	Latin America	Europe	Asia
Firm 11	15	✓	✓	-	-	-	-
Firm 12	90	✓	✓	✓	-	-	-
Firm 13	90	✓	✓	-	-	✓	-
Firm 14	40	✓	✓	-	✓	-	-
Firm 15	15	✓	✓	-	-	-	-
Firm 16	0	✓	-	-	-	-	-
Firm 17	15	✓	✓	-	-	-	-
Firm 18	95	-	✓	-	✓	✓	-
Firm 19	0	✓	-	-	-	-	-
Firm 20	73	✓	-	-	-	-	-
Firm 21	40	✓	✓	✓	-	-	-
Total	43	10	8	2	2	2	-

Source: Data from surveys.

Table 15.II.10
Location of Main Suppliers

Firm	Mexico	United States	Canada	Latin America	Europe	Asia
Firm 11	✓	✓	-	-	-	-
Firm 12	✓	✓	-	-	✓	-
Firm 13	✓	✓	-	✓	-	-
Firm 14	✓	✓	-	-	-	-
Firm 15	-	✓	-	-	-	-
Firm 16	✓	✓	-	-	-	✓
Firm 17	-	✓	✓	-	-	-
Firm 18	-	✓	-	-	✓	-
Firm 19	✓	-	-	-	-	-
Firm 20	✓	-	-	-	-	-
Firm 21	✓	✓	✓	-	-	✓
Total	8	9	2	✓	2	2

Source: Data from surveys.

16

A Network of Knowledge-Intensive Clusters for Regional Development: The Paraná W-Class Program

Carlos Quandt and Luiz Márcio Spinosa

INTRODUCTION

Competitiveness and growth in the world economy are directly linked to both the development of local innovation capacity and the ability to participate in widening networks of information and production resources. The global economy is driven by the rapid pace of technological change, the increasing importance of knowledge-intensive industries and occupations, and the globalization of production, markets, information and capital. The concept of a globalized, knowledge-based economy has led to the realization that knowledge and advanced skills are fundamental strategic resources and the essential drivers of productivity and economic performance (OECD, 1996). This context calls for the design of strategies, policies and institutions to support a new model of development based on innovation, the promotion of knowledge diffusion and the development of world-class industries, technologies and products. How to create such policies and institutions is a major challenge for Latin America and the Caribbean (LAC) and for developing areas in general. This study outlines the components of a regional development strategy that seeks to face this challenge through the creation and expansion of regional innovation clusters in LAC. It is focused on the relevance of two key concepts for SME (small and medium-sized enterprise) development—*innovation clusters* and *cooperation networks*—recognizing that both are emerging as significant tools to promote regional development, foster SME growth, reduce spatial and social inequalities, and to activate, diffuse and expand locally generated knowledge.

This approach is applied to the specific case of the *Paraná World-Class Program for Software, E-commerce and E-business*, which is being implemented in the State of Paraná, in Southern Brazil. The program, known as "W-Class", is a joint initiative of local universities, government and business associations to plan and implement actions to promote the development of software and electronic commerce in Paraná. More broadly, the study aims to generate knowledge about SME linkages, clusters and networking, particularly in the context of developing countries. The integration with public- and private-sector participants implies a great potential impact on policy formulation and implementation, as well as on the application of the resulting experience in other settings. The action-oriented nature of the project means that its ability to generate practical results will be constantly monitored and evaluated, and consequently, its ability to impact development processes at the local and regional level.

OBJECTIVES

The overall goal of this study is to generate knowledge to promote regional development by accelerating the growth of small and medium-sized software firms—increasing their capabilities to world-class levels, sharing lessons learned and leveraging their access to knowledge, skills, technology, capital and markets. More specifically, the concepts will be applied to the case of emerging innovation clusters in Paraná, aiming to produce research results that will help not only to understand, but also to accelerate the growth of regional clusters. The project also aims to understand the impact of electronic and personal networks as tools to expand and complement the capabilities of local firms and clusters by sharing knowledge with other emerging clusters and more advanced regions.

The experience of industrialized countries and the limited evidence available from the developing world show that much work is still needed to clarify the constraints and opportunities that industrializing countries face in trying to develop high-technology clusters, to master advanced technologies and to reach global markets. The broader research outcomes that are expected from this ongoing project include:

- a better understanding of the key elements of innovation clusters and how they function in this regional context;
- an assessment of the overall impact of innovation clusters on regional development, and their specific impacts with regard to several regional indicators;
- a better understanding of which activities are more effective to build and sustain innovation clusters;
- a better understanding of how some elements of innovation clusters in LAC can be supplied by virtual networks;

- an assessment of the barriers and facilitators to sharing knowledge with other emerging clusters in LAC and more developed innovation clusters in other regions;
- improved research skills and research capacity within the region, contributing to reinforce the institutional role of the local universities, and also to strengthen university-industry-government cooperation.

THE GLOBAL CONTEXT: CHALLENGES OF THE KNOWLEDGE-BASED ECONOMY

The microelectronics revolution and particularly the rapid growth of the Internet and the World Wide Web are providing extraordinary opportunities for the exchange of information and knowledge. Many governments and development agencies recognize the potential of these networking tools to enable developing countries to catch up rapidly with more developed ones. Networks are emerging as significant tools of social change, expanding opportunities for information and connectivity, and erasing the boundaries for research, education and business. Access to global information resources has become an essential condition to maintain international competitiveness and develop a knowledge-based society.

At the same time, LAC's economic structure has undergone dramatic changes, with policy measures that emphasize the role of the private sector and market mechanisms to promote economic development, in parallel with a shrinking role of the state. As a result, during the last decade most Latin American countries have experienced a strong process of industrial restructuring. After decades of protectionism, the old import-substitution industries are giving way to a new wave of foreign investment. For example, Brazil has become a major recipient of direct foreign investment, with inflows of US$16 billion in 1997 and US$28 billion in 1998. International trade in LAC has also increased quickly over the past few years, fueled by economic stabilization plans in Brazil and Argentina, the liberalization of trade and investment policies and the establishment of trading blocs such as NAFTA and Mercosur.

Despite the region's fairly strong performance in terms of trade, investment and growth, serious doubts have been raised on the sustainability of a development process that is still strongly centered on resource-based exports. This specialization in commodity exports contrasts with the fact that world trade in manufactured goods has been much more dynamic than primary exports. One major factor has been the impact of the information technology revolution on the composition of world merchandise trade. In the first half of the 1980s, office and telecom equipment accounted for 5% of world trade (only 33% of the share of agricultural products). By 1995, the share had increased to 12%, which is slightly higher than the share for all agricultural products (WTO, 1996).

The fundamental feature of international competition is becoming the mastery of skills and know-how. The link between innovation, growth and employment appears to be characteristic of the leading economies—that is, those which invest in education, training, research, innovation and new technologies.

It is apparent that the ability to achieve and sustain high growth rates will require a concerted strategy in LAC to add value to domestic products. This in turn will require efforts to improve their human capital endowments and, more generally, to improve their technological base. However, expenditures in science and technology in LAC lag far behind developed countries' expenditure levels. In addition to low levels of R&D investment, the private sector has had a limited participation in R&D efforts, leading to an inefficient utilization of such investments. This is reflected in the limited capacity of LAC countries to convert technological developments into industrial applications and commercial products.

The current situation in LAC raises many concerns about its future development prospects, because the ability to develop technologically advanced industries and products is now essential to sustain international competitiveness and to build a knowledge-based economy. The globalization process has entailed the diffusion of world-class standards of quality, productivity and efficiency, the rise of global products, process and markets, and the reduction of barriers of time and distance, with increasingly free flows of technology, capital and information. The impact of technological change on competition is closely associated with new forms of industrial organization, management of production and inter-firm relations.

At the same time, in the more complex industries, there has been a greater emphasis on nonprice factors such as technological and organizational innovations that enable the firms to reduce product development cycles and delivery times, and to increase quality and flexibility. This means that developing-country firms are in turn pressed to achieve equally efficient and flexible means of production in order to compete in the global economy.

CONCEPTUAL FRAMEWORK: CLUSTERS AND NETWORKS

The framework is centered on the relevance of two key concepts for SME development: *innovation clusters* and *cooperation networks*. In the cluster/network-based approach, both concepts are joined by a focus on interactive learning and the diffusion of different types of knowledge: tacit/codified, scientific/practical and so on in different spatial and organizational settings. That also implies a focus on the emerging field of knowledge management, that is, the explicit and systematic management of knowledge and its associated processes of creation, organization, diffusion and applications to create wealth and promote development. The research is guided by some basic assumptions:

1. *Small and medium-sized enterprises* can play a key role in triggering and sustaining economic growth and equitable development in LAC and other developing regions as well.
2. The creation of *technology-intensive firms* is essential to build local capabilities to compete in the global economy; they are also essential to strengthen academic-industry-government linkages and encourage technological innovation.

3.   The regions' development potential can be greatly enhanced by adopting a *cluster/ network-based approach* to address its development needs and spatial imbalances, searching for cooperation and partnerships among different government levels, the private sector and international organizations.

4.   In order to overcome the isolation and lack of the required institutions, skills and R&D that prevail in individual localities, *information and communication technologies (ICT)* are a key element to establish linkages within each area, among them and with strategic partners outside the region. As their individual capabilities are expanded and interlinked, they become a collective asset to sustain LAC's path to knowledge-based development.

SMEs, particularly technology-based ones, have a tremendous potential to accelerate economic growth, expand their share of exports and promote a more deconcentrated and equitable pattern of development in developing countries. However, this potential role is often not fulfilled because of their small scale. As Ceglie and Dini (1999) point out, SMEs are often unable to capture market opportunities that require a large scale of production. They are also unable to achieve economies of scale in the purchase of inputs (such as equipment, raw materials, finance, consulting services, etc.), and the creation of an internal division of labor that could foster cumulative improvements in productive capabilities and innovation.

Small size also constrains the internalization of dynamic functions such as training, market intelligence, logistics and technology innovation. Even innovative technology-based firms tend to lack key skills and resources, such as marketing or business capabilities. Berry (1997) notes that small firms' limitations typically fall in the areas of access to technological information, and guidance on quality control; access to finance; assistance in purchase of materials or equipment, in workplace organization, in financial management or in other determinants of effective performance; and market stability (security of demand over a period of time). More important, small-scale entrepreneurs in developing countries are often ill prepared to look beyond the boundaries of their firms and capture new market opportunities.

It is widely acknowledged that interfirm cooperation and linkages involving SMEs in a developing economy may have a strong impact on growth and distribution performance, as demonstrated in the successful development of East Asian countries, beginning with Japan, but also including Korea, Taiwan and others. Emilia Romagna, the Italian region most noted for its industrial districts, had the fourth largest increase in per capita income (14%) in the country between 1963 and 1984 (Pyke, 1995). The development of networks can improve the competitive position of SMEs and reduce the problems related to their size through mutual help. For example, firms may establish a localized network to become more specialized and complement each other's capabilities by sharing resources, pooling together their production capacities and purchasing power, thus achieving scale economies to conquer markets beyond their individual reach (Pyke, 1992). Some aspects of SME support (especially credit provision) have evolved considerably, but "linkage-inducing" policy remains largely a new and experimental area. As Berry (1997) notes, "the challenge for policy in this area is to understand the source of potential payoff to increased inter-firm co-operation, the contexts which facilitate it, and the potential instruments to induce it".

Horizontal cooperation and the creation of external economies among SMEs in clusters contribute to generate competitive advantages through "collective efficiency". Schmitz (1995) emphasizes that "external economies are essential to growth but not sufficient to ride out major changes in product or factor markets; that requires joint action". The advantages of cooperation among SMEs are usually connected with collective economies of scale, the benefits of dissemination of information and interfirm division of labor. These benefits tend to increase when transaction costs are low, and these in turn tend to decrease with geographic proximity and the establishment of shared infrastructure, common norms and tacit rules for cooperation.

INNOVATION CLUSTERS

The term "innovation cluster" is used to indicate a sectoral and geographical concentration of firms and other economic agents which gives rise to external economies and favors the creation of specialized technical and financial services as well as public and private local institutions to support local economic development. This type of arrangement facilitates collective learning and innovation through implicit and explicit coordination (Humprey and Schmitz, 1995). Successful clusters depend on both the private and the public sector (usually universities and research institutions), which join efforts to create innovative environments and to build synergies among agents with complementary capabilities. Their development is gradual and cumulative: over time, the region builds knowledge, skills, institutional support structures, specialized services, financing arrangements, infrastructure and collective norms of cooperation and mutual trust.

Clusters are built on linkages and relationships that integrate the isolated technological capabilities of institutions, firms and individuals into a collective, territorial asset. The establishment of mechanisms to coordinate efficiently these relationships is essential to create a supportive environment for many forms of technical interchange, cross-fertilization, risk-sharing and collective learning. This is essentially a territorially based process, as people who share the same space discover the advantages of "learning by interacting".

As Bianchi (1993) points out, the crucial characteristic of the "Marshallian" type of cluster or *milieu* is the set of competitive and collaborative linkages among agents in a socially and historically defined agglomeration, complemented by a set of collective intangible assets that belong to the production system as a whole. The cluster benefits from its complex web of interactions because innovation rarely happens in isolation. It is an experimental, trial and error activity, and each agent may draw innovation inputs from a wider matrix of institutions to take advantage of a division of labor in the generation of knowledge and skills (Metcalfe and Georghiou, 1997). In that sense, the cluster improves "dynamic" efficiency (or innovative capability) by reducing uncertainty through information sharing and screening, and by establishing a durable relational basis for the construction of competences (Camagni, 1995).

Several generic "locational ingredients" or factors are usually associated with high-tech cluster development. Two factors may be considered as necessary but not sufficient conditions for a successful cluster. The first is a "critical mass" of human resources, including entrepreneurs, scientists, engineers, technicians and skilled labor. The second is a capable scientific and technological infrastructure, or the "knowledge assets" of a region. These may include universities, public and private research labs, libraries, technological incubators, innovation centers and science parks. The main roles of these anchor institutions are to promote technology transfers and to support networking.

Other generic locational ingredients that are often associated with technology-intensive clusters include the *business infrastructure,* including institutions such as industrial associations, chambers of commerce and development agencies; *financing opportunities* through the availability of seed, venture and investment capital, in addition to grants for training and R&D, and government offices providing a wide range of business support services within the area; *physical infrastructure* such as transportation (highways, rail, airports), communications, water and power; *quality of life factors,* or the perceived benefits that certain locations offer to entrepreneurs and the upper segments of technical-scientific workers, such as pleasant residential areas, parks, recreational facilities and absence of pollution; *a diversified economic base,* comprising supplier and distribution networks, specialized services; *a favorable business climate*, usually meaning a reduced cost of doing business due to low tax levels, limited labor union activity and also other costs such as prevailing wages, housing, food and transportation. Another component is the existence of government incentives, as well as low cost of infrastructure and loans for start-ups.

A second set of high-tech cluster factors refers to less tangible governance or organizational elements, which are expressed mainly by linkages and relationships. Although most of the literature tends to highlight the physical aspects of science parks and technopoles, this second set of elements of high-technology regions points to an intangible process that is much harder to grasp: the gradual buildup of relationships, informal norms of mutual trust and cooperation, and intense exchanges of information among entrepreneurs and scientists. It comprises (Voyer and Roy, 1996; Quandt, 1997): *the existence of champions*: political and academic leaders—either individuals or regional governments—who ensure determination and tenacity in defining and pursuing objectives; *recognition of the potential* that technology-based industries offer for regional development, and also actions to identify and take advantage of regional assets; *a broad support base* for a common development goal in the region from different government levels—including procurement, research grants, regional development aid, and so on—as well as from the community, unions and other local organizations; *an entrepreneurial culture*, which is widely perceived as essential to create a dynamic business cluster; *strong linkages between the scientific and the entrepreneurial community*, and the establishment of a mutual commitment to partnerships and negotiated agreements; *information networks,* comprising formal and informal contacts as well as wider scientific, technological and business networks. Finally, there is a component of *marketing and image building:* the promotion of the region's innovative image is often seen as an important strategy to attract and retain new public and private investments as well as skilled workers and entrepreneurs.

REGIONAL AND SUPRATERRITORIAL NETWORKS

The network model is becoming increasingly dominant in modern productive sectors, not only for companies but also for institutions in the area of governance and development. The importance of long-distance linkages and on-line data exchange is growing quickly everywhere. The rise of concepts such as the networked organization and the "virtual enterprise"—which may comprise, for example, transitory teams of freelance or temporary workers organized in flexible ways to develop a specific project—has challenged directly the traditional place-bound, centralized notion of organization of production.

In a network, information is transmitted horizontally, reciprocally and iteratively, rather than following a rigid hierarchy. Hence innovation and competitiveness depend on the ability to integrate different kinds of information and to coordinate them among the different agents, types of activities and firms. As networks evolve and become more sophisticated, a *learning process* emerges through cooperation, together with increased reliability and trust. These elements constitute a shared intangible asset that helps to reduce both certain and hidden costs of the interaction among the partners as well as the probability of opportunistic behavior (Bianchi, 1993).

Cooperation networks enable firms to position themselves in the trade-off between market-related transaction costs and the high costs related to internal development of know-how. They create opportunities to reach global markets, absorb new technologies, develop joint projects and share human and material resources. Even though high costs and risks are integral aspects of the network form of organization, it is particularly suitable for coping with dynamic processes such as systemic innovation and control over future technological trajectories.

The process of clustering is similar to the network model in the sense that both are technological learning systems that help to socialize innovation-related knowledge and reduce uncertainty in the environment in which innovative agents operate. The territory of an innovative cluster is an active resource for learning through intense interaction involving a broad set of actors. A firm's ability to create knowledge is strongly related to its interaction with related firms in a process of collective learning, involving exchanges of partly codified and partly tacit knowledge.

The local system may be seen as a locus of integration of tacit (contextual) and codified knowledge, and as an instrument to connect knowledge to production. However, it should not be seen as a closed system, but as part of a global circuit of knowledge production and learning. Thus competitiveness is shaped in interaction with market structures that may range from the local to the global, with industry-specific technological trajectories, and with regionally specific resources, structures and institutions.

In this broad process of interaction, local, regional and national systems of knowledge creation retain a key role in the global economy (Nelson, 1993). Regional and national systems of innovation may be described as positive externalities that are very difficult to replicate. The sustainability of localized competitive capabilities indeed indicates the existence of strong barriers that prevent immediate or costless imitation of the institutional endowment that exists in successful regions. In addition to the factors mentioned above, the most important barrier to imitation is probably the stock of research- and experience-based knowledge, skilled people and infrastructure that some regions have already accumulated. They are better positioned to generate innovations and accumulate further knowledge than the regions that have not yet achieved such critical mass.

However, path-dependent processes of knowledge creation within closed systems can make regional capabilities deteriorate over time, as the collective assets may erode or become outdated. The process of cumulative learning within innovative milieux may require a readiness for radical change when needed to restore regional competitiveness. At the very least, they must remain open for renewal and the creation of new capabilities by absorbing inputs of external energy through the establishment of external linkages, such as through participation in wider networks.

Isolation is a great problem for most developing-country firms, particularly small ones. Technologically advanced regional systems, and industrial clusters more generally, depend on the development of territorially based networks. However, regional innovation and growth are not restricted to local sources of knowledge, capital or other factors. Supraterritorial networks, particularly with the help of advanced *information and communication technologies*, expand enormously the spatial scope and the range of opportunities for firms in any given cluster. Hence a cluster/network-based model is particularly promising for LAC precisely because few places in the region have been able to bring together a minimum set of elements to foster the creation of innovative clusters, or to keep them working efficiently.

In sum, the focus on the mobilization of local assets to achieve regional competitive advantages must be matched by a broader focus, on the ability to join increasingly wider spatial networks and to develop alliances, partnerships and opportunities with outside firms and investors as well as incubators, universities and research institutes. This represents a major shift from the established notion of the territorially based concept of innovative cluster, in which local linkages and face-to-face interactions have been always seen as vital components. However, a network of virtual linkages cannot replace the crucial role of person-to-person contacts, which are generally built on territorially defined norms and relationships of trust. Rather than being mutually exclusive, the two processes (the cluster-based personal contact and the wide network linkage) can complement each other. That is, the local system may interact with a wide range of other nodes or levels through the intermediation of actors that belong simultaneously to the cluster and to supraterritorial networks. These actors then represent a key element of the local system and, at the same time, a vital linkage to external resources and different territorial systems. Much research is still needed to explore the ways in which a cluster may take advantage of wider networks by becoming part of a "virtual region" while preserving its identity and its ability to foster a self-reinforcing process of innovation and growth.

"VIRTUAL CLUSTERS": LEARNING AND INNOVATION POLES

Gibson et al. (1998) have developed some basic principles on how to implement "virtual clusters" with the concept of the Learning & Innovation Pole (LIP). It is operationalized at the most basic level as an SME that is linked to other SMEs in a global network and also has access to a range of support activities such as training programs, workshops and mentoring activities. The application of the concept is aimed at small and mid-sized technology-based enterprises that might be considered relatively successful at the local level but are in need of assistance (e.g., talent, technology, capital and know-how) to achieve accelerated growth and global market penetration. At the moment that these virtual networks are created around the SME, it becomes a Learning & Innovation Pole. Rather than relying on a well-defined geographic area to provide all of the networks and services required for success in knowledge-based economic efforts, LIPs rely on regional and global cooperative and collaborative networks and training programs to provide service and assistance on a real-time, as-needed basis. The LIP network concept is focused on six functional objectives:

1. *Networking for Markets:* The identification of markets and successful execution of marketing strategies is a determining factor in the success and sustainability of small and medium technology-based firms.
2. *Networking for Capital:* Access to adequate financing is one of the most critical factors for the success of technology-based firms. Broad public/private/NGO partnerships could be established to offer integrated access to services such as financial planning, support for obtaining grants, opportunities for access to venture, development and seed capital.
3. *Networking for Interfirm Linkages:* A networked approach is ideal to maximize the impact of programs and projects, such as partnerships, alliances and linkages to outside suppliers. Clusters in developing countries tend to rely heavily on the local supplier base, which may become insufficient for their rapidly growing needs. Careful coordination is required to ensure that local suppliers are able to match increases in demand so that jobs may be retained and created and that other substitute supply streams can be brought online as required.
4. *Networking for Technological Support:* Electronic networks are extremely useful tools to diffuse the benefits of technological support, providing services such as technology assessment and forecasting, technology gateways (assistance on technological choices and on marketing assessment of innovative projects) and access to outside technical information. These services could also be concentrated in one or a

few centers and could be provided by public agencies, private consultants and business associations.

5. *Networking to Expand Access to Technology Transfer Opportunities:* The use of electronic networks for technology transfer is already being established in several places to stimulate investment in science and technology, research and development, technology transfer, development of commercial potential of R&D and spin-offs. Networks are necessary tools to facilitate access to technology transfer opportunities worldwide.

6. *Networking for Talent and Know-How:* SMEs often do not have and cannot afford the entire range of technical and business talents and know-how required for success in local and global markets. The process of identifying and hiring such talent and know-how on a short-term, as-needed basis is also difficult for smaller enterprises. Networks of talent would be a great asset, allowing SMEs access to the experts at affordable rates and opportune moments.

While established SME firms are the primary initial focus of LIPs, there is a critical need to develop the infrastructure and resources of the region to promote accelerated development of knowledge based firms from the bottom up as a longer term strategy. The basic need is to improve the process of knowledge transfer, acquisition, absorption and diffusion. The issues involved include basic and higher education, physical infrastructure construction and improved policy environments. For example, if knowledge acquisition—whether imported from abroad or created at home—is to lead to economic development, it must be absorbed and applied. This requires universal basic education and opportunities for lifelong learning (World Bank, 1998). The extent and economy of modern ICT greatly expand the potential for both the acquisition and the absorption of knowledge, but this can happen only after a basic level of telecommunications infrastructure is acquired. The following are potential areas for action (Quandt, 1998):

- *Creating and Strengthening Local Cluster Governance Structures:* The first step is the creation of an organizational and functional structure for the local cluster, preferably leveraging existing groups and associations. This would involve both private and public sector participants. The establishment of linkages with other clusters and incubator managers will enable a better understanding of stakeholder needs and markets and will improve organization methods. The creation of a permanent, dedicated business and technology information network would make communications more continuous and interactive, rather than sporadic exchanges that normally occur only at periodic meetings.

- *Determining Educational Needs and Offering Training:* Based on regional descriptive profiles and targeted interviews with local stakeholders, educational requirements for the LIPs and targeted companies can be ascertained. Courses could then be offered through local workshops as well as via the Internet to help improve the skills of local trainers.

- *Fostering Personnel Exchanges:* Visits of key personnel among regions in the network would greatly facilitate knowledge, technology and know-how transfer. Exchanges of students, faculty, and entrepreneurs facilitate these processes.

- *Building Local Skills and Training IT Specialists:* In order to incubate a local cluster that will depend heavily on virtual linkages, a comprehensive adaptation to the IT paradigm is crucial. The development of new types of specialists will be needed, including Technology Brokers, Research Experts, Information and Technology Guides and Animators. This type of program would benefit greatly from the experience and resources of universities and research institutes in more developed countries. At a more general level, a skilled workforce is one of the most important localization factors for technology-based companies, and a major constraint to the development of knowledge-intensive clusters in many LAC nations. This characteristic is essentially place-based, yet virtual technologies may boost the development of human resources in more remote locations through training centers, distance education, career planning, virtual job markets, and also support business development through the establishment of virtual entrepreneur schools providing all kinds of training— technical, managerial, marketing, etc.

- *Optimizing and Sharing Facilities:* For each region, the required facilities for a viable cluster could be kept to a minimum, provided they are integrated into a shared system. The operational support infrastructure could be optimized and many facilities could be shared over the network, including incubators, prototype centers, pilot plants, online library, test laboratories, and online conferencing facilities.
- *Supporting Electronic Commerce:* Electronic commerce is quickly becoming an essential business tool in an increasingly integrated world economy. The development of e-commerce capabilities for SMEs is much more feasible when implemented at a larger scale. It depends to a great extent on government policies and measures to support electronic commerce, addressing issues such as privacy, security, and consumer protection. Before it faces these obstacles, however, a successful cluster must build the skills and knowledge to enter the field.

APPLICATION: REGIONAL CLUSTER DEVELOPMENT INITIATIVE

The conceptual framework has been linked to the W-Class Program (Paraná World-Class Program for Software, E-commerce and E-business), that is being implemented in the State of Paraná, Brazil. W-Class is a set of coordinated actions from state and local governments, private enterprises and higher education institutions to promote global competitiveness among local software companies. Its actions are focused on three emerging software clusters: (1) Main Corridor (encompassing the cities of Curitiba and Ponta Grossa) which is fairly developed with some exporting companies, (2) the incipient North Corridor (Maringá and Londrina), and (3) the yet-to-be developed West-Southwest Corridor (Cascavel, Foz do Iguaçu, and Pato Branco). The Program is promoted by the State Secretariat for Science, Technology and Higher Education, and it was made possible through the establishment of broad partnerships involving state universities, several state and local agencies, private companies and business associations. W-Class comprises 30 different subprograms or lines of action that range from education, training and seminars to support for access to partnerships, capital and markets. The stated goal is to double Paraná software exports by the year 2004, and double the number of software companies in Paraná over the same period.

The W-Class program was designed with elements of the world-class model from Kanter (1995). The world class concept suggests (1) the need to meet the highest standards of competitiveness to reach a global market, and (2) the growth of a social class that is able to command assets and to operate beyond regional and national borders. The application of the model is aimed at increasing exports, creating jobs and increasing the number of firms in connection with the new business opportunities that are generated by the program.

The implementation of W-Class is focused on intangible assets—that is, world-class *concepts* (knowledge and know-how, which require investments in innovation and entrepreneurship), *competence* (with emphasis on learning and on quality and competitiveness) and *connectivity* (emphasizing the creation of linkages, networking and collaboration, in addition to access to global resources and markets). Unlike the original model, the local program also adds the need to improve access to *capital*, which is a critical obstacle to SME development in the region. Each of the model's assets have corresponding processes and specific actions associated with them. The processes are: Innovation, Entrepreneurship, Quality, Learning, Collaboration, Networking and Funding. Depending on the nature of the asset, the actions are directed to the enterprises, human resources, market, government, entities and the community.

The strategy of linking a research and evaluation component to the W-Class initiative—which was originally focused on production and commercialization—aims to strengthen the initiative and contribute to improve the capabilities of SMEs. The research focuses on issues related to building and growing clusters, based on this in-depth study of the Paraná clusters. Furthermore, the project addresses questions of resource and knowledge sharing (networking) with other Regional Innovation Clusters in Brazil and LAC and with mentoring regions elsewhere. Network links with regional SME development projects in LAC and international mentoring institutions are incorporated into the network and integrated to the processes of knowledge creation and sharing and building institutional and business relationships.

The inclusion of a component of monitoring and evaluation will also help to understand which activities are more effective to build and sustain innovation clusters, and whether the incubation of "protoclusters" can be successful. On a broader level, it will allow an assessment of the importance of innovation clusters for regional development. The networking component will contribute to clarify how Web/Internet tools can supply the elements of innovation clusters that are still lacking in LAC, and to identify the barriers and facilitators to sharing knowledge across regional clusters. The establishment of the proposed learning and innovation network will enable the participants to share lessons learned with other emerging clusters in LAC, and also to leverage access to talent, technology and capital with more developed innovation clusters.

In order to accomplish its objectives, the project has been designed as an action-research plan. In other words, the project's outcomes are expected to include not only research results that will be meaningful in scientific terms, but also actions and products that will have concrete impacts on regional development. The project will not be limited to data collection, monitoring and evaluation of the Program. It will also play an active role in the design and implementation of strategies to foster the development of SMEs, clusters and networks, as well as the creation of tools to integrate knowledge and resources. The first main component of this application of the cluster/network approach comprises a regional assessment, focused on three aspects:

- *regional development indicators* such as demographic data, income, gender-specific data, regional economic structure;
- *business indicators* such as general data (employment, founding date, spin-offs, growth) on existing small, mid-sized and large firms in the region, assessment of emerging clusters, core strengths and assets, competitive advantages in terms of regional, national and global markets;
- *University/public-sector indicators* such as R&D and technical expertise related to current and future businesses' activities in the region, assessment of existing regional innovation systems. Existing institutional data will be used, as much as possible, to conduct a benchmark/scorecard. This benchmark will focus on identifying and leveraging existing capabilities for nurturing clusters, and for networking these poles nationally and globally through the use of ICT and personal networks.

The primary and secondary data will be complemented by interviews and survey of regional leaders (academic, business and government) to facilitate regional understanding by the researchers and support by key local champions. Metrics for success will be regionally focused and will be identified and followed over time. In general these metrics will include:

- *Targeted to specific SMEs:* World-class technology and business assessment, market assessment, intellectual property creation, capital access, profit and growth rates, access to new technology and business processes, time to market, management and employee development.
- *Government and regional development oriented:* enhanced competitiveness and growth of regional economy, job creation, space utilization, capital creation, tax revenues, development of regional infrastructure, exports, impacts on income and gender inequality.
- *Academic oriented:* training of faculty and students, new curriculum development, successful student placement, enhanced relations with community, research and publications, revenue generation (royalties, license fees, etc.).

The knowledge generated in the course of the project is incorporated from its beginning into a network that SMEs and all stakeholders will use to accelerate the growth of their businesses and innovation clusters. This involves the definition of local and outside participants to collaborate on network design (physical and virtual networks), network content development for SME agents and SME end users, as well as electronic commerce design, business models and applications. The network is also designed as a tool to foster the exchange of ideas, disseminate information and promote the project.

The establishment of an operational definition of parameters to create and develop world-class companies and clusters is a basic component of the study, as well as tools to monitor these parameters. It is complemented by an evaluation of the current regional situation region with respect to the target world-class standards, and the forecasting of local needs related to capabilities, skills and employment in software and related services. This will support the identification of needed skills and tools to strengthen clusters, and the design of three related capability-building programs: for companies, for institutions and for human resources in general, targeting education and training needs including distance learning. A monitoring and evaluation program will also be developed, encompassing three subcomponents: the definition of parameters and evaluation of the results of cluster interactions and the "virtual delivery" of some elements through the network, as well as other benefits such as alliances and consortia for SMEs, other potential benefits for universities and research institutions, and for regional development more generally. Finally, a database and knowledge management tool will be developed to support the analysis, discussion and diffusion of best practices, ranging from the firm level to the cluster and the network.

REFERENCES

Berry, A. (1997), *SME Competitiveness: The Power of Networking and Subcontracting.* Washington, DC: Inter-American Development Bank, No. IFM-105.

Bianchi, P. (1993), "Industrial Districts and Industrial Policy: The New European Perspective", *Journal of Industry Studies*, Vol. 1, No. 1, October.

Camagni, R. (1995), "Global Network and Local Milieu: Towards a Theory of Economic Space". In Conti, S., E. Malecki and P. Oinas (eds.), *The Industrial Enterprise and Its Environment: Spatial Perspectives,* pp. 195-214. Aldershot: Avebury.

Ceglie, G. and Dini, M. (1999), "SME Cluster and Network Development in Developing Countries: The Experience of UNIDO", International Conference on Building a Modern and Effective Development Service Industry for Small Enterprises, Rio de Janeiro, 2-5 March.

Gibson, D., Conceição, P., Nordskog, J., Burtner, J., Tankha, S. and Quandt, C. (1998), *Incubating and Sustaining Learning and Innovation Poles in Latin America and the Caribbean.* Montevideo: IDRC.

Humphrey, J. and Schmitz, H. (1995), "Principles for Promoting Clusters and Networks of SMEs". Paper commissioned by the Small and Medium Enterprises Branch, UNIDO, Number 1.

Kanter, R. M. (1995), *World Class: Thriving Locally in the Global Economy.* New York: Simon & Schuster.

Metcalfe, J. and Georghiou, L. (1997), "Equilibrium and Evolutionary Foundations of Technology Policy". CRIC Discussion paper no. 3. University of Manchester: Centre for Research on Innovation and Competition.

Nelson, R. (1993), *National Innovation Systems: A Comparative Analysis.* New York: Oxford University Press.

OECD (1996), "The Knowledge-based Economy", OCDE/GD(96)102, excerpted from the *1996 Science, Technology and Industry Outlook.* Paris: OECD.

Pyke, F. (1992), *Industrial Development Through Small-firm Co-operation.* Geneva: International Institute for Labor Studies.

Pyke, F. (1995), "Comparing Small and Large Firms in Europe: Prospects for Incomes and Working Conditions". Paper presented on behalf of the OECD at the high-level workshop on SMEs Employment, Innovation and Growth, Washington, DC, June.

Quandt, C. (1997), "The Emerging High-technology Cluster of Campinas, Brazil". Paper prepared for International Development Research Centre, Technopolis 97 Conference, Ottawa, September 9-12.

Quandt, C. (1998), "Developing Innovation Networks for Technology-Based Clusters: The Role of Information and Communication Technologies". Paper prepared for the workshop on Tech-regiões: Ciência, tecnologia e desenvolvimento—passado, presente e futuro, Rio de Janeiro, December 6.

Schmitz, H. (1995), "Collective Efficiency: Growth Plan for Small-Scale Industry". *The Journal of Development Studies*, Vol. 31, No. 4, April.

Voyer, R. and Roy, J. (1996), "European High-Technology Clusters". In De La Mothe, John and Gilles Paquet (eds.), *Evolutionary Economics and the New International Political Economy*, pp. 220-237. London: Pinter.

World Bank (1998), *World Development Report 1998: Knowledge for Development.* Washington, DC: The World Bank.

World Trade Organization (WTO) (1996), "World Trade Expanded Strongly in 1995 for the Second Consecutive Year". WTO Press Release, 22 March.

17

The Nature of the Networks of Innovation and Technological Information Diffusion in a Region in the Initial Stages of Industrial Development

Décio Estevão do Nascimento

INTRODUCTION

The problem of regional inequality, whether it be economic, social or technological, is common to all countries independently of their stage of development. The debates surrounding this issue raise questions about differences both between and within countries. In terms of socioeconomic inequality within the same country, one of the examples cited most frequently by institutions and international organisations, because of its size and its consequences, is Brazil. In Brazil, the disparities in income distribution are reinforced by the enormous regional disparities—both economic and social. In the socioeconomic Brazilian panorama, the states in the South have a level of prosperity and equality relatively comparable to that of the developed countries. The states in the North and the Northeast on the other hand, are characterised by limited resources and a strong concentration of riches. In addition, as Droulers (1992: 412) underlines, Brazil, with this "incurable social duality finds itself confronted by a serious growth crisis which risks having durable consequences on the potential scientific-technological development of the country. The technological development does not seem susceptible to reverse the tendency of reinforcement of regional inequalities." The picture described here testifies in effect to the inexistence or the inefficiency of regional development projects. One very simple proposition—though rather simplistic—which is very pertinent is that repeated by Dominique Voynet1: "there are no condemned areas, there are only areas lacking a project".

Without a doubt, the socioeconomic development of a region is directly linked to its technological progress and to the efficiency of the local industry, which in turn "is linked in an intense way to procedure or product innovations that it can generate or absorb," as Droulers remarks (1992: 404). In this context, we can anticipate that regional development is a tributary of local competencies in the creation and/or diffusion of innovation and competencies in the mobilisation of actors and technical objects around an envisaged technological trajectory. However, we cannot forget that "in the economic development dynamic, the innovations are the motor, but it is not only the technical or organisational innovations, it is also the institutional innovations which determine and orientate development" (Corsani, 2000: 366). In other words, it is necessary to envisage organisational, interorganisational and institutional innovations, to increase the learning ability of local actors. As Lundvall (2000: 7) affirms, "we are moving into a *learning economy* where the success of individuals, firms, regions and countries will, more than anything else, reflect their capability to learn. The speed up of change reflects the rapid diffusion of information technology, the widening of the global market place, with the inclusion of new strong competitors, deregulation and less stability in market demand".

It is around the regional problem that this communication develops: the taking into account of the role of local actors in the network of competitive, cooperative, mobilised and coordinated relationships, supporting a socio-economic development project founded on the encouragement of technological development of the local industry. In a certain way, we want to launch the debate on the shape and nature of the relationships of socioeconomic actors of new industrial areas that manage to survive on the scene of exchanges in a globalised economy.

The study set out to reply to the following questions:

- From the point of view of the industrial space of a region in the initial stage of development, what is the nature of the existing networks, for the setting up of new technologies and organisational forms in a globalised economy?
- What is the role of the institutions in the technological and organisational development process of the local enterprises?
- What are the limitations of its networks from the point of view of a technological and economic development project for the region?

We base ourselves on the hypothesis that the interterritory inequalities (national, regional, etc.) of technoeconomic development are due in part to inequalities in the conception and/or coordination and execution of projects involving local enterprises. In other words, the problem of technoeconomic development can be analysed from the societal context of the local enterprise, "that is to say in their relationships with the educational space and the S&T space, but also in their relationships with different actors involved in industrial or corporate strategies or public policies, many interactions which compete to define the industrial space" (d'Iribarne et Gadille, 1998: 6). In our opinion, these different relationships between the three spaces take place in networks which we call *socio-techno-economic*[2]. These networks include the relationships between the different nodes (individuals, organisations, institutions, objects, techniques), involved directly or indirectly in research activities, development, analysis, diffusion and the adoption of technologies. In this approach, we suppose that there is a link between weak technoeconomic development and weak structuration and mobilisation of local socio-techno-economic networks. In others words, between weak technoeconomic development and weak local learning capability.

APPROACH

While aware of the complexity of the aim, using socioeconomics we studied a subject which can explain one part of the overall problem. In our opinion, a piece of research based on network analysis can explain the (dis)functioning of the local process of diffusion and adoption of technologies—essential to the development—in a territory in the initial state of industrialisation. In other words, the analysis of networks involving local actors who communicate and/or supply/acquire goods and services can be relevant for revealing part of the problem. Our approach is interdisciplinary and transdisciplinary. The problem is complex, going well beyond the field of economics. Several theories from different sciences, including sociology, philosophy, anthropology and geography, formed the bases of this study.

We wanted to centre our observations on and from the industrial space, that is to say, that "in which the enterprise is situated and which contributes to its structuration. This includes its industrial organisation, its relationships with its financial backers, its suppliers and subcontractors, customers and finally its relationships with the system of sciences and techniques, as well as with the educational system and its different spaces of training and education, which participate in the construction of its capacity for innovation" (d'Iribarne and Gadille, 1998: 6-8). In our opinion, it is unproductive to make an inventory of all the institutional mechanisms and all the existing organisations in the regional and national systems of innovation in order to understand the level of development of the enterprises. Concretely, we have tried rather to shed light on the elements of the system normally acted upon by the enterprises in the different situations of innovation, highlighting their relational and environmental contexts.

The empirical part of the study was carried out in Tocantins, a Brazilian state inserted in the region of "legal Amazonia". This new geopolitical space was created by the Brazilian constitution in 1988, from the dismembering of more than a third of the State of Goias. In a certain way, this new state can be considered as an integral part of a new frontier of industrial development in Brazil. We use here the notion of frontier in its conventional sense, that is to say, as a "fringe of pioneer development, not totally structured, and in this way a potential generator of new realities" (Becker and Egler, 1992: 382). The access to this field was due principally to our participation in a feasibility study concerning the creation of a university of technology in Tocantins, ordered by the Confederation of Brazilian Industry (CNI) from the University of Technology of Compiègne (UTC). Following this first study, we worked in Tocantins at the heart of a small structure, a *platform*, installed by the Confederation of Brazilian Industry according to the recommendation of the first study. The objective of this small organisation was to discover and develop demands and technological competencies of Tocantins, the diffusion of technological information and the collective mobilisation of the actors in the local development around innovation projects. The strategy was to improve the exploration of the region, by the platform's activities, before taking a decision concerning the pertinence of the creation of a university. In our stay in Tocantins, with the entrepreneurs/ managerial staff, we carried out an in-depth survey of the local food industry. For this chapter we will discuss the results for the milk industry (industrialisation of milk and its derivatives).

NETWORK ANALYSIS

At a time when world attention is turning towards the Internet, the "network of networks" which has radically changed the rules of the game, transforming the socioeconomic status quo around the globe, talking of the phenomenon of "network" is almost commonplace. In the last few decades of the 20th century, the passion for "the network found a very wide echo in the social sciences" (Allemand, 2000: 22). In our opinion, this approach of coordination of interdependencies is useful in the search for answers about the socioeconomic dynamic of an area. The network approach helps to understand the socioinstitutional framing of the process of diffusion and adoption of technologies or technological information in a territorial space which is theoretically not too "connected"—the case of the backward regions.

First, a fundamental notion in the network approach is that of the node. According to Knoke and Kuklinski (1991: 175), a node refers to "a set of persons, objects or events on which the network is defined. These elements possess some attribute(s) that identify them as members of the same class for purposes of determining the network of relations among them." A node also can be viewed in the singular, that is to say it can be an organisation or an individual in the network. In all cases, "each of these nodes will have a different outlook on the world with different objectives" (Swan and Watson, 1998). In the literature, we find other structural and functional representations of the network node. For example, there are the "nodes of social networks," defined as "all places where people meet in an informal way" (Bakis, 1993: 82-83). The author identifies different types of social network nodes:

- "virtually obliged nodes", which are, for example, "the enterprise canteens, the village square";
- "random node", concerning the "interpersonal communications (meeting at the market, in a queue, in corridors), and also the object-person communications (posters, magazines etc.)";
- "permanent node", in other words the "branching on the permanent networks (family, friends, members of associations)".

Often there are also places which are connected to networks not by their node but by their "hubs", which "are exchangers, the means of communication, which co-ordinate in a flexible manner the interaction of all the elements integrated in the network". Otherwise, "the nodes as well as the means are organised hierarchically according to their relative weight at the heart of the network". This hierarchy can modify itself as a function of the activities treated in the network. Certain places can even be disconnected from the network (Castells, 1998: 526). In reality, in our opinion it is not only the function of the network which is fundamental in the definition of its privileged nodes; in fact, the motivation of the nodes would perhaps explain better the existence of actors which are more *connected* and/or more sought after than others. According to Easton (1992) and Swan and Watson (1998), the links between the nodes/hubs in the networks do indeed have a transdimensional character: economic, social, technical, logistical, administrative, informational, legal and temporal. At the level of the *connectivity* of the networks, Bakis (1993: 41) affirms that this characterises the network of relationships of interdependence, "between subsystems of a territorial network, where a strong *connectivity* signifies that the relationships concern the numerous elements of the system considered".

For the concept of the actor, there exist many different definitions. Axelsson (1992: 195), for example, observes that in the field of industrial networks, the actor is frequently identified as a controller of the resources, a judicial unit, a firm or an organisation. Concerning the *autonomy* of an actor in a network, Lemieux (1999: 73-74), repeating Burt, shows that a relationship exists between *autonomy*, the quantity of structural holes[3] in the internal environment, and the quantity of structural holes in the external environment. More precisely, "*autonomy* is greater when actors find themselves in an internal environment where there are few structural holes, faced with an external environment where there are a lot of structural holes". At the same time, the *autonomy* of the actor depends also on their redundant relationships in the network. In this regard, once again Burt, repeated this time by Lazega (1998: 76), points out that the more an actor has unique relationships—nonredundant—in his network, the more chance he has of being trapped by its structure. The *prestige* of an actor is another notion inherent to network analysis. According to Lazega (1998: 46), "the number of the times one has been chosen is the easiest estimation of prestige, since it allows an appraisal of the proportion of the actors who choose the actor *i*. The bigger the index, the more the actor is common".

At the level of its structure, a network is more than a simple addition to bilateral relationships. This can be observed in the attempts to untangle the links between the elements composing a network in view of attributing them weights and values. In this sense, the notion of innovation network is compatible with the systemic approach. A system of innovation, for example, is defined as "the set of economic, social, political, organisational and institutional or other factors which influence, through their relationships the development, the diffusion and the use of innovation" (Edquist, 1997: 14). In this respect, d'Iribarne and Gadille (1998: 4) insist that the dynamics and the particularities of the systems of innovations, or the networks of innovations, are due to their confrontation by what is called "*a triple movement of structuration*:

- the specific (characteristics) linked to the activity sector or to the evolution of generic technologies cannot be neglected, notably from the point of view of the implementation of technological policies aiming at the improvement of the global performances of the system;
- the supranational or international level must be taken into account, for example due to the political and institutional influence of bodies such as the European Union on the process of and the systems of innovation, or due to the influence of the diffusion of technologies between different countries, according to the different mediations (international enterprises, brevets, co-operations, etc.);
- phenomena of structuration at the regional or local level from the policies and strategies of public actors (municipalities, regions, Chamber of commerce and industry, etc.) that private bodies (Enterprises, Associations, foundations, etc.), can generate from specific processes of innovation."

Even though cooperation and competition are two dialectical processes in a network, the network approach places an accent in particular on cooperation, complementarity and coordination. This structure, coordinated by the relationships, procures stability for networks. But a stable network does not mean a static network. The continuity of the interactions between the firms offers, on the one hand, the opportunity for innovation—the major strength in the network—and, on the other hand, the existence of a predictable and known environment in which it can be carried out (Easton, 1992: 22-24).

Lundgren brings us a relevant contribution concerning the relationship between the notions of mobilisation and coordination, and the notion of change, in the perspective of an industrial network. Lundgren (1992: 148-51) suggests the existence of two types of economic changes: continuous and discontinuous. By continuous change the author makes reference to the process of change founded in the existing activities and stabilised structure in the network, whereas discontinuous change is that of a process of change vaguely linked with the existing structure of the network. In summary, the concept of coordination is used to describe continuous changes, and mobilisation of resources is used to characterise discontinuous changes.

A basic hypothesis of networks of relationships is that one part depends on the resources controlled by another, with all winning by the pooling of resources. The "entangling strings" of reputation, friendship, interdependence and altruism become integral parts of the relationship (Powell, 1991: 272). Granovetter, with his notion of "embedded", underlines the influence of the social environment in decision making and economic actions. That is to say, actions cannot be explained uniquely by individual motivation. The author remarks that action is "embedded in ongoing network of personal relationships". In this approach, the network is interpreted as being "a regular set of contacts or similar social connections among individuals or groups," (Granovetter and Swedberg, 1992: 9).

In the same way, Llerena (1997: 369) suggests that "learning by the network of collaborations in particular, allows a sharing of the interpretation of phenomena and of information in view of constituting shared knowledge". Let us note that the "sharing of the interpretation is a consequence of a transformation of the individual representation of the environment which can lead to a modification of objectives" (Brousseau et al., 1997: 419). This constitution can make the network "a mechanism of selection of positive externalities, and elimination (at least in part) of negative externalities, or corrections of market weaknesses (Cohendet and Llerena[4]). This is because it is founded on relationships of trust" (Massard, 1997: 344).

The Actor-Network Theory (ANT), formulated by Callon, Latour and Law, has the merit of highlighting the role of the mediator/translator in the networks. According to Callon (1999: 37-38), the translators, following their objectives, consecrate a lot of energy to integrate, to adapt and to create compatibilities. Without this work, no increase in generality would be possible. The movements, the efforts, the investments and the adaptations that it involves build equivalents, trace the network and give form to the relationships and interactions. Each point remains of course irreducible to all the others, but the passage from one point to another, the action of one point on the other, what we call transmission, becomes possible: the different actors can communicate and understand each other.

In summary, a translation centre has as its mission to mobilise, negotiate, reformulate, calculate and code the set of interests, individual natures and passions, in such a way to make them compatible, understandable and satisfactory for all the other actors in the network. It involves a considerable effort of interest, enrolment, and alliance. According to Callon and Law (1989: 107), "the translation centre, by the interventions that it organises in return, is a *mobilisation centre* of the networks that it translates". In fact, "for an innovation to succeed the network must be capable of being mobilised. The resources must be available in the network and under the control of—or accessible to—the actors with an interest in the success of the innovation" (Easton, 1992: 26).

DATA FROM THE SURVEY IN TOCANTINS

The milk chain in the food industry, on a national scale and at the level of Tocantins, is very important, both for economic and social reasons. Breeding oriented to the production of milk and its industrialisation appears in all the states of the Brazilian Federation, employing workers and guaranteeing an income for a substantial part of its population. In this way, this work is limited to the survey, using a sample of 14 enterprises in the milk industrialisation sector, out of a total of 30 enterprises distributed throughout the regions of the state. The selection of the sample was made first as a function of the receptivity and availability of the enterprises and next in such a way that we could have a good representation of three principal milk basins in the state. We proceeded with semidirective interviews with owners or managerial staff—in the absence of the former.

As Table 17.1 demonstrates, the milk industrialisation sector is new in the industrial landscape of Tocantins. In 1999, 79% of the visited enterprises were between 1 and 5 years old and were in the majority very small enterprises—microenterprises—64% of the enterprises surveyed employed less than 10 employees. Yet this is a sector with a good competitive capacity given that 57% of the enterprises export their products, in particular cheeses, to the most demanding and most competitive market in Brazil, that is, the southeastern region of the country—the largest Brazilian market[5].

Table 17.1
Some *Attributes* of the 14 Enterprises Interviewed

	Creation or restart	Employees	Products	Market
1	1998	47	Cheeses, sterilised milk (plastic wrapping and *Long Life* wrapping), powdered milk	Other states of the North and Northeast Brazil
2	1993	11	Provolone, mozzarella cheeses	Northeast Brazil
3	1995	42	Provolone and mozzarella cheeses	São Paulo (Southeast)
4	1995	5	Provolone and mozzarella cheeses	São Paulo
5	1994	5	Mozzarella cheese	Tocantins
6	1997	3	Sterilised milk (plastic wrapping), mozzarella cheese	Tocantins
7	1998	7	Provolone and mozzarella cheeses	São Paulo
8	1999	48	"Prato" cheese (pressed curd cheese), mozzarella and butter	The states of Pernambuco and Ceará (Northeast Brazil)
9	1979	29	Sterilised milk (plastic wrapping), mozzarella and butter	Milk and butter for the market of Tocantins and cheese for the State of São Paulo
10	1994	7	Milk (sterilised plastic wrapping) and mozzarella cheese	Milk for the local market and the cheese for the States of Rio de Janeiro and São Paulo (Southeast)
11	1992	7	Provolone and mozzarella cheeses	During the dry period (winter) all the production is destined for Tocantins, and during the rainy period (summer), 60% Goiás and Brasilia
12	1997	2	Provolone cheese	São Paulo
13	1999	5	Mozzarella cheese	São Paulo and Ceará
14	1997	6	Mozzarella cheese	São Paulo

The survey showed that there is no establishment for the training of the workforce for the milk industry. SENAI[6] does not carry out training of workers for the food industry of Tocantins (except the profession of bakers). This provokes also the precarious aspects of offering technical services to support the technological development of enterprises. The sector demands a great deal of help and in particular technological support. The survey showed that 72% of enterprises need support in the areas of product technology and procedures. In addition, 84% of the enterprises need help for the training of personnel in techniques of the production-quality couple. Barely 37% of the enterprises have not even had a contact with the universities (placement, visit, analysis, etc.).

It is a sector which also suffers from a shortage in primary materials. In reality, this is the major problem of the sector today, according to the opinion of 61% of enterprises. One of the principal characteristics of the milk sector in Brazil is the seasonality of production. Winter in the Centre and Northern regions is characterised by a dry period. Normally, it doesn't rain during the period May–September. The wild pastures dry out, the cattle grow thin and the milk production and, as a consequence, the production of milk derivatives, drops to half in Tocantins (see, for example, Table 17.1, the experience of enterprise 11—the underproduction in winter halts the production of cheese for export to other states).

Despite these problems, and perhaps because of them, it is an innovative sector. With an open market which demands new products, with more added value, more quality and more food security, enterprises have no choice. In the case of the enterprises in the survey, 70% innovated in the period 1998–1999. The principal objective of these innovations has been, for 70% of them, the improvement in the quality and the diversification of products. The production, as Table 17.1 shows, is not very diversified. For future innovations, the same percentage affirms wanting to innovate more and in the short term.

In relation to the institutional environment, the sector is not yet united around a union or an association. In addition, enterprises feel very far from, or even abandoned by, the public institutions, at the level of the commune or public authorities or federal institutions, and by their industry federation. According to them, the public authorities make themselves present only by "controllers' visits". Yet, for 50% of the enterprises, the laws, health and environmental norms play a positive role in the processes of technological development of the enterprises. The norms are seen as a guide or a technological reference point.

Fundamentally, the survey showed that in 11 different situations of actions characterised as *innovation efforts*, the enterprises call on 21 different networks nodes. To arrive at this result we asked the entrepreneurs or managerial staff of the enterprises (nodes) who they called on in order to obtain the information and/or support for the following list of actions that we consider as innovative:

- *acquisition of equipment (AE)*
- *improvement processes (IP)*
- *development of new products (NP)*
- *enterprise creation (EC)*

- *development of new markets (NM)*
- *technical consulting services (TS)*
- *management consulting services (MS)*
- *marketing consulting services (MK)*
- *maintenance services (MT)*
- *financial support of investments (FT)*
- *employment of workers (EW)*

Table 17.2 shows us the *prestige*, or the "popularity", as well as the importance of the nodes of the local socio-techno-economic network. For example, in all of the actions considered, aiming at the implantation of technological/organisational innovations, the enterprises are directed more to the nodes: "CEO/Staff", "friends of the profession", "employees", "SEBRAE"[7] and "usual suppliers" of equipment. In summary, these are the five most prestigious nodes in the context of the local network.

In network indexes, the *density* (number of connections/maximal number of connections) indicates the intensity/the importance/the necessity of action for the enterprises in the sector. In the case of Table 17.2, the maximal number of connections is 294 (14 enterprises x 21 nodes). The table shows that the six most dense actions, in terms of mobilisation, and perhaps also in terms of importance for the local industry, are acquisition of equipment (21/294), improvement processes (20/294), technical consulting services (17/294), enterprise creation (15/294), management consulting services (14/294) and employment of workers (14/294). In terms of the analysis by the type of action, in the innovation effort of acquisition of equipment for example, we have CEO/Staff, the usual suppliers and the friends in the profession as the three privileged nodes. For the improvement of procedures, the norms and information of the competitors (directly or indirectly) play the principal role.

The *connectivity* of the network, in turn, translates the variety of nodes involved in the action. This is to say, it shows at what level of action is related to the set of network nodes. The five actions that mobilise the greatest number of different actors are: acquisition of equipment (5), technical consulting services (5), development of new services (4), improvement processes (4) and enterprise creation (4). In effect, the *connectivity* shows the problem with *the lemming effect* of unique solutions for a problem in any given context. Lundvall (2000: 12) points out that "too much copying may be problematic because it reduces diversity. Even if there were a rather clear best practice to copy from, the copying process may undermine the dynamic capabilities of the industry. One very significant insight from economic historians that has been confirmed by research in evolutionary economics is that 'diversity' in a population of firms is a key to economic evolution, innovation and economic growth."

Table 17. 2
Connection Nodes According to the Innovations Efforts

Nodes	Innovation efforts											Actor prestige
	AE	IP	NP	EC	NM	TS	MS	MK	MT	FT	EW	
CEO/Staff	6	6	4	2	4	2	6	2	-	2	-	34
Employees	-	7	-	-	-	-	-	-	-	-	9	16
Parent company	-	-	-	-	-	3	4	-	-	-	-	7
Friends in the profession	4	3	-	5	-	-	-	-	2	-	5	19
Family	-	-	-	6	-	-	-	-	-	-	-	6
Associations, federations	-	-	-	-	-	-	-	-	-	-	-	0
Fairs	3	-	-	-	-	-	-	-	-	-	-	3
Magazine	3	-	-	-	-	-	-	-	-	-	-	3
Television	-	-	-	-	-	-	-	-	-	-	-	0
Mailing	-	-	-	-	-	-	-	-	-	-	-	0
University	-	-	-	-	-	-	-	-	-	-	-	0
Technical Institute	-	-	2	-	-	-	-	-	-	-	-	2
SEBRAE	-	-	4	-	-	6	4	2	-	-	-	16
SENAI	-	-	-	-	-	-	-	-	-	-	-	0
Consultant	-	-	-	2	-	-	-	-	-	-	-	2
Public Programmes	-	-	2	-	-	4	-	-	-	-	-	6
Customer	-	-	-	-	-	-	-	-	-	-	-	0
Regular suppliers (equip.)	5	-	-	-	-	-	-	-	8	-	-	13
Commercial representative	-	-	-	-	8	-	-	-	-	-	-	8
Competitor	-	4	-	-	-	2	-	-	-	-	-	6
Bank -	-	-	-	-	-	-	-	-	-	8	-	8
Total	21	20	12	15	12	17	14	4	10	10	14	149
Action density	.071	.068	.041	.051	.041	.058	.048	.014	.034	.034	.048	*Network indicators*
Action connectivity	5	4	4	4	2	5	3	2	2	2	2	

The web of relationships of the whole set of actions is presented in Figure 17.1. With the exception of the university, SENAI, customers and the industry federation, four significant cases for our analysis, we have retained only the nodes having two or more connections per action. This said, the principal external nodes (no hierarchical nodes) which participate, or which in principle should participate in the processes of innovation of enterprises—at least in the exchange of information—are the following:

- family (FA)
- friends in the profession (FP)
- technical fairs (TF)
- public programmes (PP)
- specialised magazines (SM)
- associations, federations (AS)
- technical institutes (TI)
- commercial representatives (CR)
- technical consultants (TC)
- competitors (CO)
- banks (BA)
- regular suppliers of equipment (RS)
- SENAI (SN)
- universities (UN)
- SEBRAE (SE)

Figure 17.1
Web of Relationships and Nodes in the Process of Diffusion and Adoption of Information/Innovation

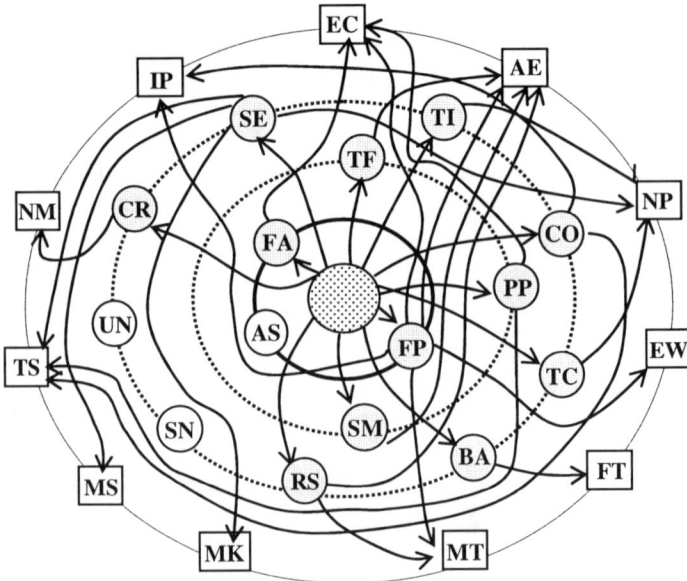

In Figure 17.1 we have placed the milk enterprises at the centre (with their internal nodes: CEO/Staff, employees and parent company). The "permanent nodes" of *family, friends in the profession and the associations* are found in a first orbit, with which the centre creates strong links. In the second orbit we have the "random nodes", intermediaries: *fairs, magazines, public programmes*. Last, the "virtually obliged" nodes of the third orbit are occupied both by the actors of the industrial space—*suppliers, banks, competitors, representatives, consultants* and *SEBRAE*—and by the actors of the science and technology and education spaces, that is to say *technical institutes, universities* and *SENAI*. For example, in the case of an action aiming at the acquisition of new equipment (AE), the enterprise will look for information and/or the opinion of friends in the profession (FP) and some regular suppliers of equipment (FE), specialised magazines (MA) and fairs (FO). The family node (FA), association node (AS), universities (UN), technical institutes (IT), SEBRAE (SE), SENAI (SE), consultant nodes (CO), public programmes (PP), commercial representatives (RC), competitors (CO) and banks (BA) are not mobilised in this case.

Figure 17.1 and Table 17.2 show the different degrees of *autonomy* of the central actor in our analysis—the enterprise—in different types of action. The enterprise is not very autonomous, for example, in the creation of enterprises (restrained to the connections to permanent nodes) or finance (restrained to the public banks nodes, given that local entrepreneurs and private banks mistrust each other regarding financing of activities inherent to production and commercialisation). Friends in the business are unavoidable for the employment of salaried workers. A last example, the consulting activities (techniques, management or marketing) are virtually monopolised by SEBRAE (due to its renowned competence, but also through lack of choice).

Another observation that we can make, from Figure 17.1, is the nonexistence of connections with training organisations for workers and technical services such as SENAI and training and research organisations such as local universities. In these nodes there is a reservoir of technological competencies— rare in this territory: teachers, researchers and students—who don't profit the local industry. In our opinion, there is a very large structural hole in this socio-techno-economic network. We cannot imagine technological development without local technological training. However, a transfer of technology does not imply solely the acquisition of new machines. The technological capacity at the interior of the enterprise is necessary not only to operate the new technology but also to select and negotiate with the suppliers and, in this way, obtain something suited to the needs and conditions of the enterprise (Sendin, 1993: 10). It is necessary to underline the absence of industrial federations in the process of encouragement of the milk industry in regard to innovation.

Looking again at the analysis of results of the survey, we emphasise that in reality the innovations set up by the enterprises, notably those linked to the improvement of quality, envisage principally the respect of health and environmental norms of production and commercialisation of milk food products. The federal norms, which apply to the products exported to other Brazilian states—

nearly all the cheese production of Tocantins requires the use of fabrication procedures similar to or the same as those used for enterprises of the most developed regions. In summary, we can affirm that the norms play a very significant role in the encouragement of technological development of industries in backward regions.

CONCLUSION

This study showed the relationship between the stage of (under)development of a given territory and the networks of relationships between agents. To arrive at this, we were interested in the system in which the agents of a region in the initial stage of industrialisation interact, communicating and/or supplying/ acquiring goods and services. Through this approach we have noted that, for the nature of the local networks, innovation or technological information propagates via socio-techno-economic networks. In a certain way, this situation of networks of social relationships, overlying business relationships, procures some means of subsistence to small enterprises. This type of network differs in relation to the *sociotechnical network* presented by the Actor-Network Theory. In this approach all things are strictly calculated. For decision making, the actors with their "technical prostheses" take into account only the calculable data or sources. The sociotechnical network approach would not explain the role of the family in the process of enterprise creation. In economic reality there are an enormous number of decisions taken in which the deciders go beyond the process of making calculations. How can we calculate the passion of an actor, his intuition, his self-confidence and trust in the system? Do agencies (involving men and technical objects) exist who are capable of doing this? On the other hand, the socio-techno-economic network is not an enormous and random network as are the *social networks*. In our study we note that friends are very important in several innovation activities. However, it is friends who know the profession. That is to say, in the structuring of the network there exists, in a certain way, a calculation, a selection of relationships, an interested decision. In short, the relationships in the networks are not as tangled as the *oversocialised* conception of man in *social networks* suggests.

Next, the survey showed that, in a region in the initial stage of development, the institutional environment is not proactive, but instead static. The milk industry in Tocantins, for example, did not have an interlocutor through whom the local institutions could express their problems, their suggestions, their objectives and participate in the creation of projects for the sector. There is no exchange relationship between the institutional environment and the enterprises, there is only a one-way channel of information, by written norms or by the visits of inspectors, whereby the institutions say directly what has to be respected. The problem of resources, which the enterprises need for this, does not appear to concern the institutions, it would seem. Despite this, the sector does not deny the value of the official norms for production and commercialisation. This institutional *intermediary* has the merit of being the carrier of technological information, very useful in a poor production and learning environment. In summary, the public authorities regulate qualitatively the production and do not feel concerned by its mobilisation or its dynamic. On the other hand, the local federation of industry, for technical, financial and political reasons, is not very prepared to support the emerging industries of Tocantins.

Finally, this study suggests that the limits of the socio-techno-economic network, in view of a local technological and economic development project, is situated at the level of both its structure and its management. In fact, development cannot result from a spontaneous evolution, nor from a simple call on goodwill. It must be articulated. For this, nodes of articulation of networks—such as *translation centres,* a designer, a motivator and a coordinator of the socio-techno-economic networks—are of primary importance. A *translation centre* procures, by the connection of nodes tailored to current and strategic needs of the local industry—reducing the number of structural holes and the nonredundant relationships—the autonomy necessary for the development of elements of the network and, as a consequence, for the whole of the network. The density of *translators* in the more advanced regions is visibly a lot more important than in the new frontiers of development.

The results of this study can be useful in the reflection aimed at the development of policies to encourage local technological development with the setting up of actions adequate to this type of social, technological and economic reality. In other words, the analysis and posterior restructuration of the local networks can reinforce the learning capabilities of individuals and organisations and contribute to the reduction of inequalities between regions and enable the backwards ones to be integrated into the learning economy.

NOTES

1. Minister of Territory Developement and Environment, France, preface from G. Benko and A. Lipietz, *La Richesse des Régions—La Nouvelle Géographie Socio-économique.* Paris: Presses Universitaires de France, 2000.

2. With the notion of a socio-techno-economic network we want to limit its application in relation to a sociotechnical network, on which the Actor Network Theory is based, or in relation to social network. We are interested only in social actors and relationships which, theoretically and/or empirically, appear in the process of innovation diffusion before an economic drop. The intention is to clearly define the raison d'être and action frontiers of the network object of our study.

3. See R. Burt, *Structural Holes. The Social Structure of Competition.* Cambridge, MA: Harvard University Press, 1992.

4. P. Cohendet and P. Llerena, "Apprentissage organisationnel et cohérence: l'importance économique de la notion de réseau", International Conference on Limitation de la rationalité et constitution du collectif, Cerisy, June 1993.

5. We consider that roughly 80% of the total production of the state be exported to the Southeast region.

6. SENAI—National Service of Industrial Training—an educative and technological branch of the Confederation of Brazilian Industry. Its mission is the reinforcement of the industry, promoting professional education, technological support, production and dissemination of information, and generation, adaptation and diffusion of technology (http://www.dn.senai.br/asplan/plano2.htm).

7. SEBRAE—Brazilian Service of Support to the Micro and Small Enterprises—is a civil society without lucrative ends, which despite functioning in synergy with the public sector, is not under the authority of the public structure. SEBRAE has as its mission to support the development of small entrepreneurial activities by incitement and diffusion of programmes and projects encouraging micro and small enterprises (http://200.252.248. 100/site/na/consb.nsf).

REFERENCES

Allemand, S., 2000: "Les réseaux: nouveau regard, nouveaux modèles", in *Sciences Humaines*, Dossier, April, no. 104.

Axelsson, B., 1992: "Network research – futures issues", in Björn Axelsson and Geoffrey Easton (eds.), *Industrial Networks—A New View of Reality*. London: Routledge.

Bakis, H., 1993: "Les Réseaux et leurs Enjeux Sociaux", *Collection que sais-je?*, Paris: PUF.

Becker, B. and Egler, C., 1992: "Geopolítica da fronteira científico-tecnológica no Brasil", in H. Drouvot, M. Humbert, J. C. Neffa and J. R. Mouroz (coords.), *Innovations technologiques et mutations industrielles*. Paris: IHEAL.

Brousseau, E., Geoffron, P. and Weinstein, O., 1997: "Confiance, connaissances et relations inter-firmes", in Bernard Guilhon, Pierre Huard, Magali Orillard and Jean-Benoît Zimmermann (dirs.), *Économie de la connaissance et organisations*. Paris: L'Harmattan.

Callon, M., 1999: "Le réseau comme forme émergente et comme modalité de coor-dination: le cas des interactions stratégiques entre firmes industrielles et laboratoires académiques", in Michel Callon, Patrick Cohendet, Nicolas Curien, Jean-Michel Dalle, François Eymard-Duvernay, Dominique Foray and Eric Schenk, *Réseau et coordination*. Paris: Economica.

Callon M. and Law J., 1989: "La protohistoire d'un laboratoire", in M. Callon (dir.), *La Science et ses Réseaux*. Paris: Editions La Découverte.

Castells, M., 1998: *La Société en réseaux*. Paris: Fayard.

Corsani, A., 2000: "Réseaux d'entreprises et territoires: la dynamique de l'innovation dans le capitalisme cognitif", Proceedings of *12th Séminaire Annuel Organisations, Innovation & International de l'Université de Technologie de Compiègne*, Compiègne, 24-27 January.

D'Iribarne A. and Gadille M., 1998: "Une composante originale de l'espace français des sciences et techniques", Colloque Changements institutionnels et dynamique de l'innovation, 2-4 December. Université Paris–Dauphine.

Droulers M., 1992: "Brésil: défi technologique et organisation spatiale", in H. Drouvot, M. Humbert, J. C. Neffa and J. R. Mouroz (coords.), *Innovations technologiques et mutations industrielles*. Paris: IHEAL.

Easton, G., 1992: "Industrial networks: a review", in Bjorn Axelsson and Geoffrey Easton (eds.), *Industrial Networks—A New View of Reality*. London: Routledge.

Edquist, C., 1997: "Systems of Innovation Approaches: Their Emergence and Charac-teristics", in Charles Edquist (ed.), *Systems of Innovation: Technologies, Institutions and Organisations*. London: Pinter.

Granovetter, M. and Swedberg, R., 1992: "Introduction", in Mark Granovetter and Richard Swedberg (eds.), *The Sociology of Economic Life*. Boulder, CO: Westview Press.

Knoke, D. and Kuklinski, H., 1991: "Network positions and strategic action: an analytical framework", in Grahame Thompson, Jennifer Frances, Rosalind Levacic and Jeremy

Mitchell (eds.), *Markets, Hierarchies & Networks—The Coordination of Social Life*. London: Sage Publications.

Lazega, E., 1998: *Réseaux sociaux et structures relationnelles*. Paris: PUF.

Lemieux, V., 1999: *Les réseaux d'acteurs sociaux*. Paris: PUF.

Llerena, D., 1997: "Coopérations cognitives et modèles mentaux collectifs: outils de création et de diffusion des connaissances", in Bernard Guilhon, Pierre Huard, Magali Orillard and Jean-Benoît Zimmermann (dirs.), *Économie de la connaissance et organisations*. Paris: L'Harmattan.

Lundgren, A., 1992: "Coordination and mobilisation processes in industrial networks", in Bjorn Axelsson and Geoffrey Easton (eds.), *Industrial Networks—A New View of Reality*. London: Routledge.

Lundvall, B.-Å., 2005: "Innovation policy and knowledge management in the learning economy", this volume, pp. 25–55.

Massard, N., 1997: "Externalités, apprentissage et dynamique technologique. Un retour sur la notion de réseau", in Bernard Guilhon, Pierre Huard, Magali Orillard and Jean-Benoît Zimmermann (dirs.), *Économie de la connaissance et organisations*. Paris: L'Harmattan.

Powell, W. W., 1991: "Neither market nor hierarchy: network forms of organization", in Grahame Thompson, Jennifer Frances, Rosalind Levacic and Jeremy Mitchell (eds.), *Markets, Hierarchies & Networks—The Coordination of Social Life*. London: Sage Publications.

Sendin, P. V., 1993: *Agroindústria: tecnologia e competitividade*. Londrina: IAPAR.

Swan, W. and Watson, I., 1998: "Knowledge Networks—A Model for Construct-IT" (http:/www.surveying.salford.ac.uk/buhu/bizfruit/1998papers/swan/swan.htm).

18

Digital Cities and the Network Society: Towards a Knowledge-Based View of the Territory?

Manuel V. Heitor and José Luiz Moutinho

The development of case studies in selected Portuguese cities and regions which have been engaged in "digital city" projects is considered in this chapter in a way to discuss main challenges, and opportunities, for mobilizing the information society in Europe, with emphasis for the conditions affecting less favourable zones. It is argued that value-based networks have the potential to make both public administration and markets more effective, which helps promoting learning trajectories for the inclusive development of society, but require effective infrastructures, incentives and adequate institutional frameworks. The analysis builds on the concept of social capital, as a relational infrastructure for collective action, but the main contribution of the chapter is presented in terms of a *knowledge-based view of the territory* to foster institutionally organized metropolitan *systems of innovation and competence building*.

Competence building is considered in terms of a dynamic and broad social and economic context associated with digital networks and the analysis suggest the need for continuous public support and monitoring, as well as for the promotion of knowledge integrated communities as drivers of larger communities of users.

INTRODUCTION

In view of the current socio-economic context, in which innovation is a key driving force for the sustainable development, which challenges are facing information-based development and cooperation, in a way to contribute for regional policies that stimulate localized learning and indigenous development within less favoured regions in Europe?

This broad question has motivated the work behind the present chapter, which considers the development of case studies in selected Portuguese cities and regions, which have been engaged in building digital networks. It is argued that value-based networks have the potential to make both public administration and markets more effective, which helps promoting learning trajectories for the inclusive development of society, but require effective infrastructures, incentives and adequate institutional frameworks promoted over time and across space (Conceição, Heitor and Veloso, 2003). Early-stage developments are shown to be particularly dependent on public funding and the necessary institutional framework, including the development of knowledge-integrated communities. The analysis builds on the need to continually adapt trajectories and foster the necessary learning capacity of increasingly diversified communities, which refers to social capital as a relational infrastructure for collective action (Conceição, Heitor and Lundvall, 2003), in a context much influenced by a dynamic of change and a necessary balance between the creation and diffusion of knowledge.

In this context, the main contribution of the chapter is presented in terms of a *knowledge-based view of the territory* to foster institutionally organized metropolitan *systems of innovation* (Fischer, Diez and Snickars, 2001) *and competence building* (Conceição, Heitor and Lundvall, 2003), which derives from observations in different Portuguese metropolitan areas with the ultimate goal of increasing regional competitiveness, by promoting public awareness and participation in decision-making processes. It is argued that the territory is a basic infrastructure that justifies and invites for the construction of several layers of information about cities and regions were people live, visit or do business. Digital city schemes should also encourage the global legibility of the information architecture of the territory and promote broad and informed participation in the decision-making process of the future of its entire influence area and not only within city limits (Tanabe, van den Besselaar and Ishida, 2002).

Although we are still in a very early and limited stage of what Mitchell (1995) called "cities of bits", it is clear that it has become a "commonplace" to discuss the diffusion of knowledge, and the "knowledge-driven economy" in general, in close association with the introduction and use of information and communication technologies (Mansell and Steinmueller, 2000; Castells, 2001). In this context, several national initiatives for the Information Society aim to achieve four broad objectives: to create a more open state, to link and make available to all the available knowledge, to promote Internet usage in education, and to support and develop digital technologies usage by firms (Tsipouri, 2002).

The evidence calls for our attention for the critical role of public funding and the dynamic adaptation and development of the context necessary to facilitate digital cities.

The work follows current discussion in Europe aiming to: (a) ensure widespread broadband access and a secure information infrastructure; and (b) services, applications and content, covering online public services and e-business (EC, 2002), but argues for the need to plan systematic actions of competence building with the ultimate goal of attracting new communities of users and to build the necessary capacity for connectivity. Community building and demand creation for digital services became the critical factor for implementing digital cities, requiring proper incentives and institutional settings.

The remainder of this chapter attempts to frame these aspects from the perspective of the challenges facing *digital cities* in Portugal. We begin by bringing empirical evidence on the Portuguese situation, as a specific case study within EU. Clearly, Portugal has significant quantitative shortcomings, but, at the same time, the country has been making good progress, in a catching-up dynamics that is well known. This combination of rapid catch up but persistent shortcomings make the Portuguese case useful to illustrate the main point of the chapter: network societies occur across time and space and require the dynamic adaptation of infrastructures, incentives and institutions, in a way that calls our attention for the need to foster learning societies. The third section, informed by the empirical evidence associated with the analysis of the Portuguese situation, discusses current evidence from specific case studies, based on specific digital city projects. The need to consider the context of social interactions and institutions that govern the behaviour of individuals and organizations is analysed in the fourth section. Thus, we discuss the necessary conditions for the establishment of knowledge driven activities and present a conceptual framework for understanding digital cities based on a knowledge view of the territory. Finally, we conclude by briefly presenting policy implications and a summary of our most important conclusions.

BUILDING THE NETWORK SOCIETY ACROSS DIVERSITY: PORTUGAL IN EUROPE

Focusing our attention to information and communication technologies (ICT), Figure 18.1 presents the intensity of ICT expenditure in 1997 against the growth rate of this intensity from 1992 to 1997. Following recent analysis for knowledge-based industries (Conceição and Heitor, 2004), the results show that Portugal was the leading OECD country in the growth rate of ICT expenditure from 1992 to 1997, with a growth rate of more than 10%, and mainly accounted for by increases in expenditures in telecommunications (about 9%). Expenditures in IT services and software are particularly low, below 1%, and only Turkey, Greece and Poland have shares of expenditure on IT software and services below the Portuguese value. The growth in this category has been equally dismal, below 2% a year.

In terms of our analysis, we would like to argue that Figure 18.1 shows large variations associated with countries characterized by small absolute values, exhibiting patterns typical of latecomer industrialization for Portugal. In addition, the results may represent indications of the process through which late-comer countries become engaged in the new techno-economic paradigm (Freeman and Louçã, 2002). Most countries are clustered in the bottom of Figure 18.1, with growth rates below 4%. The levels, as indicated by the horizontal dis-tribution of countries, confirm the perception that the United States is a leading country. The expenditures on ICT as a percentage of GDP in the United States are about 2% above the European average. Individual countries, such as Sweden, outperform the United States, but most countries lag behind.

Figure 18.1
Information and Communication Technology (ICT) Intensity and Growth, 1992-97

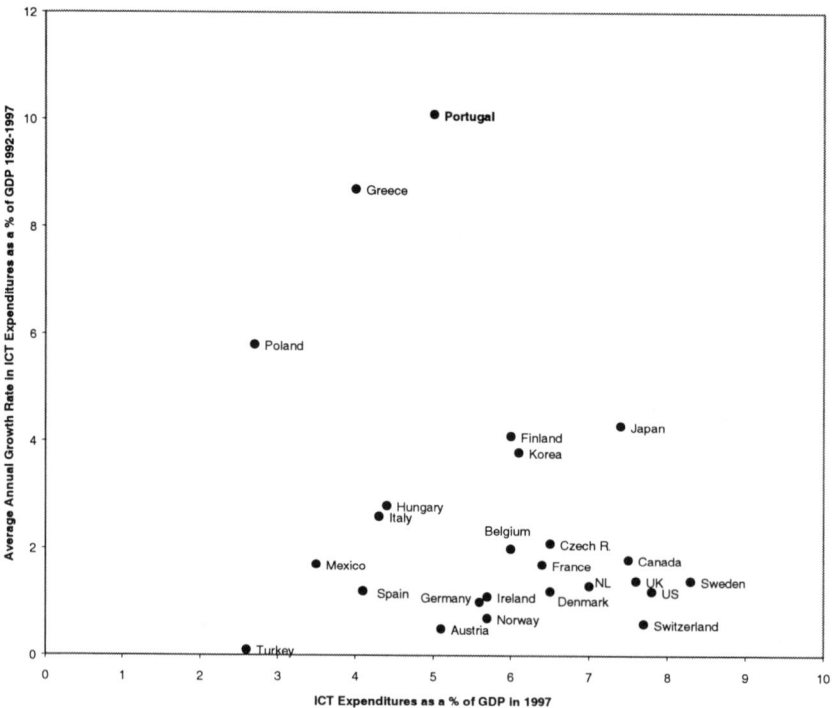

Source: OECD (2000a).

The evidence of still low absolute investments on ICT in Portugal is clearly illustrated in Table 18.1, which shows values per capita for sample European regions in the census whose programming documents indicate information society actions and that provide the necessary financial information (Tsipouri, 2002). It is clear that Table 18.1 refers, above all, to regions that have attracted

European structural funds and, on this basis, it is important to mention the wide diversity of situations and framework conditions for attracting these funds, which clearly influence any analysis to be considered. But for the purposes of our analysis, it is interesting to attempt defining the extent to which the performance of digital networks and cities would depend exclusively on the limitations of funds, as well as from the capacity to attract them.

Table 18.1
Expected ICT Expenditure Per Capita for selected European Regions, 2000-2006

Border Midland and Western Region	357.8 €
La Rioja	357.8 €
South Aegean	269.4 €
Ionian Islands	241.4 €
Baleares Islands	238.2 €
Western Greece	151.1 €
Açores	**117.9 €**
Highlands & Islands	98.4 €
Epirus	83.4 €
Alentejo	**44.5 €**
Peloponese	43.1 €
Continental Greece	42.8 €
Algarve	**42.5 €**
Centro	**29.9 €**
Norte	**13.3 €**
Southern Scotland	9.2 €
Lisboa e Vale do Tejo	**6.8 €**
Liguria	2.2 €

Source: Tsipouri (2002).

Besides large growth rates in ICT investments, the extent to which the Portuguese society is engaged in the knowledge economy comparatively to other nations can be analysed making use of the recently established systematic assessment by the World Economic Forum in collaboration with INSEAD and the World Bank's *info*Dev programme through the "networked readiness", as represented in Figure 18.2 for 2002 (Dutta and Jain, 2003). This indicator offers an aggregated idea of "the degree of preparation of a nation to participate in and benefit from ICT developments" and illustrates the still weak position of Portugal in the European context, only above Greece. The main point to note is that the results for Portugal and for most of the OECD countries appears to be dependent from other than the country's overall wealth (as measured in terms of GDP per capita). Considering the partial log regression plot included in Figure 18.2, Portugal is in fact entering the cluster of countries where the effect of increasing GDP on network readiness is less pronounced and other factors, namely at institutional and contextual level, have been shown to particularly influence country's competitiveness (Conceição and Heitor, 2004).

Figure 18.2
Network Readiness Index versus GDP (PPP) Per Capita, for 2002,
with Partial Log Regression

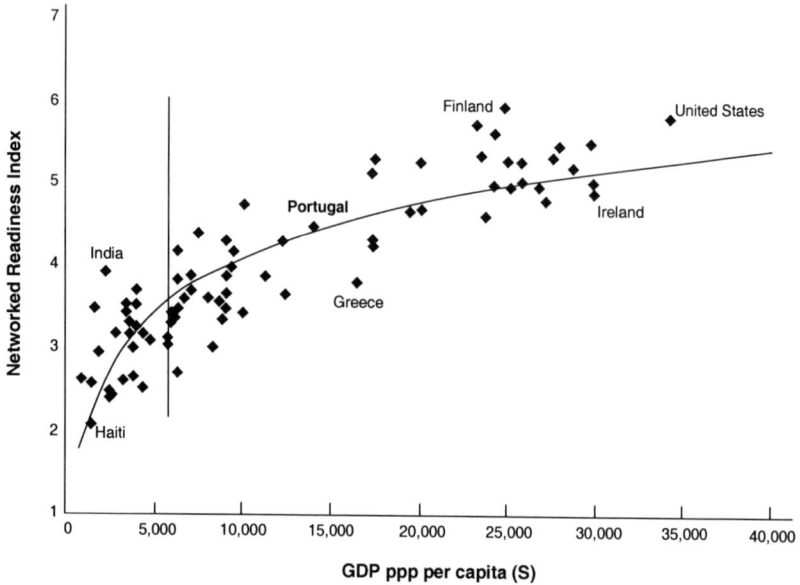

Source: Dutta and Jain (2003).

The pattern of small absolute values regarding the mobilization of information society, but large variations, can be further analysed making use of a number of typical indicators to characterize the penetration of ICT's in a country and, for example, Figures 18.3 and 18.4 show values for internet penetration rates, with Portugal getting the highest position in southern Europe for 2002 (Portugal 42%, while Spain 42%, Italy 40%, Greece 18%, with an EU average of 51%), although far away from typical north European penetration rates (Eurobarometer, 2003). A similar picture can be obtained making use of Internet access in the household, with Portuguese rates of 31%, as compared with 29% for Spain and 9% for Greece, while 40% for the EU average and 74% for the United States, although Portugal exhibits growth rates between 2000 and 2002 considerably larger than the European average (namely 72% for Portugal, with 81% for Spain and 89% for France, as compared with 43% for EU average) (EOS Gallup Europe, 2002). Turning to the type of telecom infrastructure, the country follows typical average EU trends, with standard telephone lines as the most frequent connection to the Internet access at home (Portugal 74%, EU average 72%), followed by cable modem (Portugal 12%, EU average 7%). ISDN, ADSL and Wireless connections are still relatively low.

Figure 18.3
Internet Penetration Rates in Europe, 2000-2002

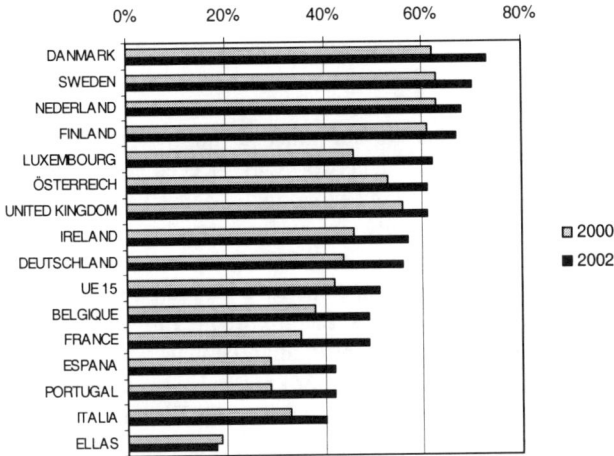

Source: Eurobarometer (2003).

Figure 18.4
Internet Access Type, 2000-2002

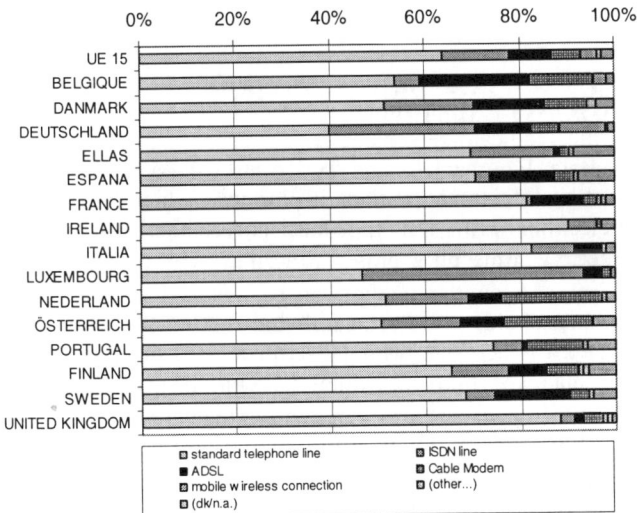

Source: Eurobarometer (2003).

For this brief analysis of the Portuguese context for the use and application of ICT's in an international perspective, it should also be mentioned that the country has one of the lowest European usage rates of Internet for on-line purchases of products or services (9%, but only 1% frequently) and the third-lowest percentage of Internet users that have already contacted the public administration (Eurobarometer, 2003). These figures are important to set the context of information networks and clearly call our attention for the need to consider contextual levels beyond pure infrastructural issues, when considering measures to foster information networks.

But Figures 18.3 and 18.4 should be further explored in terms of the main point of this chapter, in that we are aimed to improve our understating of the conditions necessary for digital networks to succeed. Learning from the conceptualization of knowledge-based economies (Mansell and Steinmueller, 2000), it can be said that, fundamentally, the performance in knowledge-rich competitive environments in terms of innovative performance depend on the quality of human resources (their skills, competencies, education level, learning capability) and on the activities and incentives that are oriented towards the generation and diffusion of knowledge. But beyond human capital, which corresponds to the aggregation of an individual capacity for knowledge accumulation, developing a collective capacity for learning—as suggested by Wright (1999) in the context of the United States—is as, if not more important, than individual learning. Instead of individual or even aggregated human capital, a further important concept for learning seems to be social capital, as analysed by Conceição et al. (2001), among others.

Before further exploring social capabilities and related issues associated with the development of digital networks, we present further evidence on the penetration of ICT's in Portugal through the analysis of specific projects aimed to build digital cities, namely making use of European structural funds.

INSTITUTIONALIZING "DIGITAL CITIES": EVIDENCE FROM PORTUGAL

The evidence presented in this section is built on the analysis of sample projects for digital cities and regions in Portugal, which have been structured around the electronic provisioning of local government administrative services complemented by some pilot projects in areas such as e-business and telemedicine.

The first experiences in Portugal with digital cities started in 1998 through a program funded by the Portuguese Government and the European Union and involving 5 small and mid-sized cities (Aveiro, Bragança, Guarda, Marinha Grande, Castelo Branco) and 2 rural regions (Trás-os-Montes and Alentejo), as identified in Figure 18.5. The main objectives of the program were to (a) improve the quality of life in cities; (b) contribute to development of peripheral areas; (c) improve local economy and employment; and (d) fight info-exclusion and help citizens with special needs (MCT, 1997).

The project sites were chosen for reasons which are out of the scope of this chapter and we concentrate our analysis on issues associated with their effective implementation. Alentejo and Trás-os-Montes are remote agricultural regions, among the least developed in Portugal and Europe, sparsely inhabited by an aging population. Both projects were designed to create new opportunities for the local population, mitigate social and economic disparities and promote regional networking. Aveiro is developing a true innovative and entrepreneurial image, in particular connection with the local university and the local branch of Portugal Telecom, which includes important research and development activities. On the other hand, Marinha Grande is particularly engaged in traditional, labour-intensive industries and the digital city project has been particularly promoted through the industrial network associated with the local moulds industry. Both these two projects invested mainly on local competitiveness and competence building. Bragança, Guarda and Castelo Branco are peripheral cities with relative regional significance. Their approach was to support the adoption of information and communication technologies by individuals, firms, associations and local government and other public organizations.

Figure 18.5
Identification of Main Projects for the specific Development of Digital Cities and Regions, established over the Period 1998-2001, making use of European Structural Funds

Source: Adapted from MCT (1997).

In terms of regional penetration, Table 18.2 shows that the previous projects listed covered about 11,30 % of the total Portuguese population (10,44% of the population under 15 years of age) and about 42% of the total surface of Portugal. All projects involved a broad range of relevant actors and change agents within each one of the territories being nonetheless always leaded by local municipalities. Local higher education institutions were particularly involved only in a limited number of projects (Aveiro, Bragança, Trás-os-Montes).

It should be noted that, at least for the initial projects analysed here, the institutional framework established by the central government was quite flexible and fostering local voluntary initiatives. It was based on the simple provision of guidelines focused on providing content and services related to local public administration and to specific activities with social implications (e.g., healthcare), economic impact (e.g., business-driven corporate networks for regional competitiveness), and aimed to promote cultural contents (Aveiro, 1998, 2001; Bragança, 2001). Initiatives to mobilize and promote the adoption of the Information Society were part of various applications, although not always considered at the required level, at least beyond that given to the implementation of infrastructures (Aveiro, 2001).

Table 18.2
Characterization of Digital City Projects, in terms of Population
and Area Considered in each City/Region

Municipality	Pop.	(%)	Pop. < 15	(%)	Area (Km²)	(%)
Aveiro	69.560	0,67%	12.160	0,73%	208	0,23%
Bragança	32.440	0,31%	4.760	0,29%	1.138	1,23%
Castelo Branco	54.260	0,52%	7.440	0,45%	1.440	1,56%
Guarda	38.560	0,37%	6.230	0,38%	709	0,77%
Marinha Grande	33.370	0,32%	5.050	0,30%	186	0,20%
Alentejo	510.690	4,93%	71.930	4,34%	27.227	29,55%
Trás-os-Montes	431.540	4,17%	65.450	3,95%	11.122	12,07%
Total	1.170.420	11,30%	173.020	10,44%	42.030	45,61%

Source: INE, 2001.

Table 18.3 shows sample data in terms of public funds made available to the seven projects mentioned, illustrating diversified situations with levels of funds per capita raging from low to moderate when compared with those considered within the overall usage of European structural funds (Tsipouri, 2002). Within the broad range of digital city projects considered at international level (Tanabe, van den Besselaar and Ishida, 2002), *Aveiro Digital* represents an interesting case study in that it has comprised diversified initiatives promoted and coordinated by an autonomous organization formed among the local government, the local University and the incumbent Telecommunication operator, PT Telecom. It represented the result of a long preparation effort and provided the opportunity to evaluate concepts and dynamically testing ideas, involving a limited but well informed group of people (Aveiro, 1998).

Bragança Digital focused on creating basic ICT infrastructures and wireless networking environment for local government buildings, health institutions, educational institutions, and local employment agency to provide information and services to local citizens. Other initiatives included the provision of local products (www.rural.net), health, educational and e-business activities (Bragança, 2001).

Guarda Digital was promoted by and organization formed by the municipality, local educational institutes, associations" and the incumbent telecommunication operator. It included pilot projects in healthcare e-business, teleworking and educational initiatives (http://www.ipg.pt/adsi/).

Castelo Branco Digital aimed to connect all public institutions (municipality, social security and health institutions) and local associations (sports, culture and business) to provide an integrated information network to citizens and tourists. For example, it has included the provision in rich media of old Portuguese theatre contents (http://www.cm-castelobranco.pt/cb_digital/).

Marinha Grande Digital, as managed by the local municipality and the Technological Centre associated with the moulds and plastic injection industries, focused on creating an Extranet to provide business-related (mould, plastics and glass) content and services and on facilitating communication among companies and clients. Other initiatives included a centre of advanced telecommunications to promote the use of the Internet (http://www.marinhagrandedigital.com/).

Trás-os-Montes Digital included regionally-based web contents (i.e., www. espigueiro.pt), managed by the local University, that aggregates content and services of 31 municipalities. The portal is managed by the local university and includes business and employment opportunities, geo-referenced information, healthcare facilities and technologies to coordinate medical services in rural areas (http://www.espigueiro.pt/).

Alentejo Digital brought together 47 municipalities and 3 regional agencies to create a regional information network to provide services and territory-related content to citizens and local firms through regional web-based contents. The main objective was to enable local government teams to learn, use and promote new technologies, namely computer network management and digital content production and publishing. An Intranet was set up linking all municipalities and regional agencies to enable the necessary collaborative work environment. About 50 people were recruited, mostly from local unemployment lists, to work on the project that lasted until July 2001 (http://www.alentejodigital.pt/).

Table 18.3
Public Funds Expenditure Per Capita in the first Phase of the Digital Cities
Program, 1998-2000

Project	Pop. (a)	Total investment	Investment p.c.
Aveiro	69.560	5,590,000 € (b)	80,39 €
Marinha Grande	33.370	1,200,000 € (c)	35,96 €
Bragança	32.440	1,044,000 € (d)	32,18 €
Castelo Branco	54.260	1,082,000 € (e)	19,94 €
Guarda	38.560	350,000 € (f)	9,08 €
Trás-os-Montes	431.540	1,735,000 € (g)	4,02 €
Alentejo	510.690	1,500,000 € (h)	2,94 €

Sources: (a) INE, 2001; (b) Aveiro, 2001; (c) http://www.marinhagrandedigital.com; (d) Bragança, 2001; (e) personal communication, http://www.dpp.pt/pdf/info52.pdf; (f) http://www.freipedro.pt/tb/110698/guarda3.htm; (g) personal communication; (h) http://home.telepac.pt/telepac/net/13/regionalismo_2.html.

BUILDING A CONCEPTUAL FRAMEWORK

We now turn to the analysis of the empirical evidence provided and start by discussing necessary framework conditions for the success of digital cities. Then, we argue that knowledge-integrated communities are drivers of larger communities of users. Under this context, we continue our analysis by identifying elements that resembles a *knowledge based view of the territory* in order to discuss a conceptual framework for digital cities. We conclude our analysis by presenting main policy implications derived from our conceptual understanding of digital cities.

But before turning to our analysis, it should be mentioned that our emphasis is on the conditions favouring the mobilization of the information society in less favourable zones in Europe, LFRs, which have been shown to lag behind the adoption of measures as rapidly or intensively as were the core regions of Europe (Tsipouri, 2000). In fact the type of structural funds used to support the projects discussed before derive from increasing awareness of that growing disparity in the European scenario, based on three arguments, namely (1) LFRs tend to get little new hardware and applications because of the weakness of their markets (lack of scale and agglomeration economies); (2) most LFRs have no track record of intensive interaction leading to innovation or new ways of learning and, therefore, most LFRs put their efforts into catching up, as opposed to proactive capacity building for IS; and (3) although the deregulation process and much of the hardware infrastructure may be national jurisdiction, applications and content are vital in regional terms.

Building Infrastructures, Capacity and Connectivity across Time and Space

The evidence provided in the previous paragraphs may be interpreted in light of the framework identified in Figure 18.6, which considers the relative importance of the processes of making available and implementing infrastructures, together with the increasingly important role of incentives for mobilizing communities of practice and dynamically adapting institutions, and the overall context, towards innovative societies (Conceição, Heitor and Veloso, 2003). Under this conceptual approach, we may refer to three main levels of analysis, including *infrastructures*, *contents* and *context*, although there is a growing body of opinion that balanced information society depends on five main aspects, namely: infrastructure, access, application and services, digital content development, and ICT skills development. In fact, the evidence provided by Tsipouri (2002) throughout Europe leaves us to jointly consider the first two levels under infrastructure, as well as to join application and services and digital content developments into a single level of analysis. In addition, we broaden the scope of the so-called ICT skills development to include other contextual issues and local characteristics of communities of practice.

The methodology uses the socio-technical approach to information systems (Laudon and Laudon, 2002), where the dynamic mutual adjustment of both technology and users determines its final performance. The contextual analysis provides to both content (existing, needed or planned) and communities bring their significance as nodes of information. Their relationships—information flows—must be mapped to the technological roadmap to determine the amount of change needed in the organization, community or society. Technology path-dependency and usage permanence must be well understood before designing, implementing or exploiting information systems. As a result, knowledge creation, accumulation and diffusion play a critical role in this process.

Looking first at infrastructures in general, in the neoclassical view, they are related with the existing amount of labour, capital, and natural resources. The new growth theories bring to stage other important factor inputs, in particular human capital, and R&D expertise embodied in firms, universities, and laboratories (Conceição, Heitor and Veloso, 2003; Conceição, Heitor and Lundvall, 2003). Thus, infrastructure will encompass, in addition to labor and capital, what we call technology infrastructure, or technostructure. Tassey (1991) has proposed a definition that suits our discussion: technostructure consists of science, engineering, and technical knowledge embodied in human and organizational forms. In the context of our analysis, we consider these three types of infrastructures in two main terms, namely physical infrastructures and web-based contents (or non-physical infrastructures), on which most of the sample projects considered in this chapter have concentrated their investments. Table 18.4 provides the results of a brief analysis of main activities considered in the various digital city projects discussed and list typical infrastructures that have been considered.

Figure 18.6
Framework for the Analysis of Digital Cities, C³ITIES, Including the Consideration of the Overall Context and Communities of Practice, with a Diversified Network of Infrastructures and the Processes of Their Implementation

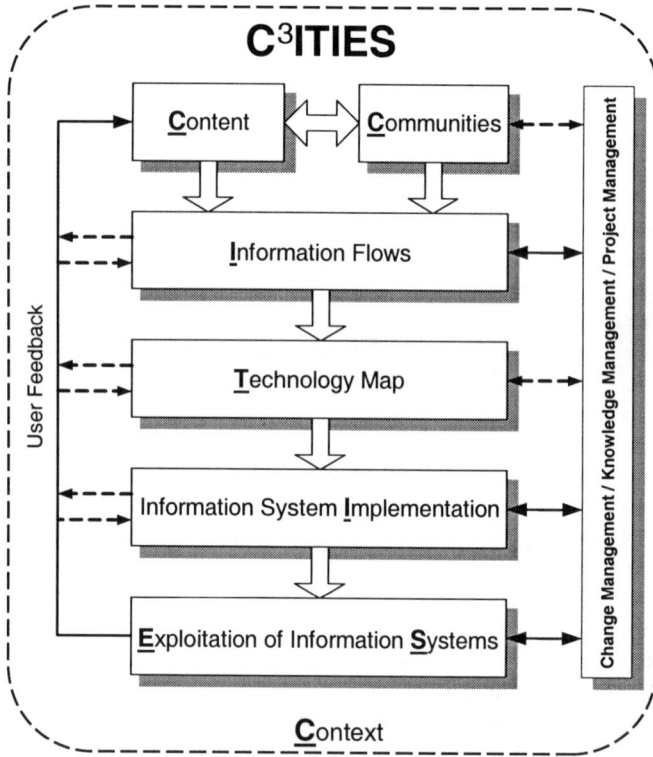

Turning to incentives, current understanding of knowledge-driven activities based on endogenous growth theories are based on the existence of dynamic externalities and imperfect markets, and require a careful understanding of the structure of competition. On the one hand, because of the nature of knowledge, investment of private agents often fails to acknowledge spillover effects, or may not be able to anticipate the full extent to which there is further learning potential in a new technology. On the other hand, incentives to invest in new knowledge depend on the existence of some degree of monopolistic rents. These rents may not exist in latecomer countries exposed to international competition, if they are solely adopting foreign technology.

As a result, Conceição, Heitor and Veloso (2003) call our attention that private investment levels (which result from the incentive structure provided by the market to economic agents) in activities with learning or spillover potential tend to be lower than the social optimum, and may even generate what is known in the literature as "low-level equilibrium traps" (Azariadis and Drazen, 1990; Aghion and Howitt, 1998). In principle, these shortcomings of the market mechanism call for some sort of government intervention. Governments are concerned with making sure that societal costs and benefits are endogenized in the decisions of private firms. In a learning environment this may mean subsidizing specific activities, investing in education, or protecting infant industries (Shapiro and Taylor, 1990; Chandler and Hikino, 1996). But government intervention must balance the potential distortions on competition that may come from intervention with the needs to "correct market failures": artificial restraints on competition can also divert profits to activities other than building technological capabilities.

Against the background of the conditions described, it is clear that digital cities cannot be promoted independently of an innovation policy fostering capacity and connectivity and that in turn innovation determines and is determined by the market. However, it is also clear that it will require an effective mix of public support mechanisms that take a relatively long-term perspective, taking into consideration specific regional and thematic aspects, thus promoting a diversified environment.

But still focusing on the issues of incentives and looking at their relation with the operational effectiveness of digital infrastructures, applications and services, the evidence is that the most important web contents associated with the digital city projects discussed before and summarized in Table 18.4 were available to the public domain only for the time public support was also available. Besides the notable exception of the Aveiro Digital Project, this result may be obvious for the local promoters of those projects, but should be acknowledged as a major issue for public policies fostering the information society. We will argue that early stage developments, as those we are considering throughout this chapter, do require continuous support, together with adequate monitoring and evaluation procedures, in order to acquire the necessary strength for their sustainable development. Market mechanisms do not necessarily work at the level of the issues associated with digital cities, namely in less favourable zones, where incentives structures should be effectively designed and adapt over time.

Although incentives and infrastructure greatly inform our understanding of economic development, they do not tell the whole story about the differences across the various projects or even across the countries discussed earlier. This is because both incentives and infrastructure do not operate in a vacuum, being shaped by and shaping the particular context where they operate. In the scope of our analysis, the city or region must have embedded a set of social capabilities that define the context under which digital cities evolve. Consideration of contextual issues in building-up network societies have not always been considered in many different situation throughout the world, as acknowledged by Castells (2001), among others, and Table 18.4 shows that specific measures to promote adequate contexts in the projects considered in this chapter have also been scarce.

Table 18.4
Main initiatives developed within "Digital Cities" in Portugal, 1998-2000

Projects	Physical Infrastructures		Content (non-physical infrastructures)		Context[e]
	Networking and Connectivity[a]	Information Systems[b]	Information Services[c]	Interactive Services[d]	
Aveiro	Local health institutions communication network; Internet access in public schools; People with special needs	Local public administration management information systems; Justice court Intranet; GIS	City guide; Entertainment, Arts and culture initiatives; Local government website	e-business, Agriculture; Job opportunities; Environment; Teleworking	Community building based on city metaphors
Bragança	Municipality communication network; Internet access in public schools	Municipality management information systems; GIS	City guide; Local government website	e-business; Telemedicine; Agriculture	
Guarda	Internet access in public schools		Local government website	e-business; Telemedicine; Teleworking	

Table 18.4
(continued)

	(a)	(b)	(c)	(d)	(e)
Marinha Grande	Advanced tele-communication demonstration centre; Internet access in public schools	Local industries Knowledge network (Glass, moulding and plastics)			Mobilization of firms and public institutions for the use of ICT
Castelo Branco	Municipality communication network; Internet access in public schools		City guide; Local government website; Art and culture		
Trás-os-Montes Digital	Internet access in public schools	Content management platform	Regional Portal	Telemedicine; Agricultural Network; Job opportunities	
Alentejo Digital	Intranet for 47 municipalities	Content management platform	Regional Portal	Job opportunities	

(a) Networking and connectivity includes communication networks and Internet access.
(b) Information Systems includes technological components that store and process data like data bases, electronic mail, ERPs, management information systems, content management, application serves and business intelligence software.
(c) On-line presence or downloadable forms.   (d) Electronic form submission or interaction through the web.
(e) Mobilization and context building initiatives.

If one considers the broad social and economic context under which digital cities may be facilitated, we must consider the conditions for integrated learning processes. This has led Conceição, Heitor and Lundvall (2003) to build on Lundvall and Johnson's learning economy (1994) and to discuss the learning society in terms of innovation and competence building with social cohesion. They view innovation as the key process that characterizes a knowledge economy understood from a dynamic perspective, while competence is the foundation from which innovation emerges, and which allows many innovations to be enjoyed. In other words, it contributes both to the "generation" of innovations (on the supply side of the knowledge economy) and to the "utilization" of innovations (on the consumptions side of the knowledge economy). Conceptually, the foundations for the relationship between learning and economic growth have been addressed in the recent literature (Bruton, 1998), with learning being reflected in improved skills in people and in the generation, diffusion, and usage of new ideas (Conceição and Heitor, 2002).

Further, the ability to learn seems to be the main driver of long-term growth, but learning can occur at different levels. Individual people, firms and organizations, and countries all are dependent of learning for development. Lamoreaux, Raff and Temin (1999) write: "more than any other factor, the ability to collect and use information effectively determines whether firms, industry groups, and even nations will succeed or fail." There are also different ways through which people, firms, and countries can learn. Learning can be an unintended consequence of experience and augmentation of scale, as formalized at the firm and then country level by Arrow (1962). On the contrary, formalized and intentional learning methods such as education, training or R&D is often the result of an utility maximization rational decision from the point of view of the firms. It is clear that previous evidence on digital cities (Tanabe, van den Besselaar and Ishida, 2002), shows that we refer to complex networks of formal and informal learning processes leading to wealth creation and shared prosperity, but also to forms of inclusive development.

As we emphasized earlier, learning can occur in many shapes and forms, some of which are informal, some formal. As described before, the institutional framework that comprise the national and regional systems of innovation formalize the technological infrastructure critical to generate the learning processes for individuals, firms, and nations, that ultimately lead to long-term development. Thus, looking at a particular set of organizations, their capabilities and related institutions, provides important lessons for development. This is the reason we argue for the need to combine adequate infrastructures and incentives with institutions, to foster the necessary context for digital cities to succeed.

Knowledge Integrated Communities (KICs) and Beyond

The previous analysis is broad in scope and considers network societies as wide social and economic processes, which we argue occur across time and space and require the dynamic adaptation of infrastructures, incentives and institutions, in a way that calls our attention for the need to foster learning societies.

However, the evidence of the projects discussed in this chapter show that we must extend our analysis to other aspects of the learning society. This is because the experience of projects such as those developed in the cities of *Marinha Grande* and *Aveiro* clearly shows the important mutual relationships that specific project-based communities have on the facilitation of network societies, but also the fact that the implementation of digital cities may significantly improve the efficiency of those communities. In the following paragraphs, we extend this evidence and argue that the success of digital cities rely on the specific development of knowledge-integrated communities (KICs).

We refer to project-based communities, oriented to specific social and economic goals, that will benefit, and gain from, digital networks if particularly challenges by knowledge-based activities. In the case of *Marinha Grande* the evidence is that economically-oriented networks based on mould-forming companies has particularly launch business networks, which still require long-term processes and continuous funding, as well an adequate institutional setting. In this case, it should be noted the role of the related industrial association and technology centre in promoting the necessary links and networking facilities, which again support our previous analysis of the need to consider basic framework conditions.

In a different scale, but also using relatively reduced level of incentives, namely at an international scale, the evidence provided by the *RuralNet Project* developed in the city of *Bragança* also shows the critical importance of project-based mechanism to support and sustain digital cities. But of specific interest in our context, are some of the activities developed in *Aveiro*, in that knowledge-based activities could promote and sustain digital networks well beyond the period under which public incentives were made available.

The reason why knowledge-based activities are particularly prone to foster and sustain digital networks is because they will increasingly rely on "distributed knowledge bases", as a systematically coherent set of knowledge, maintained across an economically and/or socially integrated set of agents and institutions, as discussed by Smith (2000) and Conceição, Heitor and Veloso (2003), among others. The relevance of considering distributed knowledge bases across economically and/or socially integrated set of agents and institutions leads us to the concept of social capital. In the broadest sense, social capital is associated with the "social capabilities" (Lamoreaux, Raff and Temin, 1999) that allow a country or region to move forward in the process of development. In a more sophisticated treatment, Coleman (1988) states that social capital is "a variety of different entities, with two elements in common: they all consist of some aspect of social infrastructure, and they facilitate certain actions of actors— whether personal or corporate actors—within the structure." The relationship of social capital for the economic performance of nations was recognized by Olson (1982) and North (1990), in broad descriptions of the process of development.

Referring again to the evidence provided by some of the projects discussed in this chapter, namely those at Aveiro, the role of higher education institutions appear to be particularly important in fostering network activities, namely in the form of knowledge-based communities. Following the analysis of Castells and Hall (1994), "it takes a very special kind of university, and a very specific set of linkages to industrial and commercial development, for a university to be able to play a role it often claims to play in the information-based economy". Definitely, those technical universities that are pure teaching factories, or work under a bureaucratic structure, are unlikely to act as generators of advanced technological milieu. Again, this recalls our attention to the role of institutions in planning digital cities and promoting their impact.

Still in this context, Mitchell (2004) argues that the most obvious advantage of digital networking is that it provides an efficient way of "aggregating specialized expertise" through "common access to project databases, compatible software tools, and advanced telecommunication capabilities". But he emphasizes that "it does little about the problems of creating trust and confidence, and of building intellectual and social capital for the long term", requiring the development and maintenance over weeks and months of "project-based learning communities" looking at a common and complex target. Long-term collaborations can provide a more permanent framework of online resource-sharing, and examples of such an initiative, shows the need to bring scale and diversity, beyond time. Based on this example, Mitchell concludes that we should look beyond the popular idea of learning communities and seek to produce communities that "motivate and sustain creative discourse yielding original intellectual products such as architectural and engineering designs", the so-called "creative communities".

A final remark associated with the form and role KICs may play in the process fostering university-based network societies, should be discussed in terms of the evidence provided by the Program "Ciência Viva" in Portugal, namely in association with some of the projects discussed (http://www.ucv.mct.pt/home/). It refers to specific networks formed among basic and secondary schools with university groups and research centres through project-based activities oriented to promote a culture of learning. Beyond the critically important role of this type of activities, as explained by Ziman (2000), among others, taking Pine and Gilmore's (1999) contentions about what they termed "the experience economy" and the role experiences play in building stronger and more personal relationships in the corporate world, our argument is that schools, and universities in particular, must deliver authentic experiences to build and encourage sustainable and entrepreneurial growth. Pine and Gilmore explore the idea of experiences as a fourth economic offering, as distinct from services as services are from goods, but one that has until now gone largely unrecognised. While services may be considered as a set of intangible activities carried out on behalf of a person, experiences are memorable events that engage that person in an individual way, so that they determine and guide transformations. Experiencing entrepreneurial processes at the school (and the university, in particular) thus sets the stage for the societal transformations required to progress successfully towards innovative societies.

From this analysis, it is clear that knowledge-integrated communities may develop over different institutional, thematic and social frameworks and Table 18.5 summarizes the evidence provided by the various projects analysed.

Table 18.5
Typical Experiences fostering KICs as identified in the various Digital City Projects analysed and other Sample Initiatives

Driving factor	Sample Experiences	Remarks
Scientific	*Biorede*—Biology knowledge network about local biodiversity, molecular biology and estuary ecosystems launched at Aveiro (www.biorede.pt)	Website developed and managed by Research Centre
Education/ Training	"Engineering in Portugal", providing historical data and information for Basic and Secondary Schools, as well as university students (http://www.engenharia.com.pt/)	Learning materials and information exchange between experts, teachers and students; Website managed by Research Centre
Public Health	Health information and communication network of the Bragança Digital City extension services (www.espigueiro.pt/servico_cooperativo/ servico_coop_puh.html)	Portable computers and Internet access to foster the communication and information exchange between doctors and patients
Managing Public Risks	Water quality monitoring and public diffusion system (www.simoqua.pt)	Raise public awareness about water quality, flooding and other public risks
Corporate strategy and competitiveness	Marinha Grande local-industry (moulding, plastics and glass) network (www.marinhagrandedigital.com/)	Extranet managed by Technological Centre

Striving for a Conceptual Framework:
Fostering a Knowledge-Based View of the Territory

The previous paragraphs provide empirical evidence on specific digital city projects developed in Portugal in the recent past, as well as on particular framework conditions for their success, but now we turn to the discussion of a conceptual framework required to improve our understanding of digital cities and networks.

It is clear that focusing on digital cities, we must consider the conditions that foster innovation and the related processes of knowledge sharing in local contexts. Traditional neoclassical approaches in industrial economics have emphasized the analysis of the microeconomic behaviour of firms and built theories specialized in the American and Anglo-Saxon systems and related market dynamics. Following the analysis of Conceição, Heitor and Veloso (2003), it provides an excellent context to understand incentive structures and outcomes, but ignores most of the remaining issues associated with learning discussed above. Evolutionary economics focuses on routines and capabilities rather than incentives to improve our understanding of learning processes and the role of institutions in economic development. Nevertheless, they have not addressed the specific historical context of any region, namely those characterized by late industrialization (Cooke and Morgan, 1998). Building on the evolutionary approaches and in system theory, the concept of "national system of innovation" (Nelson, 1993; Lundvall, 1992; Edquist, 1997) has led to numerous studies of individual countries, but there is still a long way to go in order to assess the specificity of metropolitan systems or late industrialized regions.

The importance of the learning dynamics of firms and regions has been increasingly considered as key to the processes of knowledge accumulation, innovation and growth (Nelson and Winter, 1982). In this respect, "firm competencies" affect the ability of firms to innovate and shape their technology trajectories. Building on this idea, Conceição, Heitor and Lundvall (2003) discuss the need to consider the systemic nature of innovation together with processes of competence building.

At the same time, the spatial patterns of innovation and the related geographical dimension of economic and social development have witnessed a renewed and increasing interest in the literature (Cooke and Morgan, 1998; Storper, 1998), but attention is to be focused on the ability to build social capital, including interactive learning, local externalities, and networks among institutions (Swann, Prevezer and Stout, 1998). This focus on relational assets is part of the "institutional turn" in regional development studies, as a result of the relative failure of classical approaches which sought to privilege either "state-led" or "market-driven" processes, regardless of time, space, and milieu.

In conceptual terms, we attempt to explore features in the seminal work of Nelson and Winter (1982), for which organizations know how to do things through simple rules and procedures (routines) which represent the knowledge memory of the organization. Even firms in the same industry differ in the sense that they accumulate and develop idiosyncratic routines, which form the basis of the firms' distinctive capabilities. Fundamental to the idea of skills and routines is that they are constituted essentially by tacit knowledge and are thus not easily replicated. Replication of routines is thus possible only as a costly, time-consuming process of copying an existing pattern of productive activity. The dynamics in the theory is brought about by the processes of searching for new routines and creating variety and mutations amongst firms, which are then subject to selection processes. The combined interaction of search and selection processes form the basis of the evolutionary approach and relate Nelson and Winter's approach to the theories of organizational learning and population ecology respectively. The routines are thus seen as the knowledge genes of the organization, being transformed by organizational learning and innovation.

Although Nelson and Winter's work provided a conceptual foundation for a knowledge-based view of the firm, an essential development was a deeper understanding of what constitutes knowledge, which we attempt to extend for a territory bases. Figure 18.7 attempts to provide a schematic representation of a possible framework of analysis considering main functions to satisfy the knowledge-based view presented here.

Figure 18.7
Schematic Representation of a Possible Framework
for the Development of Digital Cities

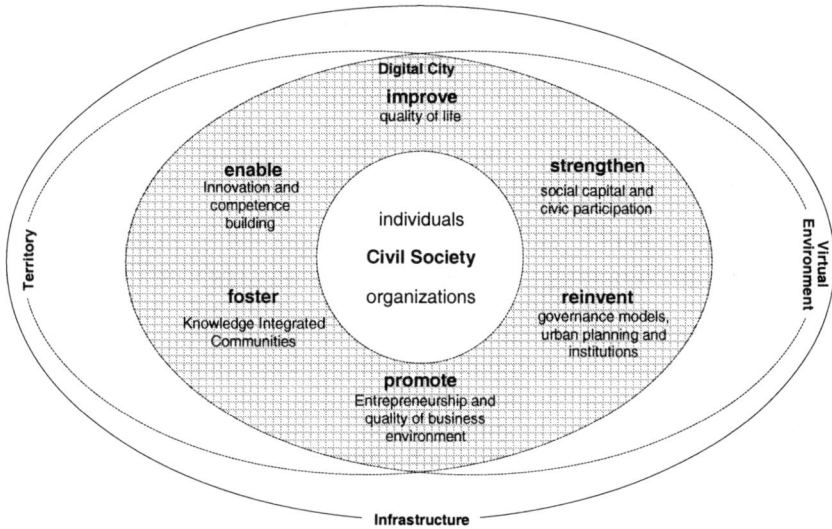

In this context, a knowledge-based view of the territory assumes that individual, firms and organizations operate in dynamic environments, where markets and technology are changing fast and in unpredictable ways, as discussed by Eisenhardt and Santos (2002) for the open discussion of emerging theories of strategy. It also assumes a highly competitive setting, with those agents operating within ecologies of learning, interacting and adapting to the environment. In this framework, organizational capabilities or competencies are understood as clusters of knowledge sets and routines that are translated into distinctive activities. Dynamic capabilities are those that enable individuals and firms to build, integrate and reconfigure internal and external competencies. The knowledge that is embedded in capabilities is a complex and dynamic combination of tacit and explicit knowledge. Individuals operate within organizational contexts in order to be able to share and use their specialized knowledge. As a result, digital cities should consider *communities of users* and build a context favourable to their increasing participation over time.

Following the analysis of Santos and Heitor (2003), a determinant of the effectiveness of those communities of users is the level of the absorptive capacity, as identified by Cohen and Levinthal (1990) as the ability of a firm to recognize the value of new external information, assimilate it, and apply it to commercial ends. Extending this concept for the collective dimension typical of the territory, the absorptive capacity should be largely a function of the level of the regions' prior knowledge (which emphasizes the cumulative nature of knowledge) and is also history or path dependent (which emphasizes the importance of earlier decisions). The level of absorptive capacity is heavily dependent on the level of absorptive capacity of each actor (individuals and organizations) in the territory, but is different form the sum of the parts. It not only depends on the interface with the environment but also involves the internal transfer and communication of knowledge. This concept calls attention to the internal channels of communication, to the diffusion of knowledge in the region, and to the pattern of investments.

Which Policy Implications?

Knowledge creation and competence building at the regional level, when considered as a dynamic process of learning, can mitigate the uncertainty about the future of urban areas. Metropolitan areas must be liveable and attractive to all citizens, most of them needing urgent sustainable and innovative solutions to overcrowding, pollution, traffic jams, insecurity, social inequalities, unemployment, and population aging. City authorities will have to invent new models of participative governance and learn to manage change. Information rich networks can provide the main resources to overcome physical barriers to share interests and experiences to prop up creativity and entrepreneurship and to diminish the pressure on urban areas. We conclude by recommending elements and components for policy making and design of digital cities in Portugal, arguing for the need to promote *regional systems of innovation and competence building.*

Referring to recent work within the framework of the OECD International Futures Program (OECD 2000b, 2001), two broad policy-related conclusions apply not only to OECD countries in general, but to a large extent also to the case of Portuguese regions. The first is that if one is to build on the opportunities offered by the considerable progress that has been made in key technological sectors, if one is to reap to the full the economic benefits of rapidly integrating markets and the emerging knowledge society; and if solutions are to be found to tackling the challenges that the management of such rapidly changing world raises, then what is needed are innovative, creative societies. The second is that in achieving that higher degree of innovativeness and creativity, policy will matter. The way ahead does not necessarily mean less government, not less policy but—certainly in some key areas—different policy.

Just because we are headed into a rapidly changing world in the coming decades does not mean that we have to throw out all policies and make a completely fresh start. Indeed, some policies that have proved their worth in the past may well continue to do so in the future. However, it is clear that in other policy areas at least incremental adjustments are called for, and in yet others some radical new thinking is required. This provides, in fact, a simple but convenient framework for looking at the role of general policies in the future and their implications for innovation: (1) policy continuity, (2) policy reform, (3) policy breakthroughs.

In Portugal, most of the complex social, economic and political advances towards the Information Society are governed by public decisions. The evidence provided in this chapter shows that investments in ICT infrastructures, although very necessary, haven't been sufficient to create a sustainable knowledge-based living and working environment. Consistent public policies, innovative regulatory frameworks and strong incentives are thus needed to create over time the conditions to catch up with more developed societies and mitigate the uncertainty associated with the adjustment process (Mansell and Steinmueller, 2000).

Within this perspective, our analysis calls for policies that consider long-term approaches of dynamic environments, which require to be continuously monitored and evaluated. Specific incentives for infrastructures should continue, but articulated with the need to foster knowledge integrated communities as drivers of larger communities of users. This requires a continuous pubic effort, but also a better understanding of the effectiveness of the mix of public support mechanisms and private incentives for the development of digital cities.

SUMMARY AND MAIN CONCLUSIONS

The main contribution of the chapter is presented in terms of a knowledge-based view of the territory to foster institutionally organized metropolitan systems of innovation and competence building. The analysis is based on observations in different Portuguese metropolitan areas and regions with the ultimate goal of increasing regional competitiveness, by promoting public awareness and participation in decision-making processes. It is argued that the territory is a basic infrastructure that justifies and invites for the construction of several layers of information, but above all for communication infrastructures and digital contents, but well arranged with local contexts. It is suggested that knowledge driven communities are important drivers of larger communities of users and different types of KICs are identified. Particular attention is suggested for those established among basic and secondary schools with university and research groups and evidence is provided from sample case studies in Portugal.

Our analysis led us to suggest that while the role of institutions needs to be re-examined, the variety of demands and the continuously changing social and economic environment is calling for diversified systems able to cope with the need to produce policies that nurture and enhance the learning society. We refer to the need for individuals, firms and organizations to operate in dynamic environments, where markets and technology are changing fast and in unpredictable ways. This calls for the need to combine adequate infrastructures and incentives with institutions, to foster the necessary context for digital cities to succeed.

The institutional framework should be dynamically considered in order to foster local conditions over time, and this does not necessarily mean less government, but rather continuous public support and monitoring.

ACKNOWLEDGMENTS

We acknowledge the collaboration and many useful conversations with the administration of the Program for the Information Society in Portugal, POSI. The collaboration of staff at *Aveiro Digital City* and the *Council of Aveiro* is also recognized, as well as the information provided by staff of the Bragança's Polytechnic Institute and the *RuralNet* Project.

REFERENCES

Aghion, P. and Howitt, P. (1998), *Endogenous Growth Theory*. Cambridge, MA: MIT Press.
Arrow, K. (1962), "The Economic Implications of Learning by Doing", *Review of Economic Studies*, 28: 155-173.
Aveiro (1998), "Programa Aveiro Cidade Digital", Municipality of Aveiro, University of Aveiro and Portugal Telecom, S.A. Available from http://www.aveiro-digital.pt/.
Aveiro (2001), "Relatório Global da 1ª Fase do Pgrgrama Aveiro Cidade Digital 1998-2000", Aveiro Digital City Coordination Team. Available from http://www.aveiro-digital.pt/.
Bragança (2001), "Relatório Final Bragança Cidade Digital", Associação para o Desenvolvimento de Bragança Cidade Digital. Available from http://www.braganca-digital.pt/.
Azariadis, C. and Drazen, A. (1990), "Threshold Externalities in Economic Development", *Quarterly Journal of Economics*, 105(2): 501-526.
Bruton, H. J. (1998), "A Reconsideration of Import Substitution", *Journal of Economic Literature*, 36(2): 903-936.
Castells, M. (2001), *The Internet Galaxy—Reflections on the Internet, Business, and Society*. New York: Oxford University Press.
Castells, M. and Hall, P. (1994), *Technopoles of the World: The Making of 21st Century Industrial Complexes*. London: Routledge.
Chandler, A. and Hikino, T. (1996), "The Large Industrial Enterprise and the Dynamics of Modern Economic Growth", in Chandler, A., Amatori, F. and Hikino. T. (eds.), *Big Business and the Wealth of the Nations*. New York: Cambridge University Press.
Cohen, W. and Levinthal, D. (1990), "Absorptive Capacity: A New Perspective on Learning and Innovation", *Administrative Science Quarterly*, 35(1): 128-152.
Coleman, J. (1988), "Social Capital in the Creation of Human Capital", *American Journal of Sociology*, 94: S95-S120.
Conceição, P. and Heitor, M. V. (2002), "Knowledge Interaction Towards Inclusive Learning: Promoting Systems of Innovation and Competence Building", *Technological Forecasting and Social Change*, 69(7): 641-651.
Conceição, P. and Heitor, M. V. (2004), *Innovation and Competence Building: Learning from the Portuguese Path in the European Context*. Westport, CT: Praeger.
Conceição, P., Gibson, D. V., Heitor, M. V. and Sirilli, G. (2001), "Knowledge for Inclusive Development: The Challenge of Globally Integrated Learning and Implica-

tions for Science and Technology Policy", *Technological Forecasting and Social Change*, 66(1): 1-29.

Conceição, P., Heitor, M. V. and Lundvall, B.-Å. (eds.) (2003), *Innovation, Competence Building, and Social Cohesion in Europe—Towards a Learning Society*. London: Edward Elgar.

Conceição, P., Heitor, M. V. and Veloso, F. (2003), "Infrastructures, Incentives and Institutions: Fostering Distributed Knowledge Bases for the Learning Society", *Technological Forecasting and Social Change*, 70(7): 583-617.

Cooke, P. and Morgan, K. (1998), *The Associational Economy*. New York: Oxford University Press.

Dutta, S. and Jain, A. (2003), "The Networked Readiness of Nations", in Dutta, S., Lanvin, B. and Paua, F. (eds.), *The Global Information Technology Report 2002-2003: Readiness for the Networked World*, pp. 2-25. New York: Oxford University Press. Available at http://www.weforum.org/pdf/Global_Competitiveness_Reports/ Reports/GITR_2002_2003/Networked_Readiness_Nations.pdf

EC (2002), *e-Europe 2005: An information society for all*, COM(2002) 263. Brussels: Commission of the European Communities.

Edquist, C. (1997), *Systems of Innovation: Technologies, Institutions and Organizations*. London: Pinter Publishers.

Eisenhardt, K. and Santos, F. (2002), "Knowledge-Based View: A New Theory of Strategy?", in Pettigrew, A., Thomas, H. and Whittington, R. (eds.), *Handbook of Strategy and Management*, pp. 139-164. London: Sage Publications.

EOS Gallup Europe (2002), "Flash Eurobarometer 125—Internet and the public at large", Report. Brussels: European Commission (Directorate General "Information Society"). Available at http://europa.eu.int/comm/public_opinion/flash/fl125_en.pdf

Eurobarometer (2003), http://europa.eu.int/comm/research/press/2003/pdf/cc-report_en.pdf.

Fischer, M., Diez, J. and Snickars, F. (2001), *Metropolitan Innovation Systems: Theory and Evidence from Three Metropolitan Regions in Europe*. Berlin: Springer-Verlag.

Freeman, C. and Louçã, F. (2002), *As Time Goes By: From the Industrial Revolutions to the Information Revolution*. Oxford: Oxford University Press.

INE (2001), *Recenseamento Geral da População—2001*. Lisbon: Instituto Nacional de Estatística.

Lamoreaux, N., Raff, D. M. G. and Temin, P. (eds.) (1999), *Learning by Doing in Markets, Firms, and Countries*. Chicago: The University of Chicago Press.

Laudon, K. and Laudon, J. (2002), *Management Information Systems: Managing the Digital Firm*. Upper Saddle River, NJ: Prentice-Hall.

Lundvall, B.-Å. (1992), *National System of Innovation: Towards a Theory of Innovation and Interactive Learning*. London: Pinter Publishers.

Lundvall, B.-Å. and Johnson, B. (1994), "The Learning Economy", *Journal of Industry Studies*, 1/2: 23-42.

Mansell, R. and Steinmueller, W. E. (2000), *Mobilizing the Information Society: Strategies for Growth and Opportunity*. Oxford: Oxford University Press.

MCT (1997), "Iniciativa Internet". Lisbon: Portuguese Ministry of Science and Technology (in portuguese).

Mitchell, W. J. (1995), *The City of Bits*. Cambridge, MA: MIT Press.

Mitchell, W. J. (2004), "Challenges and Opportunities for Remote Collaborative Design", in Bento, J., Duarte, J. P., Heitor, M. V. and Mitchell, W. J. (eds.), *Collaborative Design and Learning: Competence Building for Innovation*. Westport, CT: Praeger (forthcoming).

Nelson, R. (1993), *National Innovation Systems*. Oxford: Oxford University Press.

Nelson, R. R. and Winter, S. G. (1982), *An Evolutionary Theory of Economic Change.* Cambridge, MA: The Belknap Press of Harvard University Press.

North, D. (1990), *Institutions, Institutional Change and Economic Performance.* Cambridge, UK: Cambridge University Press.

OECD (2000a), *Information Technology Outlook.* Paris: OECD.

OECD (2000b), "The Creative Society of the 21st Century", *OECD Future Studies.* Paris: OECD.

OECD (2001), *Governance in the 21st Century.* Paris: OECD.

Olson, M. (1982), *The Rise and Decline of Nations—Economic Growth, Stagflation, and Social Rigidities.* New Haven, CT: Yale University Press.

Pine, B. J. and Gilmore, J. H. (1999), *The Experience Economy.* Boston: Harvard Business School Press.

Santos, F. and Heitor, M. (2003), "The Cognocratic Organization: Toward a Knowledge Theory of the Firm", in Gibson, D., Stolp, C., Conceição, P. and Heitor, M.V. (eds.), *Systems and Policies for the Global Learning Economy*, pp. 465-481. Westport, CT: Praeger.

Shapiro, H. and Taylor, L. (1990), "The State and Industrial Strategy", *World Development*, 18(6): 861-875.

Smith, K. (2000), "What is the 'knowledge economy'? Knowledge-intensive industries and distributed knowledge bases", paper presented to DRUID Summer Conference on The Learning Economy—Firms, Regions and Nation Specific Institutions, Rebild (Denmark), June 15-17.

Storper, M. (1998), *The Regional World—Territorial World in a Global Economy.* New York: Guilford Press.

Swann, G. M. P., Prevezer, M. J. and Stout, D. K. (1998), *The Dynamics of Industrial Clustering.* Oxford: Oxford University Press.

Tanabe, M., van den Besselaar and Ishida, T. (2002), *Digital Cities II—Computational and Sociological Approaches.* Berlin: Springer-Verlag.

Tassey, G. (1991), "The Functions of Technology Infrastructure in a Competitive Economy", *Research Policy*, 20(4): 345-361.

Tsipouri, L. (2000), "Europe and the Information Society: Problems and Challenges for Supranational Intervention", *Journal of Comparative Policy Analysis*, 2(3): 301-319.

Tsipouri, L. (2002), *Final Report for the Thematic Evaluation of the Information Society.* Brighton: Technopolis Ltd.

Wright, G. (1999), "Can a Nation Learn? American Technology as a Network Phenomenon", in Lamoreaux, N., Raff, D. M. G. and Temin, P. (eds.), *Learning by Doing in Markets, Firms, and Countries.* Chicago: The University of Chicago Press.

Ziman, J. (2000), *Real Science: What It Is, and What It Means.* New York: Cambridge University Press.

PART III:
LANGUAGE, DEVELOPMENT AND POLICY

Introductory Note

Lawrence S. Graham

When planning took place for the fourth International Conference on Technology Policy and Innovation in Curitiba, Brazil, early on the issue of language arose. The practice for this particular network of individuals and organizations interested in technology policy and innovation has generally been to use English as the working language for communication. Yet the hosting of this particular conference in southern Brazil in the city of Curitiba and the state of Paraná in August 2000 led a number of the organizers to insist that as matter of courtesy Portuguese be used, especially in the opening session, since both those hosting the meeting and funding the local venue were drawn mostly from the public and private sectors in Brazil and Portugal. Hitherto and subsequently, keynoters in Austin and The Hague have used English. Hence, when in the Curitiba opening session, several speakers preferred to speak Portuguese a block of non-native-speakers of English protested on the grounds that such a practice if continued would rupture the attempts to sustain an international forum meeting annually in different high tech centers on a worldwide basis.

The four chapters in Part III look at four very different issues in the response to globalization and how language and culture relate to technology policy and innovations linked to the use of computer-based technologies.

One current, and clearly the one most closely tied to the focus on technology policy and innovation in this conference group, looks to the development of a Universal Networking Language (UNL). As Tarcisio Della Senta and M. G. K. Menon write, the core of this project is centered in the development of an electronic language which will enable "computers to recognize information and represent it in digital format" through a common set of symbols and codes. This chapter discusses how the participants in this United Nations University program in Tokyo, Japan, have been developing a system for representing information and expressing knowledge in digital codes that "can be reproduced using Roman characters and others symbols available on any standard computer keyboard."

The chapter lays out the formatting involved in this particular project and reports the results of the first phase in the effort to create a common electronic representation of the ten major languages in the world, measured in terms of the numbers of speakers using these languages. As this project moves ahead, the goal is to expand gradually the number of languages covered to reflect more comprehensively the linguistic and cultural diversity of the contemporary world and to establish computer-based networks to transmit more rapidly new scientific and technical knowledge by converting words and symbols into a common UNL representation.

But for now, the testing and refining of the encoding in process is focused on producing greater accuracy in the UNL "Enconverter" software, which simultaneously converts all inputs into the program through a common UNL representation. What is missing in this written document, however, is the discussion that the Power Point presentation of this project generated, when the preliminary results of this project were presented in terms of computer-generated translations. The workshop format adopted for this session generated a lively debate over issues related to language as the vehicle through which communication occurs cross-culturally and the shift in technology markets from production-oriented "drivers" to smaller and medium-sized firms centered on knowledge-based initiatives. Generally speaking, the conclusions reached by the participants in this workshop was that the utility of this project is correlated with the creation of a more effective base for communicating more rapidly advanced scientific and technology-based research. Where the system is least satisfactory is centered in that writing and conceptual formation which is linked more closely with abstract ideas and symbols, such as those represented in a poem or an essay linked to aesthetics.

The next chapter, by Mercedes Lynn de Uriarte, takes a position that is diametrically opposed to the conceptualization represented in the United Nations UNL project. Her stance is that rapid technological advances in the communications industry are producing a digital divide that is singularly insensitive to the cultural diversity that one finds in the contemporary world within existing national communities and transnationally. Translated into the images and concepts used to produce a common mass-media-based language that is increasingly being "globalized," her charge is that the symbols and images used are increasingly insensitive to the color of voice and the resonance of language as vehicles for expressing the diversity of human experience, on a worldwide basis as well as within individual national communities. Seen from this perspective, her chapter emphasizes the forces that are leading to patterns of communication which are producing opposite effects to what is the goal of the UNL project, one that aims at democratizing the Internet through increasing the access of individuals all over the globe to new scientific and technologically oriented research. Uriarte's charge is that the movement in media-oriented language is toward a common information highway in which there is "the absence of color, voice and ultimately accuracy within [the] media." Taking the case of the United States, she asserts that the correlation becomes the following in the use of the Internet and mass-media-generated images: "the higher the income, the more likely [one is] to be coasting along the information super highway."

The third chapter focuses on issues in language instruction pertaining to how language is used, in this case to prepare non-native-speakers of Portuguese to be able to communicate more effectively in areas related to conducting business in Portuguese-speaking countries. The theme here is very different from either Della Senta and Menon or Uriarte in that the object of Orlando Kelm's chapter is "to demonstrate how changes in technology provide us with newer and more effective ways of learning." While this applies to all the top 10 or 12 worldwide languages, the case in point is focused on how to globalize the use of Portuguese, when the outside world—especially Spanish-language-based initiatives—discount the fact that Portuguese is one of the major world languages in terms of those who speak and use it as their primary vehicle for communication. In this regard, the conflicting values represented in the first two chapters, when reduced to the level of training people with greater skills in communicating across the language divide, is transcended in this document by centering the discussion on looking "at how technology is used within society in general, and then see[-ing] how these situations can be modified for language instruction."

In drawing on the results of Kelm's project to devise a more effective way to train businesspeople who as nonnative speakers of Portuguese need to be able to conduct their business in Portuguese-speaking countries, the author arrives at a conclusion that is perhaps more instructive than presenting the reader only with an either-or situation, of arguing for computer approaches to language instruction versus those that are more traditional. The more important concept here, he argues, is that we focus our attention on "the sorts of communication patterns [that] can also be modified for language teaching purposes" by taking advantage of newer computer-based technologies. Kelm goes on to write that as a consequence "our challenge is not to use or not use technology, but to look for innovative ways to communicate and expose our students to language and language practice." There is, in short, both a place for computer-generated methods to improve the ways in which we communicate new scientific and technologically oriented research and for increasing the number of users of the major language. For, all it takes, at a time in which individuals invest enormous amounts of time in accompanying the technical shifts in computer-based technologies, is that these same individuals interested in globalizing computer-based technologies also meet the minimal requirement that they learn at least one major language other than that to which they are accustomed, to access more effectively human diversity, since language also gives expression to differing culturally based perceptions of reality. From this follows the fact that there is not a one way for us to engage in the pursuit of knowledge through the use of new computer-based technologies, but instead many, many different ways to approach basic issues and concepts and ways of thinking that cannot be reduced to a common symbolic representation that is universally valid for all places and at all times.

Herein lies the contribution of the fourth chapter, by Regina Dell'Isola. A leader in her field in developing programs for the teaching of Portuguese as a second language in Brazil, within the context of a country that is increasingly the "driver" of regional economic development in South America, she is part of a small but important new language network, if one will take seriously that this country which is larger than the continental United States constitutes not an English but a Portuguese-speaking world of its own. This particular project is focused on developing approaches for the teaching of Portuguese in Portuguese-language countries which will replicate there the long-standing practice of teaching English as a second language (designated as ESL) in Canada and the United States, in the countless number of ESL offerings to be found in North American colleges and universities. Along with this process goes the development of a national network in Brazil for certifying people trained in the teaching of Portuguese as a second language by arriving at a commonly agreed-upon set of standards for certifying those who will teach Portuguese to foreigners in Portuguese-speaking countries.

All in all, these four chapters sample the diversity to be found in approaching the issue of language as a significant part of the current shift going on in the high-tech field, as we move toward restructuring and recovery in world technology markets centered around the concept of how best to situate ourselves in the shift toward knowledge-based industries. At the macro level, attention is being focused on how to arrive at a universal computer-based language which will facilitate the more rapid transfer of new knowledge. At the same time, at the micro level, when one studies how Portuguese is now being approached by native speakers of Portuguese engaged in developing a greater capacity to teach Portuguese as a second language, one can tap into another domain of innovation and experimentation that is occurring in foreign-language instruction, to ensure that at least among the 10 to 12 major languages there are more viable and increasingly competitive options to the use of English as a global language.

19

The Universal Network Language:
An Electronic Esperanto?
No, a Language for Computers

Tarcisio Della Senta and Mambillikalathil G. K. Menon

INTRODUCTION

This chapter introduces the UNL as a "language for computers" that will help machines to assist humans in multilingual communication. The second section of the chapter provides a general explanation on UNL and how it works. The final section discusses its possible application to the United Nations and the potential benefits it can bring to humankind in the current globalisation trend.[1]

UNL, AN ELECTRONIC LANGUAGE FOR COMPUTERS

Universal Networking Language, an initiative of the Institute of Advanced Studies of the United Nations University,[2] is an electronic language enabling computers to recognise information and represent it in digital format. Thanks to the UNL, it is possible to store and build up a vast reservoir of knowledge already available, or in the process of being generated by different peoples. To achieve this, it is necessary to equip computers with a language they can understand.

Without going into lengthy scientific explanations on linguistics or technical discussions on language engineering, let me introduce the general concept of the

UNL. The fact that it is defined as an "electronic language" suggests two important features.

The first one is that the UNL represents information and expresses knowledge in sets of digital codes that can be recognised by any computer. These codes can be reproduced using Roman characters and other symbols available on any standard computer keyboard. A sample of UNL expressions is shown in Figure 19.1. A sentence in Spanish is codified in UNL according to predefined specifications.

Figure 19.1
A Sample of UNL Expressions

THE UNL EXPRESSION

```
                        Hace tiempo, en la ciudad de Babilonia,
                        la gente comenzo a construir una torre
[S:2]                   enorme que parecia alcanzar los cielos.
{org:es}
{/org}
{unl}
tim(begin(icl>do).@entry.@past, long ago(icl>ago))
mod(city(icl>region).@def, Babylon(icl>city))
plc(begin(icl>do).@entry.@past, city (icl>region).@def)
agt(begin(icl>do).@entry.@past, people(icl>person).@def)
obj(begin(icl>do).@entry.@past, build(icl>do))
agt(build(icl>do), people.@def)
obj(build(icl>do), tower(icl>building).@indef)
aoj(huge(icl>big), tower(icl>building).@indef)
aoj(seem(icl>be).@past, tower(icl>building).@indef)
obj(seem(icl>be).@past, reach(icl>come).@begin.@soon)
obj(reach(icl>come).@begin-soon, tower(icl>building).@indef)
gol(reach(icl>come).@begin-soon, heaven(icl>region).@def.@pl)
{/unl}
[/S]
```

With such a code system, it is possible to express the meaning of words from any natural language, written in any kind of script. Conversely, such digital expressions can be decodified into any natural language when linked to the same code system. The process of codifying a human language for computers to understand is known as "Enconversion", while "Deconversion" exchanges the digital codes into normal words that humans can understand. Both Enconversion and Deconversion are tasks for computers to serve the human mind.

The second meaning of "electronic language" is that encoding into UNL expressions, and back into native languages, is a task for computers and not for human beings. The human author just writes the text in his or her natural language. While typing his or her thoughts, the machine automatically enconverts the text into the UNL, and provides the author with its understanding of what is being typed by deconverting from UNL expressions into the author's language. Any discrepancy between the author's thinking and the machine's understanding is indicated on the screen so that the author can revise the text until the computer represents it correctly in the UNL.

THE UNL "LANGUAGE"

The UNL is an "electronic" language, but it is still a "language". There is a vast literature about human and artificial languages that reflects the advances made in understanding the different ways minds and machines work. However, it is not the purpose of this chapter to enter into the discussions of scholars on this topic[3]. The main point here is to stress that the UNL works as an intermediary language to enable people—with the help of a computer—to cross the boundaries of natural languages.

Technical explanations of the UNL Language, as well as its relevant software, can be found in the book *The UNL, A Gift for a Millennium*[4]. This chapter borrows the basic concepts from this book. The UNL language, as in any natural language, is composed of "words" to represent concepts, and a "grammar" to establish the correct relationships between the words, thus enabling the author to express what he or she wants to convey.

Universal Words

In the UNL, words are called "Universal Words" (UW), which represent the meaning of words used in native languages—that is, *concepts*. UWs can represent the image of a concrete thing—for example, *book*—or something very abstract—for example, *kindness*. Such concepts are expressed by different words and are written in different characters from one language to another. For instance, the English word *book* has a corresponding French, Arabic or Japanese word to indicate the same object, which is represented in Roman, Arabic or "Kanji" scripts. Furthermore, the concept of *book* or the feeling of *kindness* exists in all languages even though they are written in different characters. They are "universal concepts", though under multiple and different written representations. The UNL represents such concepts in digital format (UWs) so that computers can immediately recognise their correct meaning from the UW "script"; and then they can deconvert them back into the scripts of any natural language linked to the UNL.

The UNL Word Dictionary stores information about concepts represented in UWs, as well as about the syntactical features of each word. As this information is in digital format, computers can recognise the exact meaning of a given word from a natural language, and also express that meaning in many other natural languages. Each UW entry defines the correspondence between concepts from a native language and the various functions they may have in a given sentence. The collection of these entries constitutes the UNL Master Dictionary, which serves as a reference for developing the corresponding UNL representation of the meaning of the words in native language dictionaries. These dictionaries, therefore, contain the UW concepts and define their corresponding meaning in the words of the native language.

The primary task of the UNL developers, therefore, is to create the UWs of the UNL, to build the Master Dictionary of universal concepts and then link them with the corresponding Word Dictionary that expresses the concepts in each native language.

The UNL vocabulary is an immense task, as it must contain a very great number of UWs. It is also a never-ending task, as languages create new words every day. To ensure that a UW is truly universal—that is, that it represents the same meaning in all languages—meticulous work is required to define the precise concept expressed by a given UW. Sets of specifications and norms have been defined for this purpose. These will enable all languages to contribute towards building a constantly expanding UNL vocabulary. To facilitate this work, the English dictionary is used as a common reference, but the UWs should be language-independent. As the English vocabulary often contains ambiguous words, or words with multiple meanings, further efforts are needed to create UWs that express precise concepts.

UNL Relations and Attributes

The UNL syntax is governed by "relations" and "attributes", which determine the functions of UWs in sentences and the subjectivity underlying the statements.

To write a sentence in UNL, as in native languages, words are grouped into an appropriate structure to convey the intended meaning of the author. The intended meaning determines the role played by each word (subject, verb, object, etc.) and the structural relationships between them. Likewise, in UNL sentences, the relationship between one UW and the others has to be clearly defined in order to make meaningful statements. Relations, therefore, build the syntactic structure to link concepts in UNL expressions. A "label" represents the conceptual relationship between two UWs. These relation labels denote the specific function or role that UWs may have in a sentence (agent, beneficiary, conjunction, etc.).

To cope with the wide variety of sentence structures in different native languages, the UNL has defined a comprehensive set of relation labels to cover the different roles they may play in sentences. There are 42 different types of relation labels listed, each one identified by three characters (ex. "agent" = *agt*). Very clear specifications have been established for all of them. The specifications are provided by the UNL Centre and adopted by all the UNL developers. This is a basic condition for ensuring the homogeneous matching between the UNL and native languages.

Another important feature of the UNL is the treatment of attributes (adjectives and adverbs), words that modify the meaning of UWs or make their meaning in a sentence more precise. Attributes in a given statement express the perception of the author regarding the things described by UWs. Attributes denote quality, mode, tense and so on, in relation to a given concept. In that case, it can be said that an attribute manifests the author's subjectivity underlying a statement. In the UNL, subjectivity is represented by "labels" expressing the author's attitude, perception, viewpoint and other information that may enrich the meaning of UWs.

Here again, the work of the UNL developers consists in drawing up inventories of attributes in the native language, and then matching them with the UNL representation specified by the UNL Centre for all languages. While carrying out this task, they must follow the rules of representation specified for each type of attribute.

Knowledge Base

Words and sentences in any natural language derive their full and correct meaning when referred to the context in which they are used. For example, in the sentence *As I got closer to the bank, I realised how big it was*, it is only in the context of the whole paragraph, or of the whole text, that it becomes clear that the word *bank* means the *bank of a river* and not a *bank where money is managed*. Human intellect has the background knowledge of both meanings of the word *bank*, a kind of "semantic background" that enables a person to identify which meaning fits the sentence when additional information is provided. Similarly, in UNL sentences, in addition to correctly grouping UWs, relations and attributes, a knowledge structure is provided by the computer, within which information is clearly expressed and shared, and sentences understood. The knowledge structure constitutes the "semantic background" to give correct meaning to the UWs and the entire UNL expressions.

In natural languages, the "semantic background" is the cumulative result of each person's experience in his or her cultural context, and the formal knowledge acquired. This renders words and sentences meaningful, and the concepts they convey can be understood without much ambiguity. The wider the cultural and scientific background of the person, the greater and more accurate will be his or her understanding, hence the importance of education and learning. For instance, almost every human being knows the meaning of the word *horse*, but a biologist has a "scientific" knowledge of all the characteristics of the animal, while a child in an urban setting would probably have only an intuitive image from a picture of a horse. Both, however, have enough semantic background to distinguish between a horse and a cat, or any other animal.

In the case of the UNL, computers provide the semantic background in a format that can be recognized by other computers. Knowledge is expressed in sentences, built on an appropriate grouping of UWs, relations and attributes, with the help of the "Knowledge Base". In essence, the Knowledge Base consists of information on the way of mastering the binary relations between two UWs. This constitutes a vast and complex system of knowledge that in theory contains all the concepts represented in UWs, all the meaningful modifications that may occur between them, as well as all the meaningful modifications that they may have from the perspective of the author of a given sentence. The Knowledge Base is, to a certain extent, a "replica" of what exists in the mind of person with advanced knowledge of his or her native language.

Computers with a Knowledge Base, therefore, can generate any kind of information and universal expressions of knowledge and thus make this available in native language words understood by human beings. This first requires human knowledge from any language to be expressed in the UNL and stored in a "reservoir" accessible to users through the Internet. The UNL Centre has created the basic design, technical specifications and technologies for building up the Knowledge Base. Following these specifications, developers in any language can take part in the huge and never-ending endeavour to build up the reservoir of global knowledge, an endeavour that is carried out with the use of UNL technologies.

THE UNL SYSTEM

The architectural design of the UNL system has two components: the "UNL Language Server" (UNL-LS) unit, which consists of a set of software programmes and files, and the network of UNL-LS units, interlinked throughout the Internet.

The UNL Language Server

There is at least one UNL-LS unit for each native language containing a set of four basic components: the twin Enconverter (EnCO) and Deconverter (DeCO) software, and the twin Editor and Viewer tools. These operate in conjunction with the UNL Language Dictionaries, Knowledge Base and language-specific rules, as shown in Figure 19.2.

Figure 19.2
The UNL Language Server Unit

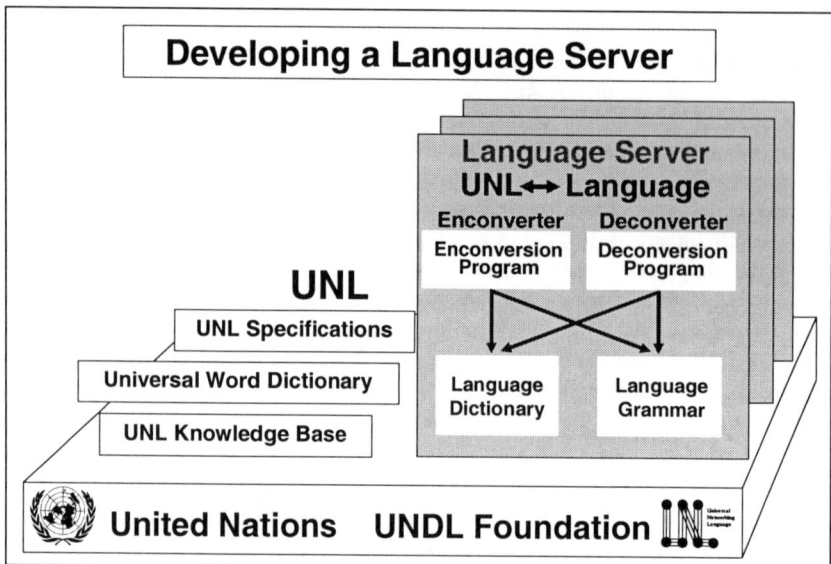

EnCO: Enconverting Information into UNL. The EnCO and DeCO UNL software programmes play a central role in making interlingual communication possible. Together they enable communication from a given native language into many other native languages. The generation of a UNL document starts by the enconversion of a text from a native language into the UNL with the help of the EnCO programme software. The EnCO works in conjunction with a UWs Dictionary, the UNL Knowledge Base and the conversion rules of each native language. It performs morphological, syntactic and semantic analyses synchronously. The end result is a UNL document generated from natural language texts. This is done automatically and through an "author-computer interactive process", without any knowledge of the UNL. As it is different from Esperanto, users do not need to learn the UNL in order to take advantage of its utility.

The EnCO is a language-independent software, but it can be applied to any languages. It has been created by the UNL Centre and provided to partners so that they can develop the necessary conversion rules specific to each natural language.

DeCO: Deconverting from UNL into Native Languages. The DeCO software automatically deconverts UNL into native languages. It works in conjunction with the UWs Dictionary and the conversion rules specific to each language, and carries out deconversion from UNL into a native language. It can achieve high standards and also correct results.

The DeCO is also a language-independent software, but it is applied to each native language to make interlingual communication possible. The UNL Centre has created the DeCO software and distributed it to all members of the Languages Centres and the UNL Society to ensure that all languages are treated with the same standards of quality and precision.

UNL Tools

Each UNL-LS unit operates with several computer tools, but two are essential to enable users to work with the UNL. One is the UNL Editor, which assists users in enconverting texts into the UNL, and the second is the UNL Viewer, which enables users to read texts deconverted in their own native language.

The UNL Editor. The UNL Editor functions as a sort of word processor for managing the enconversion process. It enables users to input a text in their native language and simultaneously have it enconverted into UNL expressions. The EnCO programme, with the help of the UW Dictionary, the Knowledge Base and the grammar rules of each language, carries out this enconversion process. As the author inputs text, the Editor automatically generates the corresponding UNL expressions, and instantly displays on the screen the text in the author's native language after being processed by the EnCO. With this feedback, the author can check whether the enconversion is correct by comparing the intended meaning of the original text with the enconverted sentences. As in the "spelling-check" of any word processor, the author can then revise his or her original phrase word-by-word and sentence-by-sentence until 100% accuracy is achieved in the UNL expressions. This process is automatic but it may require some pre-editing if the set UW Dictionary and the UNL grammar rules of a given language have not yet been well developed.

The UNL Viewer. The UNL Viewer is a tool for users to access and read UNL documents in their native language. It deconverts UNL expressions into native language words and sentences, and displays them on the screen in the desired script (Roman, "Kanji", Cyrillic, Arabic, etc.). To generate a text in the reader's desired language, the Viewer functions are interlinked with the DeCO programme, which in turn accesses the UWs Dictionary, the Knowledge Base and the UNL grammar rules of that language. The quality of the text generated depends on the correctness of the enconversion of the original text, and on the quality of the UW Dictionary and grammar rules of the target native language.

The UNL System: Connecting to Internet

The United Nations is working towards the development of the UNL for this purpose. The UNL will provide a common communication environment for different native languages. Furthermore, the UNL will expand information, education and business opportunities around the world. Mutual understanding among different cultures is one of its ultimate goals.

Understanding How the UNL System Works. Figure 19.3 helps to understand how the UNL System operates by showing the activities taking place in three interlinked blocks.

In block 1, the user writes a document in his or her native language using a PC. The UNL Editor tool, in connection with the UNL Language Server, enables him or her to write it in the UNL. As the user types word by word, and sentence after sentence, a full paragraph or a complete document, the UNL Enconverter software (EnCO) simultaneously enconverts all inputs into a UNL representation. This can be done through an interactive process between the writer and the computer. The UNL Viewer shows the writer the document as it is enconverted from the UNL, which represents how the system understands the original document being produced by the writer, thus allowing the latter to check the accuracy of the enconversion. In such an interactive process, the writer can produce a UNL document as precisely as desired. Users do not need to know the UNL, nor how the EnCO and DeCO programmes operate. All that is required is to input the correct sentence in their native language with the help of the UNL Editor, like a word processor.

Figure 19.3
How the UNL System Works

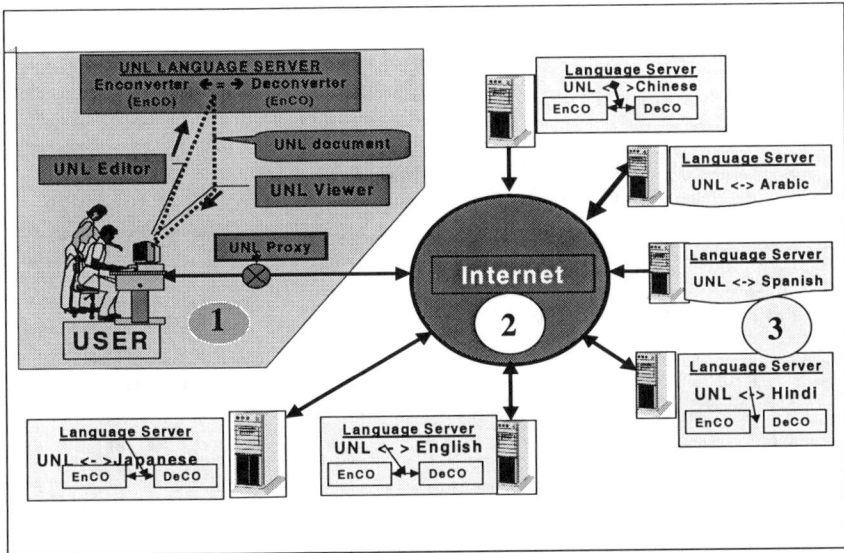

In Block 2, the UNL document is placed on the Internet through a UNL Proxy Server. Information, text, documents, Web pages written in the UNL can be stored in archives or downloaded and shared throughout the Internet with multiple users in all native languages equipped with the UNL Language Server set. UNL documents can be processed in individual computers, exchanged through local networks (LAN), or distributed through WWW servers. They can also be forwarded through file transfer programmes. UNL documents received in a network terminal can be deconverted into each native language and read by anyone on a browser equipped with the UNL Language Server set. This is one of the outstanding features of the UNL system, for it allows for the synchronous and asynchronous operation of multiple language servers simultaneously.

Block 3 shows that each native language has its own UNL Language Server. This allows any user to interact with others in his or her own language, while the others use their own language through the Internet. The number of languages that can be supported is unlimited. The Language Servers are all equipped with the same set of software as in Block 1—the EnCO and DeCO programmes, as well as the Editor and Viewer tools, which are connected to the Master Dictionary of UWs and the UNL Knowledge Base. Users can therefore write, read and exchange UNL documents in any language that has developed its UNL Language Server. They can also improve the existing UWs Dictionary, or create one where it does not exist, and expand the Knowledge Base indefinitely. The necessary tools, specifications, instructions and manuals are available on the Web to carry out these tasks.

POTENTIAL BENEFICIARIES AND BENEFITS OF THE UNL

Communication is essential for understanding among peoples to bring about world peace. Although the Internet holds out the promise of sharing information with all societies and nations, language barriers limit their access to the wealth and power that information and knowledge can provide. Some figures exemplify such a situation. For instance, there are 18 official languages used in India and national government documents should be written in all these languages. Table 19.1 illustrates dramatic situation from another angle. The database "Ethnologue" lists the ten most spoken languages according to the primary country location.

Table 19.1
Top Ten Languages by Population

RANK LANGUAGE	POPULATION
1. CHINESE (MANDARIN)	885,000,000
2. SPANISH	332,000,000
3. ENGLISH	322,000,000
4. BENGALI	189,000,000
5. HINDI	182,000,000
6. ARABIC (ALL COUNTRIES)	177,000,000
7. PORTUGUESE	170,000,000
8. RUSSIAN	170,000,000
9. JAPANESE	125,000,000
10. GERMAN, STANDARD	98,000,000

Source: Grimes, B. (ed.) (1996), *Ethnologue: Languages of the World*, 13th Edition. Dallas, TX: SIL International.

Although the networking technology of the Internet is now available to link all peoples, a high percentage of the world's population cannot read or write the languages of advanced societies. Language barriers continue to prevent effective communications. By allowing people to participate in global information exchanges in their native languages, the UNL will fulfil the promise of equal opportunities for all, a commitment of the United Nations and of governments professing democracy.

Potential beneficiaries of the UNL include citizens, scholars, business circles, the media, as well as government and nongovernmental organisations in a variety of contexts. For example, the UNL could facilitate the dissemination of research findings and educational material among scholars. Visionary educational institutions are already creating "virtual universities", in which students can participate in degree programmes without physically attending classes, through courses offered on the Internet. The UNL expands this concept to include students from anywhere in the world attending a university of their choice, regardless of geographical location or language considerations.

In the business world, the use of the UNL can be applied to all fields of activities covered by the Internet. UNL applications in certain fields would be attractive even at the commercial level. These include a variety of activities ranging from e-commerce to interactive television. For instance, telecommunication services, in particular mobile telephony, and media information providers could use the UNL for mass circulation of information. Multinational organisations could resort to it for internal communications, collaborative research, staff training and market information between headquarters and overseas operations in many different languages simultaneously.

Put more broadly, the UNL is designed to benefit individuals, peoples and nations. As implemented on the Internet, the UNL will enable:

- citizens to produce and share information around the world in their native languages;
- institutions to circulate information, collaborate and undertake joint ventures in the fields of research, education, business and industry, thus providing everyone with equal opportunities for individual, industrial and economic development;
- international organisations, whether governmental or not, to disseminate information about their missions and activities, as well as to promote collaborative programmes across cultural and language barriers;
- governments and states to eliminate language barriers within countries, improve civil citizenship education and social cohesion, and promote global understanding.

THE WORK AHEAD

The UNL is a long-term endeavour, and will evolve continuously towards attaining an increasingly higher quality. When launched in 1996[5], it was estimated that a period of ten years would be required to achieve full coverage of the languages of the Member States of the United Nations. The initial phase (1996-2000) was dedicated to creating the UNL language and developing the architecture of the UNL System. The UNL language server for 15 native languages was developed simultaneously. In its second phase (2000-2006), it is planned to extend access to the UNL to cover the natural languages of all UN Member States. In this way, approximately 90% of the world's population will have the opportunity to communicate with each other through the Internet in their native language.

The structures of the UNL language, as well as the architectural design of the UNL system and its core technology, have now been completed. Although the conceptual basis and the core technology, with the relevant software programmes and tools, are already in place, improvements of the core UNL itself always will be needed. Furthermore, the development of language servers will continue as long as there are nations that wish to participate. The mission of the Universal Networking Digital Language (UNDL) Foundation, a newly established entity, is to carry on with this task.

Since January 2001, therefore, the development and application of the UNL is being guided by the UNDL Foundation, under the auspices of the United

Nations. While the UNDL Foundation is responsible for creating and improving the UNL language and its operational system, as well as providing prototype enconversion and deconversion software, international research institutes and industrial partners will develop the individual modules for their native languages. The UNDL Foundation manages the overall R&D programme and provides the set of UNL language specifications, software and tools. It is also responsible for the development of language server units for each native language interested in being part of the UNL System. For this purpose, the UNDL Foundation has set up an infrastructure of computer servers, making it possible to collaborate and share the progress and experience of the international scholarly community that is developing the UNL. To facilitate interaction among the different players, the Foundation organises multidisciplinary conferences, workshops and training courses, drawing on both existing practices and the creativity offered by partners operating in any language.

The UNDL Foundation welcomes cooperation at all levels. When all is said and done, the UNL initiative is the result of close collaboration between the academic community and industry, and it invites the sponsorship of corporations and foundations.

CONCLUSION

Sharing information among countries is a crucial part of the work of international organisations such as the United Nations, UNESCO, the European Union (EU) and many others. These organisations are concerned with providing equal opportunities and access to these benefits for all their Member States. Many nations, however, have fewer opportunities to learn about the activities of international organisations, despite the great significance these activities have for them. At present, most information material on science, technology and education is written in English or in a limited number of other languages. While this benefits millions of people, millions more are deprived of access to such information material because they do not speak the required languages.

Although the UNL is still at an early stage, its infrastructure and architectural design are available for collaborative work between scholars, developers and providers working in any language. Our common and clearly defined strategy is to work together towards a goal that will be beneficial to everyone. Achieving this goal will depend on the vast intellectual resources available in all languages. Thus, the development of the UNL offers a unique opportunity for genuine global solidarity. This is an invitation for a universal win-win collaboration that meets the ideals of the UN, UNESCO, the EU, and many other international organisations striving to ensure equal opportunities for all peoples. It is a call for a global partnership to achieve a precisely targeted, attainable, tangible and long-lasting communication facility. The UNDL Foundation provides the means to accomplish such an ambitious collaborative win-win endeavour.

NOTES

1. This chapter was prepared with the assistance of the UNDL Foundation.

2. The Institute of Advanced Studies of the United Nations University (UNU/IAS) was created in 1995 and is located in Tokyo, Japan. It is an advanced research and education institution with a flexible and multithematic programme orientation focusing on the interaction between social and natural systems. The United Nations University, headquartered in Tokyo, is an international academic organisation that brings together the world's leading scholars to address pressing global problems, through multidisciplinary research, postgraduate education and information dissemination.

3. Interested readers can see, for instance, G. Varile and A. Zampolli, eds., "Survey of the State of the Art in Human Language Technology," in *Linguistica Computazionale* Vols. XII-XIII (Pisa: Giardini Editori e stapatori, 1997). See also a previous analysis on "Current Issues in Computational Linguistics: in Honor of Don Walker," edited by Antonio Zampolli, Nicoletta Calzolari and Martha Palmer in *Linguistica Computazionale* Vols. IX-X (Pisa: Giardini Editori e stapatori, June 1994).

4. Hiroshi Uchida, Meiying Zhu and Tarcisio Della Senta, *The UNL, A Gift for a Millennium* (Tokyo: Institute of Advanced Studies, November 1999).

5. The Institute of Advanced Studies at United Nations University launched the UNL in April 1996. Its mission is to eliminate language barriers among the peoples of the world by creating a medium of information exchange that permits individuals to communicate in their own native languages. After the development of the basic architecture, the UNU/IAS created a new entity, the UNDL Foundation, fully dedicated to carrying out R&D and promoting the UNL for public use worldwide. The Foundation is continuing the initiative in cooperation with international partners.

20

The Color of Voice,
the Resonance of Language:
Precedents of the Digital Divide

Mercedes Lynn de Uriarte

INTRODUCTION

Early in August, 2000, *Washington Post* Reporter Michael Fletcher reintroduced a topic that has long agitated media scholars, professionals and audiences. He reported that the color of Spanish-language television in the United States is white. This was not surprising. In fact, this formula is found throughout the Western Hemisphere, distributing more generously images that mirror those of power holders.

Popular electronic culture from entertainment to news has always had a bleached aspect, reflecting as it does a certain hegemony that has emerged from discourse born in the 1700s and anchored in the Enlightenment and liberalism. It is perhaps the most discernible but certainly only one way in which populations lose voice.

Fletcher recounted the latest experiences of Malin Falu, who "has long been among New York's most beloved Spanish-language radio personalities. The region's huge Latino community," he reports, "has showered her with awards and has compared her to Barbara Walters and Oprah Winfrey". All three women are known for powerful voices and civic language. Walters and Winfrey are the two most powerful women in U.S. mainstream television. And although Winfrey is African-American, she remains a singular black representation in the role of a major media figure. Her accomplishments may have been aided by her earlier success in film, where her blackness first emerged in appropriate dramatic roles. Audience acceptance there allowed for a transition to mainstream television. But such trajectories are rare. Indeed, there is so little brown or black on the U.S. small screen that in 1999 a coalition of minority organizations, led by the National Association for the Advancement of Colored People, urged minority supporters to boycott the medium. Before long all major networks were negotiating small increased inclusion. Images that do appear tend to be negative stereotypes. "In television drama Latino representation primarily takes place through peripheral, nonrecurring roles frequently depicting youths as involved in drugs and/or gangs and depicting adults as immigrants and undocumented workers. These images are reinforced in news broadcasts."[1] This happens even though statistics prove that these are not the usual experiences.

For Falu the difficulty lies in making her contributions visible rather than audible. Brown-skinned Falu is African-Latina. She represents an invisible aspect of the multicultural environment both North and South Americans confront both domestically and in globalization. She personifies the matter of communication access. Who may easily navigate these highways of exchange? Her experience is a microcosmic reflection of the conflict manifested in communication history, where contemporary issues now revolve around the digital divide, carved to exclude certain sectors of race and class.

The 1999 report by the U.S. Department of Commerce, "Falling Through the Net: Defining the Digital Divide," documented the absence of poor and especially people of color in the increasingly wired U.S. society.[2] Others peg the participation in Latin America at less than 5% of the population, with the same populations excluded. In fact, as the new millennium opened, "barely 2% of the world was connected in one way or another to a world information network."[3] Nor is the communication technology balance of power apt to change given history and the current distribution of research and development resources: 40% by North America, 28% by Western Europe and a little over 18% by Japan and the newly industrialized nations.[4] These stark communication gaps continue not only a history of technological development but the promotional advocacy that accompanies change.

Despite claims of communication democratization, which have been reinvigorated in the age of the Internet, a selective inclusion in this dialogue is evident. History repeats itself.

"The communication network is an eternal promise symbolizing a world that is better because it is united. From road and rail to information highways, this belief has been revived with each technological generation, yet the networks have never ceased to be at the center for struggles for control of the world."[5] But technology emerges with a fanfare of promise. At the 1893 Chicago World's Fair, for instance, "promoters presented the railroad as a major civilizing force, one bringing propriety, communication, and understanding to the world.... America's most significant gifts to the twentieth century were advanced technology and mass culture."[6] By 1893 the United States was well into liberal developmentalism, a vision of high technology and mass consumption, much anticipating the ideology of the current era.

Liberal developmentalism advocated five major positions: "(1) the belief that others nations could and should replicate America's own developmental experience; (2) faith in private enterprise; (3) support for free open access for trade and investment; (4) promotion of free flow and cultures; and (5) growing acceptance of governmental activity to protect private enterprise and to stimulate and regulate American participation in international and cultural exchange."[7] These ideas sparked growing interest in internationalization, much as similar ideas push globalization. Between 1897 and 1914 American direct investments abroad more than quadrupled, rising from an estimated $634 million to $2.3 billion. This direct investment provided the basis of many huge international corporations. Eventually these ideas led to multinational and transnational corporations. A larger market was perceived as a means to level barriers between nations and class, to move toward an age without scarcity. The American Dream, historian Daniel Boorstin would later observe, became "a democracy of things." Today we speak of globalization in much the same terms.

But early technology developed within an intellectual era that fostered ideals of equality, the perfectibility of human society and thus saw free exchange as a human right. By 1763 it was established as a principle of human rights, free, borderless communication of thoughts and opinions. Soon thereafter, in its push toward internal unification, France sought to establish both a national and a universal identity. In 1789 it called for "One nation, one law, one language." The objective—to absorb differences and to break "barriers of specific local characteristics inherited from the times of feudalism and absolute monarchy"[8]—has since manifested itself in a quest of economic dominance. The arguments that promote these movements have changed little. From its earliest stage, "long-distance communication technology was promoted as a guarantee of the revival of democracy."[9] But dreams of empire soon captured that potential; by the mid-1800s, the United States sought both commercial and territorial expansion. In 1842 one of the first graphic ads for agricultural technology aimed at large-scale farmers appeared in Mexico—an equipment sketch embedded in culturally irrelevant images. Eventually, media also exported culturally hollow and ideologically driven content. These trends continue virtually unmodified.

Less than six years after the farm equipment ad, the United States had stripped Mexico of half its territory. The U.S. invasion—which initiated U.S. foreign media coverage and motivated the naming of the first correspondents—spurred communications development as war so often does. The 1846 U.S. war with Mexico marked the first use of electric communiqué in the history of war. It also created some of the earliest and most enduring stereotypes about Mexicans in what was to become a system of limited-access communication—where exclusion from news reporting on any significant scale was limited to white men. The press took early advantage of the technological pioneers that became the first corporations. By the early 1900s telegraph was laying the foundations of what would later become a communications empire.

Each technological development brought confrontation in geocultural territories where borders have been repeatedly redefined in struggles revolving around language, physiognomy and, ultimately, consciousness. "The homogenization of societies is inherent in the unification of the economic sphere; their fragmentation is the collary."[10] Thus the color of mass media and news, including the color of consciousness, which this chapter addresses, is both predictable and disappointing. It is the color of power.

For on the one hand, as each technological advance promises liberation, so it continues to restrict in ways that compress control. This chapter does not address the outright dictatorial control of press content (although that, of course, inappropriately continues in some parts of the world). This chapter is concerned with the forms in which communication structures reality and transmits cultural values embedded in political intent, and power which retains technological and mercantile mediation and whose devices—the fragmentation that dislocates and decenters, the flow that globalizes and compresses, the connection that dematerializes and hybridizes—implement the turning of society into a market.[11]

This chapter briefly explores what Max Horkheimer and Theodor Adorno describe as "the production of culture to a commodity,"[12] and introduces concerns raised by Manuel Castells who defines "black holes of informational capitalism."[13] Because of its significant direct and indirect international role, this chapter concentrates on these matters as they are manifest in the U.S. communications industry. It provides brief commentary on exported imagery in advertising, commercial ambassadors that convey values and a selection of examples from the companion construction of reality that is news. For it is within the latter constructed space that the rationalization for actions by the powerful in defense of those commercial values take place. Historically, communication technology has served to exclude from view most of the world's inhabitants, to market outside their sphere of need, to convey agendas of those who define political interest absent their participation in preliminary discourse. Nevertheless, communication power holders once again argue that the emergence of new technologies will foster change as multiculturalism becomes integral to globalization.

Today, five communication corporations within the United States—Disney, Westinghouse, GE, Murdoch's News Corporation and Time-Warner-Turner—control virtually everything seen, heard or read. Nine major entities share this role worldwide. It is a media system that "plays a central role in the development of 'neo-liberal' democracy; that is a political system based on the formal right to vote, but in which political and economic power is resolutely maintained in the hands of the wealthy few."[14] These entities speak the language of class interest, a language often reflected in news reports which serve to construct a reality sometimes blatantly contradicting documented events, as for example, the disinformation campaign to which some news reports contributed from Central America during the 1980s, when the United States both overtly and covertly intervened in revolutionary struggles there. Throughout this period U.S. news reports were censored both by omission of critical information or by outright distortion of facts that contradicted the U.S. State Department version of events.

Perhaps the most absurd of these were reports covering the 1990 elections in El Salvador. First the United States staged and financed much of the process, then it objected to the outcome and subsequently it negotiated an arrangement whereby someone not even on the ballot was inaugurated as president. Television news anchors and other reporters hailed the new president as a "triumph of democracy."[15]

Equally serious is censorship by exclusion: the distribution of information within certain population configurations, which decrease in circulation size relative to audience access to power. Thus decision makers, the smallest group, draw upon the broadest range of sources, while mass media addresses a larger, class-defined, but power-conscribed group, leaving the most numerous population to fall into information voids or "black holes." Today, according to 1999 reports by the U.S. Department of Commerce, whites and Asian-Americans use the Internet most often—37.7% and 35.9% respectively. They also have more home access. Only 19% of African-Americans and 16.6% of Latinos use the Internet, usually at a site outside their homes. Only 12% of the lowest income level use the Internet, while almost 60% of those earning $75,000 a year or more do so; 15 million high-income, U.S. households are online compared to only 3 million of the poorest households.

Thus the digital divide reflects the well-being divide. As is often the case, race and ethnicity matter less than class. But the distribution of class places poverty less frequently within white populations. For example, almost half of all black and Latino children in the United States live below the poverty line; the figure is higher for American-Indian children. This class fracture manifests even within those societies whose computer development is strongest. The digital divide in the United States means that the higher the income, the more likely to be navigating the information superhighway. The U.S. Department of Commerce reports that more than one-third of all whites use the Internet, while fewer than one-fifth of blacks and Latinos do. However, two-thirds of those on line indicate that e-mail was their main motive for getting wired.

The digital divide simply continues media divisions already manifest in the other sectors of the communications industry. Similar problems exist in U.S. newsrooms, where for more than a quarter century, the American Society of Newspaper Editors (ASNE) urged racial and ethnic parity by the year 2000. ASNE set this goal a decade after the National Advisory Commission on Civil Disorders condemned the de facto segregation in U.S. newsrooms. The Commission was appointed by President Lyndon Johnson after more than a hundred race riots wracked the United States in 1967. Among the questions they were asked by the president to explore was the role of the press in race relations. At the time 1% of all journalists were black. When ASNE began taking an annual newsroom census in 1978, 4% of all print journalists were minorities (African-, Asian-, Native-American or Latino). In 1998, when they announced that parity would not be reached, almost 12% of the print press was minority. Figures for radio (10%) and television (19%) integration fall within the same range, with television closest to parity. Minorities account for more than 22% of the U.S. population. However, these populations are the fastest growing in the nation, where it is estimated that no group will comprise a majority within the next 30 years. The goal of parity is sought to ensure voice so that the civic interests of all constituents, which the press is ethically bound to address in a democracy, may be heard.

Thus two pertinent questions become: What past use—and by whom—have commerce and news media made of communications technology? What evidence is there of a change in newsroom culture?

EARLY PATTERNS

"The contrast is striking between utopian discourse of promises for a better world due to technology, and the reality of struggles for control of communication devices and hegemony over norms and system."[16] Cycles of resistance and minor accommodations are also part of this history. Such a pattern within the Western Hemisphere traces recurring confrontations between the United States and various Latin American nations, often driven by U.S. expansion motives. An early example is provided by the era of "manifest destiny," a concept originally coined in 1845 by John O'Sullivan, editor of the *Democracy Review*, which rationalized armed aggression with supremacist arguments also rooted in concepts of a better world. This period shaped both cultural perception and that of the news industry in ways that are still being felt.[17] "It is our manifest destiny," wrote O'Sullivan, "to overspread the continent allotted by Providence for the free development of our yearly multiplying millions."

Expansion campaigns always involve communication technology advances —beginning with the first offensive U.S. war, which had major impact on journalism. The 1846 U.S. invasion of Mexico initiated four major press developments: (1) It pioneered the use of electronic communiqué in the service of war. President Polk's message to Congress calling for war was telegraphed to the *Baltimore Sun*, for its exclusive use, marking the first time the two-year old device had been used in this way and opening a new communications era. (2) It initiated U.S. foreign coverage; the first foreign correspondents were named to cover this war. (3) It fostered press use of state-of-the-art technology; a communications patchwork of Pony Express, stagecoach, telegraph and rail carried press dispatches to home newsrooms. (4) It motivated cooperation between news competitors; because communications technology was too expensive for any one newspaper to support, five organizations joined together to cover costs, becoming the Associated Press. So proficient did journalists become that they often beat the military in advising President Polk of battle results.

Correspondents traveled with U.S. troops recording the battles. U.S. victories against poorly prepared opponents in a bankrupt nation independent barely 25 years gave the U.S. military and the press an arrogant sense of superiority, which echoed in press accounts and was reinforced by politicians. It provided an opportunity to appeal to exaggerated patriotism.

However, even with shared expenses, cost remained a consideration. Compressed accounts were encouraged, with direct impact on language, especially as it was used to construct a reality in the service of motive more than as a tool for an accurate chronicle. The birth of the sound bite began with print bites. The pyramid structure of news articles, which put the most important aspects of the story at the top, and shorthand terminology became standard practices. This was language in the service of agenda—in this case, of military and political agenda. Ultimately, the press provided stereotypes of Mexico and Mexicans that remain entrenched.[18]

Communication technology led also to gross production of inexpensive books and the rise of a mass audience. Soon perceptions by the press contributed to the tapestry of observations about the southern neighbours.

The Mexican War boosted the popularity of the new paperbound books. Soldiers bought them in quantity and carried them to Mexico in their knapsacks. They became the standard reading fare on the long sea voyages from eastern ports and broke the monotony of camp. Publishers, with their eyes on sales, solicited authors and tailored their lists with the military market in mind. The war provided settings and plots for scores of works that, like the penny press, carried the conflict into households throughout the country. "This war literature has circulated through the newspapers and cheap works over the land," moaned a critic of the war. "It is so diffused, that it enters every crook and corner of the land."[19]

These events initiated a pattern in which those nations with the greatest communication technology become the constructors of reality in the press, popular culture and early education. These patterns are embedded in those of commerce, which also began during the same historical period.

Equally significant, the war and the efficiency of communication technology laid the cornerstone for future U.S. interpretation of Mexico (and eventually other Latin American nations) to others. This was the first foreign war for the relatively new northern nation. It would be used to define North Americans and to construct both a national and international identity. It was fought almost entirely in a distant exotic land few of them knew. "The Mexican War," wrote one historian, "was a window through which Americans were able to view another land, another people, and thus overcome limiting parochialism."[20] But accounts of the clash were mostly chronicles of U.S. superiority. No voice was provided for Mexico's version of events. Thus began a pattern protected by control of communication technology in which U.S. voice conveyed the sole descriptions of reality circulating within the United States. That technology quickly became the ally of commerce.

The 1893 Chicago World Fair took place less than 50 years after the United States had absorbed Mexico's northwest. By then, the United States was well into liberal developmentalism, a vision of high technology and mass consumption, in which much of the current era's motivation is anchored. These ideas sparked growing interest in internationalization. Between 1897 and 1914 American direct investments abroad more than quadrupled, rising from an estimated $634 million to $2.3 billion. This direct investment provided the basis of many huge international corporations. Eventually, these ideas led to multinationals and transnationals. A larger market was advocated as a means to level barriers between nations and class, to move toward an age without scarcity. Glowing reports of modernization claimed that improved communication technology would become the handmaiden of democracy. Today we speak of globalization in much the same terms.

Direct investment allowed North Americans, the British and other Western European nations to move into control of many foreign resources in less industrialized nations. In Mexico, under dictator Porfirio Diaz, Mexico turned over its mineral resources, wide expanses of arable land and a cheap labor force to foreign advantage. President Taft wrote: "Immediately before us at exactly the right time, just as we are ready for it, great opportunities for peaceful, commercial and industrial expansion to the south are presented.... In many respects, the people of the two continents are complimentary to each other; the South American is polite, refined, cultivated, fond of literature and of expression of graces and the charms of life, while the North American is strenuous, intense, utilitarian. Where we accumulate, they spend."[21]

In 1909 the first international syndicate—the International News Service— was established by Randolph Hearst, offering a wide variety of articles and other media content for reproduction and translation to nations with less developed media reach. "The nineteenth century hailed communication as an agent of civilization." The concept of interdependence was seen as pushing the world "toward cultural unification."[22] In 1912 Taft called for expansion of U.S. foreign investments and for the establishment of U.S. banks and newspapers in other countries. Between 1900 and 1920 the United States vacillated between dollar diplomacy and moral diplomacy, between foreign policy of self-interest and one of idealism. This period was accompanied by what some called the "export invasion" as low-priced, high-quality U.S. goods entered foreign markets. American industrialists pioneered assembly-line technology, reducing the cost of goods by using interchangeable parts, mechanization and specialization. The use of image as commodity multiplies soon afterwards. Here is seen an early push for physiological conformity and early messages of technological superiority— familiarly called "Yankee know-how."

The expansion of international trade was promoted within liberal tenets of "free enterprise, free trade, free men," bolstering the myth of American exceptionalism.[23] On keeping with what would become a standard mark of value, a 1899 Mexican tailor's ad legitimizes his skill by attributing to U.S. training. But directions of value were unidirectional. They left unchallenged an early myopia, an invisibility of the unfamiliar. A willful ignorance of the consumer target framed exported commercial messages.

By 1912, as the Mexican Revolution entered one of its most violent phases, patent medicine faced regulation in the United States, so setting a trend that continues today, the controversial product was marketed in Mexico. Ads for Lydia Pinkham copy described women as fragile and easily incapacitated by illness, unable to do household chores, but the women then most likely to be doing such work were unable to read, were fired if unable to meet labor expectations and enjoyed no socially secure position. In fact, during that period the domestic worker often became a *soldadera* or female participant in revolutionary forces.

In 1927, before Mexico emerged from the revolutionary era, Ford Motors advertised snappy two-door roadsters for the Mexican woman driver. They exported what had been a successful marketing campaign in the United States aimed at the newly "liberated" suffragette, whom they had just discovered as a new U.S. consumer pool, without information about the southern nation. The car, which had only one seat, was an extremely expensive luxury item meant only for the elite. To purchase such an item during a period of recurrent violence would be foolhardy. For wealthy women to purchase such an item at any time would risk social censure. Wealthy women neither drove, nor shared a seat with their chauffeurs. Throughout the 20th century a steady stream of ethnocentric, naive images marched south. "The upward curve of U.S. investments abroad was accompanied by the establishment of advertising agencies, which became the bridgehead of commercial culture."[24] U.S. lifestyle and predominant images were celebrated with minor adjustments. One in particular symbolizes the superficial approach to difference. When Gerber, the maker of baby foods, was criticized during the 1970s civil rights era for its uniformly white babies in marketing messages, it eventually air brushed earrings on the ads meant for Latino market distribution. More recently, *Cosmopolitan,* a U.S. magazine with translated and tailored counterparts, displayed its globalized version of female beauty: four slighter darker-skinned versions of women with Anglo features and a certain slightly exotic shape to the eye.

PRESS COUNTERPARTS

This myopia can also be found in news coverage then and now. The narrow, ethnocentric interpretation of Mexico, begun in the 1840s, framed accounts of the 1910 Mexican Revolution, the 1930s rebellion in El Salvador and the nationalism of 1930 Nicaragua. One of the most consistent oversights is U.S. press failure to provide context to crisis. It also fails to report the intellectual foundation to resistance and revolution or to report regularly, consistently or comprehensively during the periods of repression that precede internal strife. Despite advanced technology that assists research and makes experts more easily located, the flaw remains, leading to serious misrepresentation, which is often embedded in stereotype. For example, both Farabundo Marti, who led the 1932 Salvadoran rebellion of peasants forced off their lands by large coffee growers, and Augusto Cesar Sandino, leader of the Nicaraguan troops resisting U.S. military occupation, were influenced by the intellectual framework of the Mexican Revolution, while they were in that nation during the early days of the upheaval. Both brought ideas home, where they helped shape domestic response to similar problems. But during the 1980s, when revolutionary forces rose again in Central America to confront those long-unresolved issues, U.S. media attributed the conflict to Soviet influence. That analysis supported the Reagan Administration's anti-Communist foreign policy, U.S. covert action in the area and economic warfare. But it failed to enlighten audiences to the real problems in these countries. More seriously, press complicity with those policies violated both international and domestic press ethics.

The Hutchins Commission crafted the first U.S. press ethics codes after World War II. During the conflict, in the interest of national security, the press both participated in propaganda campaigns and allowed censorship. One of the most intense cases involves U.S. manipulation of Latin American press. Nelson Rockefeller, who served as Coordinator of Inter-American Affairs (CIAA) a component of the State Department, used the international crisis to capture Latin American press independence. With a $140 million budget over five years, a staff of more than 1,200 journalists, advertising experts, public opinion analysts and the latest technology, Rockefeller "mounted a propaganda effort virtually unprecedented in the annals of American history."[25] With assistance from the U.S. Treasury, he sought to control news content. U.S. corporations were provided tax write-offs for ads placed in Latin American media; Latin American media was provided the advertising revenue if they agreed to accept content guidance from Rockefeller's office.

The Rockefeller office provided not only "canned" editorials, photographs, exclusives, feature stories and other such news material, but also manufactured its own mass-circulation magazines, supplements, pamphlets and newsreels. To ensure understanding of the issues being advanced in Latin America, the office sent 13,000 carefully selected opinion leaders a weekly newsletter, which was to help them clarify the issues of the day. The CIAA also arranged trips to the United States for the most influential editors in Latin America (and later scholarships for their children). More than 1,200 newspapers and 200 radio stations, which survived the economic warfare, were fed a daily diet of some 30,000 words of news in Spanish and Portuguese, which was disseminated by cooperating news agencies and radio networks in the United States to their clients in Latin America.[26]

By the end of the war, 75% of news content south of the U.S. border was provided by Washington. The process established not only a media elite, which soon organized the Inter-American Press Association, but one that benefited by internalizing U.S. perspective.

World War II also provided a window of opportunity for U.S. news agencies and wire services. Absent significant competition from English or French wires services while these nations were under military attack, North American news agencies expanded. Before the war ended the American Society of Newspaper Editors organized an international campaign to retain that new dominance using air transportation provided by the U.S. military to sell support for a concept called "free flow," which promoted the idea that all information should flow freely without interference of any kind. The concept was appealing after years of war propaganda, but it soon was recognized as a one-way, U.S. communication flow. Despite repeated efforts of resistance, that pattern remains pervasive today.

But other media leaders and scholars worried about the postwar state of the U.S. press. In the United States, Henry Luce, a co-founder of *Time Magazine*, provided Robert Hutchins, chancellor of the University of Chicago, $200,000 to head a study on the role of the press in a democracy. He and a dozen scholars and statesman interviewed 58 journalists and editors and 225 individuals who, as businessmen, government or private agencies, regularly interacted with the press. They headed 17 two- and three-day meetings and examined 176 documents. The project initiated an interactive examination of the press which drew upon idealism, philosophy, social reality and established media performance. Then drawing on Enlightenment ideas espoused by founders of the Constitution, the Hutchins Commission set down five standards for a free press, which at the time was radio and print. They believed that citizens in a democracy had the right to expect the press to provide:

- a truthful, comprehensive account of the day's events in a context that gives them meaning;
- a forum for the exchange of comment and criticism;
- the projection of a representative picture of constituent groups in the society;
- the presentation and clarification of the goals and values of the society; and
- full access to the day's intelligence.

Besides a deep commitment to freedom of expression, their report reflects a firm respect for intellectual diversity.

Civilized society is a working system of ideas. It lives and changes by the consumption of ideas. Therefore it must make sure that as many as possible of the ideas which its members have are available for examination. It must guarantee freedom of expression, to the end that all adventitious hindrances to the flow of ideas shall be removed. Moreover, a significant innovation in the realm of ideas is likely to arouse resistance. Valuable ideas may be put forth first in forms that are crude, indefensible or even dangerous. They need the chance to develop through free criticism as well as the chance to survive on the basis of their ultimate worth. Hence the man who publishes ideas requires special protection.[27] The Commission stressed throughout its report the central importance of broad inclusion—of people, ideas and representation.

As they deliberated, the United Nations Educational, Scientific and Cultural Organization (UNESCO) began examining communication issues, first in 1945, at the urging of the United States, which was seeking support of its free-flow campaign. But the concern soon passed to small, recently independent nations that feared for their cultural survival as foreign communications products—especially movies, television and news—overwhelmed their own. These concerns would eventually lead to the appointment of the 16-member McBride Commission to examine these issues.

In 1948 the Universal Declaration of Human Rights, adopted and proclaimed by the United Nations, defined freedom of expression as a basic human right, thereby embracing inclusion. Article 19 stated: "Everyone has the right to freedom of opinion and expression; this right includes freedom to hold opinions without interference and to seek, receive and impart information and ideas through any media regardless of frontiers."

In 1968, after more than a hundred racial conflicts in 1967 had disrupted the social fabric of the United States, President Lyndon Banes Johnson appointed a National Advisory Commission on Civil Disorders headed by then-Governor Otto Kerner of Illinois. Among the five questions Johnson asked the Commission to answer he included "What effect do mass media have on the riots?" The result was a sweeping study of race relations, and a comprehensive analysis of press performance, including that of the newest technology, television. Echoing much that the Hutchins Commission had concluded about the responsibility of a democratic press, the Kerner Commission minced no words in condemning the coverage divide between white and black communities.

The media report and write from the standpoint of a white man's world. The ills of the ghetto, the difficulties of life there, the Negro's burning sense of grievance, are seldom conveyed. Sights and indignities are part of the Negro's daily life and many of them come from what he now calls "the white press"—a press that repeatedly, if unconsciously, reflects the bases, the paternalism, the indifference of white America. This may be understandable, but it is not excusable in an institution that has the mission to inform and educate the whole of our society.[28]

Foreshadowing later findings by media sociologists that the press was an example of institutionalized racism, the Kerner Commission observed that "By failing to portray the Negro as a matter of routine and in the context of the total society, the news media have, we believe, contributed to the black-white schism in this country."[29]

To meet their civic responsibilities, the Commission believed that newsroom culture must change. The media, said the Commission, "must make a reality of integration—in both their product and their personnel."[30] Both U.S. and international commissions firmly favored the elimination of information and participation divides.

In 1976 UNESCO named Irish statesman and journalist Sean McBride, who had won both the Nobel Peace Prize and the Lenin Peace Prize, to head a group of 16 delegates from the United States, the Soviet Union and less industrialized nations to examine conflicted communication issues. They began what would be a four-year investigation and deliberation process.

Meanwhile, meeting in Costa Rica in 1978, UNESCO adopted the Mass Media Declaration, seeking to establish an international code of ethics defining the social obligations of the press. It mandates that mass media (and by extension, communications technology) work toward an equitable world thus reaffirming the role of the press as one of social stewardship—a belief that social well-being takes precedence over individual or commercial goals. The document is a "declaration on the fundamental principles concerning the contribution of the mass media to strengthening peace and international understanding, the promotion of human rights and to countering racialism, apartheid and incitement to war." It recognizes that "the mass media throughout the world, by reason of their role, contribute effectively to promoting human rights, in particular by giving expression to oppressed peoples who struggle against colonialism, neo-colonialism, foreign occupation and all forms of racial discrimination and oppression and who are unable to make their voices heard within their own territories." These precepts are of key importance both in the United States and in Latin America, where populations of color are routinely excluded from discourse and today find themselves on the debit side of the digital divide.

In 1980 the McBride Commission concluded its work and called for a New World Information Order, which sought (1) greater communication freedom and a "horizontal" information exchange; (2) more balance in the directional flow of news and information; and (3) greater diversity of content and wider participation of voice. They noted with concern that the information balance was intricately entwined with the economic structures. Their report called upon technologically more developed nations to assist those less developed to acquire the tools necessary to assure full participation in global discourse. "The right to inform and to be informed makes possible the exercise of all other rights."[31] They Commission specifically also addressed the plight of minorities, noting the situation of those who are denied voice within their own territories, those who live in isolated geographic areas, who live in difficult material conditions, who live on the margins of society as a whole or those where multilingualism has fractured their participation. Their focus foreshadowed the current communication crisis. Today many of these segments of world population are found on the disadvantaged side of the digital divide.

They proposed that financial and technical assistance be provided to close the communication disparity, or divide, which they believed was becoming greater and more serious. "Communication reflects the disparities which characterize the entire international scene and therefore stands in need of equally far reaching changes."[32] Commission chair McBride spoke of the danger inherent in the imbalance in the flow of information. "It was not only the imbalance in news about which the Commission complained, but also the imbalance in the flow of all kinds of information, especially through instruments of advanced technology."[33]

Clearly, all these documents call for the elimination of conditions that contribute to the digital divide. Indeed, the digital divide is a new conceptualization of old problems. These documents provide a precedent for addressing the growing digital divide, not only between nations, but within nations along race and class lines. Moreover, they identify a number of structures through which information distribution is provided that contribute to the problem, especially those that are embedded within economic or commercial determinants. Tensions within nations over the exclusion of whole populations from information exchange continue. Standards set by these commissions seek to avoid a deepening of social and political disparity. Communication technology can lead to increased information flow and to enhanced services dependent upon such flows. But they are too often one-way activities. Smart cards, plastic identity cards within which computer chips are embedded, may speed delivery of health care or consumer services, but they do not equal participation in civil discourse. Today, when nine corporate entities provide virtually everything seen, heard or read, the world faces a global media system that devalues the democratic process. "Survey after survey shows that by 1998 the commercial broadcasters had reduced, almost eliminated, any meaningful coverage of electoral campaigns in their newscasts."[34] Media scholars make similar observations about print press. All evidence indicates that the digital divide contributes toward the establishment of such a system.

IMPLICATIONS OF THE DIGITAL DIVIDE

These patterns do not bode well for the democratization of the Internet, which was not originally conceived as a consumer service, without any responsibility beyond profit concerns or enhanced government functions. Originally developed with taxpayer funds, its commercial potential moved it out of the public-service sphere and increasingly into the control of big business. There are no universal ethical guidelines by which to close the information gap or to widen the distribution of knowledge, let alone to diversify the definitions of content. Today, those who navigate the information highway represent a minuscule number of the world's population. This fracture is manifest even within those societies whose computer development is strongest. In the United States, for instance, the higher one's income, the more likely one is to be coasting along the information superhighway.

Consider the case of Malin Falu, whose case so symbolizes the absence of color, voice and ultimately accuracy within media. U.S. Latinos are virtually invisible in all forms of television fare, including newscasts. This problem is documented every year in the annual Brown Out report compiled by the National Association of Hispanic journalists. For example, in its 1996 study of 12,000 news stories airing on the three major networks—ABC, CBS and NBC—only 121 or 1% focused on Latinos or Latino issues. Yet within about a decade, Latinos will be the largest minority population—a population that is itself racially diverse.

But inclusion cannot be conceived as necessary simply across lines of race and ethnicity. Within those groups are diverse perspectives that must also be drawn into discourse. The struggle over consciousness is often disguised by assumptions of genetic homogenization. That assumption operates as if there is no diversity within groups. Thus the inclusion of those who appear to be different, but who echo established truths, is invited, while intellectual diversity is rejected. This process serves to extend and intensify hegemony, while promoting the appearance of change. As the case of Supreme Court Justice Clarence Thomas so well proves, a carefully selected minority can be used by power holders to undermine others from the same ethnic or racial group who seek inclusion, but bring different, perhaps controversial or resistant, points of view.

In the United States, Internet use falls sharply as income level drops. Figures vary on the number of Latin Americans online from reports of fewer than 1% to almost 3% claimed by NAZCA, the French communications research organization. Experts from the Massachusetts Institute of Technology (MIT) state that fewer than 2% of the world's population is connected in one way or another to a world information network.[35] Nevertheless, by all accounts the figure is minute and resource driven. "Networks, embedded as they are in the international division of labor, organize space hierarchically and lead to an ever widening gap between power centers and peripheral loci."[36] In the process, populations either become identified as markets, or disappear from digitized reality, their cultures and progress assigned to the noncommercial void. Realities are constructed to the convenience of the powerful, with full participation of communication corporations who control technology. "What we are witnessing now, is the formation and reinforcement of powerful multimedia conglomerates that control, at their convenience, in some cases the interested defense of the protection of national cultural production and in other cases the defense of transnational flows."[37]

These capacities shape not only the ability to insert voice into discourse, but they affect the impact of participating languages. Indeed, in celebration of digital communication, silence is overlooked or ignored. Today only 250 languages in the entire world still have at least a million speakers. This is the number considered necessary as a safety level of preservation, as globalization seeks to homogenize the world—every nation, every village, no matter how remote. It is unlikely that those without state sponsorship, including those in the Amazon, will survive. Increasingly English dominates the world, conveying as it does inclusion in the world. "It is the lingua franca of science, the Internet, the movies, rock and roll television and even sports."[38] Where then is *mestizaje*? As Jesus Martin-Barbero notes: "The recognition of the *mestizaje* that constitutes Latin America does not refer to something that happened in the past, but what we are today. *Mestizaje* is not simply a racial fact, but an explanation of our existence in the web of times and places, memories and imagination which, up to now, have been expressed only at a literary level."[39] Martin-Barbero likens the current cultural condition to that of a palimpsest: "The territories and discourses of identity today have the fragile texture of a palimpsest, a text in which an erased past emerges tenaciously, although blurred, between the lines that write the present."[40]

From this perspective, he writes, Latin American authenticity has become a distant form of being and instead has become a decentering suffered by modernity, which is a subject of study for sociologists and anthropologists: "its having less to do with erudite doctrines and lettered aesthetics than with the massification of schooling and the expansion of the cultural industries and therefore with the formation of a cultural market in which the sources of cultural production, instead of the dynamic of communities or authority of the church, are the logic of industry and specialized apparatuses that replace the traditional ways of life with lifestyles formed by publicity and consumption, secularize and internationalize symbolic worlds and segment the people into publics constructed by the market...modernity becomes the collective experience of Latin American majorities because of social and perceptual dislocations of a clearly post modern imprint effecting strong displacements of the compartments and exclusions that modernity has instituted in the course of more than a century."[41]

In his recent book, *Networking the World 1794-2000*, Armand Mattelart acknowledges that Asia and Latin America have adapted technology and "have taken advantage of it, both to perform in the world market and to link up with social and scientific networks... (but) these new sources of modernity coexist— as the second side of the coin—with a galloping process of impoverishment and exclusion of large sections of the population."[42]

As has always been the case, class issues are intersected by race, ethnicity and gender. Distribution of well being placed poverty less frequently within white populations. For example, in the United States, according to census figures, almost half of all black and Latino children live below the poverty level. The figure is greatest for American Indian children. In the richest country in the world, 13.5 million children live in poverty, 5.8 million live in extreme poverty. Latin America has a much greater number. The exclusion of those impoverished from economic opportunity is compounded by literacy limitations. According to the U.S. Department of Education, 44 million North American adults, nearly one out of four, are functionally illiterate. Another 50 million suffer limited literacy. Globally, literacy remains a major problem even within many nations, like Brazil, where elite now enjoy easy Internet access. Literacy and education provide the foundation for participation in the electronic information age. Today, as an increasing number of jobs globally require computer skills, proportionally a decreasing number of individuals can perform them.

Additional concern should be raised by the limitation of deliberation related to inclusion taken by those who decide the direction of technological expansion. For example, "competitiveness was the only argument used to legitimize the historic decision taken at the G7 summit of Brussels in 1995, to rely on the market alone to favor the expansion of information highways." At the same time, "organizers of this summit devoted to the global information society refused to include on the agenda the subject of 'content,' that is cultural diversity, because it was seen as too controversial by nature."[43]

The rhetoric of diversity, multiculturalism and globalism masks continuation and intensification of established patterns of exchange. "Perhaps only in literature does mestizaje cease to be an abstract theme and become a living actor who speaks with a distinctive way of perceiving, narrating and being aware of the world."[44] Thus Falu is included in media, but contained within manageable boundaries of participation; she is heard, but not seen. A disembodied voice of *mestizaje*. "The dimension of ethnic affirmation is not so easy to perceive or to interpret as are the external pressures and mediations through which hegemony operates."[45] Thus although the popularity of Falu represents ethnic affirmation, the pressures of hegemony deny her visibility. Carlos Monsiváis writes of the revolutionary role of Mexican muralists who transformed the masses and customs of daily life into affirmations of Mexican nationalism by making them visible as people, thus socially legitimizing "gestures, customs and manners of speech that were until then, widely rejected or repressed by Mexican society."[46] Falu, in her Latino blackness, inserts a definition of Latino rejected by both

English- and Spanish-language television. In their terms, her exclusion makes sense to power. Monsiváis explains it thus: "Much more important than the 'expressions' of popular in different national contexts was the recognition of the *meaning* these expressions acquired—the masses making themselves socially visible, 'configuring their hunger to attain visibility by having a social space of their own.'"[47]

This definition also helps interpret current newsroom and academic tension between minorities for whom space is made for the appearance of diversity and in response to affirmative action—and those minorities who manage within the space provided to conform to that which remains unchanged or minimally adjusted as a result of their inclusion. Their contributions become "externally picturesque, but increasingly hollow." (Words used by Martin-Barbero to describe 1950s popular U.S. culture are appropriate here.)

So it is that although ASNE mounted a newsroom integration campaign almost a quarter of a century ago, minority participation figures have grown merely 8% in print and not much more in broadcast. In fact, according to a 1999 study by the Freedom Forum, over the course of the previous five years, newspapers hired an average of 500 minority journalists a year and about 440 a year had left the business. Exit interviews cited hostile environment or isolation as a contributing factor. In that context, it is perhaps more significant to consider how many, if spread evenly across the mediums, could be the average number of minorities on the staff of each: 6.2 for all local TV newsrooms and .66 for radio. If distributed across the number of print newsrooms, minorities would average fewer than 4.0. In fact, without the demographic measurements by ASNE and more recently, the Radio and Television News Directors Association, minority participation is largely invisible.

In academia minority participation is minimal. Consider the communications field, where only 10% participate according to 1999 figures from the Association for Education in Journalism and Mass Communication. Moreover, although intellectuals have historically provided critiques of society, that space is increasingly eroded. "The mobilization of energies around competitiveness precipitates the encounter between places in which knowledge is traditionally produced and diffused, such as universities, and the needs economic actors.... Their aim is to enroll geography, history, ethnography, psychoanalysis, sociology and linguistics in the service of corporate performance."[48] The corporatization of universities, which have become increasingly dependent on the business sector for the funding of research, converges with the corporatization of information technology and the compression of media ownership. Simultaneously, corporate marketing seeks uniformity of image, but not universal democracy.

The forging of macroregional markets (NAFTA, GATT, etc.) had hardly been announced when communication groups and international TV channels (such as CNN) or regional ones (pan-American, pan-Arabic, pan-Asian or pan-European) inaugurated a search for cultural universals. Helped along by the integration of communication operations, the third generation of advertising networks, so-called global networks, followed suit responding to the trend of interconnection of markets. One of the axioms in the search for global common denominator is the cultural convergence of consumers.[49]

This continues the trend described earlier, which began in the early 1900s with the internationalization of markets and the exportation of myopia. The cultural convergence of consumers, says Mattelart, is "the product of years of influence of mass culture on the collective imaginary of consumers of diverse cultures. As natural mediums for universality, the culture industries in the United States still appear to be playing an excessively predominant role in defining the parameters of globality".[50] Although he notes the emergence of Brazil's El Globo and Mexico's Televisa, it must be noted that these media corporations, like their U.S. counterparts, include even their own populations selectively.

"Today's technology provides the possibility of a renewed manifest destiny—a perspective by those who control advanced communication technology to native peoples, to museums, to make them as dead as yesterday... Of all the arts and sciences made by man, none equals language. For only language in its living entirety can describe a unique and irreplaceable world."[51]

In his recent essay, Earle Schorris describes the arrival of a blue butterfly, which settled nearby as he sat in a Mayan forest. The color, he says, "was a blue unlike any I had ever seen, hue and intensity beyond naming, a test for the metaphor.... There are nine different words in Maya for the color blue in the comprehensive Porrua Spanish-Maya Dictionary, but just three in Spanish, proving beyond a doubt that when a language dies six butterflies disappear from the consciousness of earth."[52] His observations return this discourse to the plight of those who absent space and image of inclusion disappear, together with history and intellect, from the technological construction of reality.

CONCLUSION

This chapter addresses the color of voice in mass media, including the color of consciousness, now increasingly technologically determined. For although each technological advance promises liberation, it serves mostly to restrict options. This chapter does not address outright authoritarian control of the press (although that inappropriately continues in some parts of the world). This chapter is concerned with the forms in which communication structures reality and transmits cultural values. This chapter is concerned with what Martin-Barbero describes as the burden of "discursive opacity and the political ambiguity introduced by technological and mercantile mediation and whose devices—the fragmentation that dislocates and decenters, the flow that globalizes and compresses, the connection that dematerializes and hybridizes—implement the turning of society into a market".[53]

For, although historically communication technology has served to exclude from view most of the world's inhabitants, to market outside their sphere of need, some once again argue that the emergence of new technologies will foster change as multiculturalism becomes integral to globalization. This argument ignores the repetion of history found today in the serious implications of the digital divide.

NOTES

1. Cristina Bodinger-de Uriarte and Mercedes Lynn de Uriarte, "Falling Through the Media Grid: Secondary Narratives in the Defintion of Latino Families," *Till Death Us Do Part: A Multicultural Anthology on Marriage: Contemporary Studies in Sociology,* Vol. 14 (Stamford, CT: JAI Press, 1999), 282.

2. Jessica Brown, "Digital Divide Provides Opportunities for Corporate Spin," *Extra!*, March/April 2000, 16-17.

3. Armand Mattelart, *Networking the World, 1794-2000* (Minneapolis: University of Minnesota Press, 1996), 100.

4. Ibid.

5. Ibid., viii.

6. Emily Rosenberg, *Spreading the American Dream: American Economic and Cultural Expansion, 1890-1945* (New York: Hill and Wang, 1982), 4.

7. Ibid., 7.

8. Mattelart, 3.

9. Ibid., 2-4.

10. Ibid., viii.

11. Jesus Martin-Barbero, *Communicacion, Culture and Hegemony: From Media to Mediations* (London: Sage Publications, 1993), 28.

12. In the 1940s, Max Horkheimer and Theodor Adorno developed the concept of the culture industry in which they described art as having been "reduced to the status of ornamenting everyday life as a result of the rise of the culture industry." These concepts are discussed by Armand Mattelart and Michele Mattelart, *Theories of Communication* (London and Thousand Oaks, CA: Sage Publications, 1995), 58-65.

13. Manuel Castells, "End of the Millennium", Vol. 3, *The Information Age: Economy, Society and Culture* (London: Oxford and Malden, MA: Blackwell Publishers, 1998), 162.

14. Robert McChesney, *Rich Media, Poor Democracy* (Urbana, IL: University of Illinois Press, 1999), 79.

15. A decade of news distortion is documented by a number of scholarly works, including *War Stories* by Michael Pedley (London: Routledge, 1995), *Manufacturing Consent* by Edward Herman and Noam Chomsky (New York: Pantheon Books, 1988). The Salvadoran elections were covered in many publications and television reports. News treatment of the elections is detailed in *Making News Fit*, a documentary by California Reel.

16. Mattelart, 21.

17. Mercedes Lynn de Uriarte, "Crossed Wires: U.S. Newspaper Construction of 'Outside Others'—the Case of Latinos." Unpublished dissertation for Yale University, 1996 UMI Dissertation Services, 90-129.

18. Ibid.

19. Robert Johannsen, *To the Halls of Montezumas: The Mexican War in the American Imagination* (New York: Oxford University Press, 1985), 176-177.

20. Howard Zinn, *A People's History of the United States* (New York: Harper & Row, 1980), 153.

21. William Appleman Williams, *The Contours of American History* (Chicago: Quadrangle, 1966), 472.

22. Mattelart, 33.

23. Rosenberg, 23.

24. Mattelart, 42.

25. Jeremy Tunstall, *The Media Are American: Anglo American Media in the World* (New York: Columbia University Press, 1977), 139.

26. Ibid., 140.

27. Commission on Freedom of the Press, *A Free and Responsible Press* (Chicago: University of Chicago Press, 1947), 6.

28. Commission on Freedom of the Press, *The Report of the National Advisory Commission on Civil Disorders* (New York: Bantam Books, 1968), 366.

29. Ibid., 386.

30. Ibid., 389.

31. Excerpts from the McBride Commission Report as condensed in "Communication Problems Today," in *Crisis in International News: Policies and Prospects,* eds. Jim Richstad and Michael H. Anderson (New York: Columbia University Press, 1981), 3.

32. Herbert Altschull, *Agents of Power: The Media and Public Policy* (New York: Longman, 1995), 316-333.

33. Ibid., 319.

34. McChesney, 264.

35. Mattelart, 100.

36. Ibid., 98.

37. Jesus Martin-Barbero, "Transformation in the Map: Identities and Culture Industries," trans. Zilkia Janer, *Latin American Perspectives*, May 2000, 42.

38. Earl Shorris, "The Last Word: Can the world's smallest languages be saved?" *Harper's Magazine*, Vol. 301, No. 1803, August 2000, 36.

39. Martin-Barbero, *Communication*, 188.

40. Martin-Barbero, "Transformations," 32.

41. Ibid.

42. Mattelart, 107.

43. Ibid., 118.

44. Martin-Barbero, *Communication*, 188.

45. Ibid., 191.

46. Carlos Monsiváis (1978:101; also 1977:319) as referenced in Martin-Barbero, *Communication*, 194.

47. Carlos Monsiváis (1978:101; also 1977:319) as referenced in Martin-Barbero, *Communication*, 194.

48. Mattelart, 82.

49. Ibid., 33.

50. Ibid., 36.

51. Shorris, 38.

52. Ibid.

53. Martin-Barbero, "Translations," 28.

21

Foreign Language Materials for Business Portuguese: The Role of Technology in the Development of Foreign Language Curricula

Orlando R. Kelm

INTRODUCTION

It is interesting to note that an international conference on technology policy and innovation would address the issue of language instruction, and more specifically the role of technology in the area of foreign language instruction. One need only observe the structure of this conference to see that language is part of the overall issue of development. This international conference, for example, is being held in Curitiba, Brazil, and the official language of the conference, as stated in the printed program, is English. As such, a decision regarding language was central to the very initial planning of the conference. As part of this planning process, conference organizers had to determine what language would be used and what translation and interpretation services would be provided to help presenters and participants get the most out of the conference. So, not even the conference itself could be held without addressing the issue of language, let alone promoting the objectives behind the conference, namely, the use of science and technology in fostering socioeconomic development. Clearly, as we analyze the barriers which affect the transfer and dissemination of knowledge, communication and language play central roles. The objective of this brief report is twofold: demonstrate some of the Portuguese language materials that we have created at the University of Texas at Austin, and show how technology has come into play in order to develop and implement the materials.

BUSINESS PORTUGUESE LANGUAGE MATERIALS AT THE UNIVERSITY OF TEXAS AT AUSTIN

This subheading is, by design, a little misleading because all of the materials that we demonstrate are available not only at the University of Texas, but also online to anyone and everyone. As such, we already see one of the changes that technology brings both in development and implementation of language materials, namely, access. All of the following materials are accessible through my course homepage or through the foreign language homepage of the Center for International Business Education and Research (CIBER) at the University of Texas at Austin (Orlando R. Kelm Course Homepage: http://www.sp.utexas.edu./ork/-kelm.html; CIBER Spanish and Portuguese Language Homepage: http://bevo2.bus.utexas.edu/ork/ciberspanish.htm).

CIBER has been supportive of the development and implementation of pedagogical materials for the teaching of business language. Through the efforts of CIBER, a series of courses has been developed and implemented for both graduate and undergraduate students of business, in response to the needs of these business students. The ability to provide increased access to language materials is not a trivial issue. Especially in the case of the less commonly taught languages, such as Portuguese, increased access provides instructors with a means of sharing pedagogical materials which do not go through traditional publishing channels. For example, based on correspondence that we have received from interested parties, the materials presented in this chapter, and others that we have developed at CIBER, have been utilized by language instructors throughout the United States, Canada, Mexico, Brazil, Iceland, Sweden, Germany, Portugal, France, Spain and Japan. Some are individual learners or private tutors, while others are instructors of structured courses. But in each case, increased access to materials has been one of the benefits of newer technologies. What follows is a description of two different sets of materials that we use to teach business Portuguese. After a brief sampling of these materials, I will touch on some of the ways in which technology has affected their creation and implementation.

FALANDO DE NEGÓCIOS: PORTUGUÊS COMERCIAL (SPEAKING OF BUSINESS: COMMERCIAL PORTUGUESE)

The online materials for this course contain brief videotaped interviews from 27 Brazilians who discuss a wide variety of topics related to business and professional activities. The objective of the videos is to provide intermediate- and advanced-level students of Portuguese with a sample of technical vocabulary within the context of business topics. In these interviews, the Brazilians discuss the following topics: Accounting, Advertising, Banking, Economy, Finance, Human Resources, Insurance, Investment, Organizational Structure, Sales, Stock Market, Trade and Unions. In all there are nearly 90 video clips.

The premise behind these materials is threefold: First, offer students of Portuguese with context-rich information from the point of view of real business professionals. The comments are not edited, changed or elaborated; they simply contain ideas and opinions from real people who say what they really think about the business topics. It bears mentioning that these interviews focus on oral speech. As such, it is sometimes easier to understand the oral interviews than it is to read the written transcript. Second, the interviews provide a context from which students can see the use of specific related vocabulary. The objective was not to provide a complete list of business vocabulary, but to be able to show students how technical words are used in context. Third, although these materials were not created specifically to study grammar of the Portuguese language, the analysis of the transcripts does provide students with a way of seeing Brazilian speech patterns, both in grammar as well as in syntax and pronunciation. The majority of the interviews are from people from São Paulo, but there are also some from Minas Gerais, Rio de Janeiro, Bahia, Ceará and Rio Grande do Norte. Consequently the materials may also be used to supplement the study of grammar, especially for those who would like to study the patterns of oral speech.

Each video clip is accompanied by a Portuguese transcript of the interview, an English translation of the transcript, a vocabulary list of related business terms, a brief multiple-choice exercise to check comprehension and additional Internet links to URLs of related topics. In all there are 43 different interview categories and 2 different interviewees to discuss each category. As mentioned, there are 13 basic topics.

Chapter one deals with accounting terms and includes interviews about income statements, a description of debits and credits, assets and liabilities, and a description of depreciation. Chapter two enters into advertising and discusses the different forms of advertising, the difference between advertising and marketing, comments about meeting the needs of a consumer and ends with interviews describing trademarks and slogans. Chapter three provides vocabulary related to banking and includes interviews about checks, deposits, withdrawals, interest rates and ATM machines. Chapter four contains interviews about organizational structures of a company and explains corporations, the horizontal and vertical descriptions of companies and the characteristics of effective managers. Chapter five covers vocabulary related to the economy, Brazilian GNP, macroeconomy versus microeconomy, privatization and inflation. Chapter six interviews cover finance and include topics about different types of financing, inflation and financial planning. Human Resources is the focus of chapter seven and it has interviews about benefits, payments and the welfare of employees. Chapter eight has interviews about insurance, including comments on coverage, claims and liability. Risk, foreign exchange markets, treasury bills and treasury notes are the topics of chapter nine. Chapter ten offers ideas related to sales and includes interviews on goods and services, wholesale and retail, and some comments on price. Chapter eleven discusses the stock market, including comments on stocks and bonds, commodities and futures, and financial indicators. Chapter twelve deals with international trade, specifically dealing with letters of credit and export/import documents. Finally chapter thirteen has interviews about unions, the role of unions, the difference between mediation and arbitration and a few comments about lobbyists.

In order to give the reader an idea of what the interviews are like, what follows is the transcript and translation of one of the video clips in which César Pinotti of São Paulo discusses an income statement:

Descreva a demonstração de resultados. Quais são suas principais componentes e para que ela serve?

Describe an income statement. What are the components of an income statement, what is it used for?

Eu vou falar sobre demonstração de resultados. A demonstração de resultados é um relatório que a empresa ou banco apresenta com os resultados da companhia durante o período apresentado. Então, da mesma forma que o balanço apresenta uma foto parada da companhia naquele momento, a demonstração apresenta a performance da empresa ou banco ao longo do ano. Essa demonstração de resultados serve para mostrar ao investidor também, que tem interesse, se a empresa apresentou lucro ou prejuízo durante esse período. As principais componentes de demonstração são a receita de vendas, que mostra quanto a empresa recebeu com o total de suas vendas durante esse período que em geral é de um ano, mas também pode ser apresentado de três em três meses. Também os custos; a receita de vendas menos o custo apresenta a receita bruta. Tirando-se também a depreciação que também apresenta as demais despesas, que são as despesas operacionais, despesas administrativas, despesas comerciais, você chega finalmente ao lucro bruto. E também tirando o lucro com receitas financeiras e despesas financeiras, você vai chegar ao importante número que é o lucro líquido ou prejuízo líquido da empresa ou banco do respetivo demonstração de resultados.

I am going to talk about an income statement. An income statement is a report that a company or bank prepares about a company's income during a specific period of time. So in the same way that a balance sheet presents a photograph of a company at a given moment, the income statement shows a company's or bank's performance during the year. This income statement also serves to show interested investors whether the company had earnings or losses during that period. The main components of an income statement are the revenues from sales which show how much the company received from all of its sales during the period, which is generally one year, but it can also be quarterly. There are also expenses; the revenues minus the expenses will give you the gross receipts. Taking out depreciation and also the rest of the expenses, such as expenses like operational expenses, administrative expenses and commercial expenses, you finally arrive at your gross profit. And by taking your profit with your financial earnings and financial expenses, you will arrive at the important number which is your net profits or net losses of the company or bank doing the income statement.

Besides the video clips, the transcripts and translations, the Portuguese/ English vocabulary list of accounting terms and the multiple-choice exercise, these materials also provide learners with rich context in which to study vocabulary and grammar. For example, in the brief clip written above, students see technical words such as income statement, report, income, balance sheet, investors, earnings, losses, revenues from sales, and so on. Furthermore, there is a whole series of interesting characteristics of oral speech patterns for students to analyze. Additionally the students are also exposed to different rates of speech, regional accents and individual styles of speaking. All of these characteristics are enhanced by possibilities introduced through newer technologies, as explained below.

CULTURAL INTERVIEWS WITH BRAZILIAN EXECUTIVES

A second set of materials that is available online is *Cultural Interviews with Brazilian Executives,* which is a compilation of over 170 brief video clips in which Brazilians discuss various cultural issues that are of interest to North American executives who work or will work with Brazilians. (We also have a Latin American version of the cultural interviews in Spanish which can be accessed from the same homepages.) In these interviews over 60 native Portuguese-speaking professionals offer their opinions on questions related to these cultural differences. The interviews are subdivided into five sections: Negotiation, Language, Courtesy and Social Situations, Time and Scheduling, and A Typical Day of Work. Similar to *Falando de Negócios,* the premise of the cultural interviews is threefold: First, provide North American professionals with practical cultural information from real people. The opinions represent the opinions of individuals and not just merely "textbook" answers. Many times the answers are extremely different and may even contradict one another. However, they are designed to be a catalyst for discussion, not to provide a definitive answer about some stereotype regarding business in Brazil. Second, the interviews provide context and vocabulary in specialized areas within a professional setting. They are not designed to be a dictionary of technical vocabulary, but they do present a diverse array of vocabulary within the context of each individual's job or experience. Third, these interviews provide nonnative speakers of Portuguese with multiple examples of natural speech. These interviews contain samples of speech of people from Bahia to Rio Grande do Sul including old and young, male and female, professional and student. Consequently they can be used to supplement instruction in cultural issues or they can be used exclusively for language purposes, especially for advanced-level students.

Part 1 of *Cultural Interviews with Brazilian Executives* deals with negotiation styles. Informants answer 15 questions related to direct and indirect styles of negotiation, how the workplace is set up, what to talk about in meetings, ability versus connections, race, color and gender issues, working in groups, the role of lawyers, putting things in writing and how to avoid offending others. Part 2 addresses language issues such as how to decide which language to speak, when to use an interpreter, and whether or not it is acceptable to use Spanish in Brazil. Part 3 provides suggestions related to social situations such as how to greet people, receiving gifts, eating food, deciding which clothes to wear, addressing people correctly and understanding what *machismo* is. How Brazilians keep track of their schedules and how meals are incorporated in work-related activities is the major focus in Part 4. The description of a typical day of work in Part 5 includes samples from managers, supervisors, small business owners, bankers, sales representatives and secretaries.

One of the topics addressed in the negotiation section is that of trying not to offend others. As an example of these interviews, what follows is a transcript of one executive's comments about this topic (Rita de Cássia M. Lima from São Paulo):

A primeira coisa que a gente deve fazer para não ofender a um colega brasileiro, por exemplo empregado, é jamais chamar atenção dele na frente de alguém. E se a gente tiver de chamar atenção dele, que a gente o faça num canto separado e que a gente sempre tente ao final da chamada de atenção, tente levantar o astral dele. Fala de uma coisa boa que ele fez, por exemplo acreditar, que "Olha, OK você errou nisso mas a gente tem uma condição de você tentar fazer melhor da próxima vez. Eu vou sugerir para você que você faça isso ou faça aquilo." Ou seja, procure melhorar o astral da pessoa. Há uma dica também para quem é funcionário e quer falar com chefe, é jamais responder num tom agressivo ao chefe. E não dá uma resposta direta muito seca. Tentar sempre explicar e expor o problema de uma maneira colaborativa. As pessoas em geral no Brasil, que a gente quer, são pessoas que trabalham colaborando, que não pensam na sua individualidade, pensando somente nelas, sem pensar na empresa como um todo. Isso é muito importante para a gente.

The first thing that you should do to not offend a hired Brazilian colleague, for example, is to never chew him out in front of others. If we have to talk to him, we do it in a separate corner and after talking to him we always try to raise his spirit. We talk about something good that he has done—for example, "Look, OK, you may have made a mistake this time but we can do better next time. I would suggest that you do this or that..." That is, try to make the person feel good. There is also a suggestion for a worker who wants to talk to his boss and it is to never answer the boss in an aggressive tone. And don't give some direct dry response. Try to explain and describe the problem in a cooperative manner. People in Brazil in general want people who work together, who don't think just in their individuality, thinking only of themselves, without thinking about the company as a whole. This is very important for us.

The interviews are not designed to synthesize all of the cultural material nor to give the right answers, but they do open up discussion of the various topics and help to make students aware of the issues. As to our methodology in class, first we study specific cultural issues as presented in textbooks and in theoretical descriptions. We then compare the theoretical descriptions with the comments of the videotaped interviews. The students are then asked to go and talk to Brazilians that they know (or who are on campus or in town) to compare the theoretical issues, the interview opinions and their Brazilian contacts' opinions. Finally students come back to class and compare the issue from all three angles, adding their own opinions to the mix. This methodology has helped to present the issues, while avoiding the stereotype answers that could easily result from a discussion on cultural. Again, all of these materials are online and accessible to anyone. Additionally, *Cultural Interviews with Brazilian Executives* and *Falando de Negócios: Português Comercial* are also available in CD-ROM for those users who might not have fast access to the Internet. Having introduced these materials, let us focus our attention on the contribution of newer technologies, as related to the creation and implementation of these materials.

TECHNOLOGY AND LANGUAGE INSTRUCTION

One way of looking at the contribution of technology in language instruction is see how technology is used in society in general. Since the acquisition of a foreign language centers on one's ability to communicate with others, as we observe how society uses technology to communicate, we see ways to imitate similar activities within foreign language contexts. Consequently, as we observe how people use computers and technology in their everyday activities, instructors of foreign language are in a position to modify those activities for a language setting. The end result of this is that as we simulate our foreign language learning activities around practical everyday communication patterns, we add a sense of realism and practicality to our teaching. What follows are a few specific examples of how technology has influenced the way these materials were created and implemented.

Access

Traditionally it has not been easy to expose students to natural examples of speech, whether written, visual or oral. This is especially true for language learners who are not physically located in a country where the target language is spoken. Through emerging technologies, however, online materials provide us with the opportunity to expose students to language in ways that would have been impossible just a few years ago. Additionally, not only do students have access from distant locations, but they also have access without limitations on time. By providing access from distant sites and without limitations on when that access is made, we are altering the traditional boundaries of teacher roles and classroom settings. The whole notion of access provides us with new examples of how people get information. Those who use online resources to get information are not limited by time and location. Thanks to these new parameters, I believe that we will modify our concept of classroom dynamics and teaching methodologies, taking advantage of our ability to access information. Part of this modification will include how we utilize authentic materials for teaching purposes. For years language instructors have debated how to best use or modify authentic materials. The business Portuguese-language materials demonstrated here illustrate how materials, rich in authentic speech, have been modified for pedagogical purposes and how they can then be used to send students on to new authentic locations or on to other natural experiences for language use and practice.

Nonlinearity

Implementation from emerging technology also helps to eliminate the limitations that linearity brings to education processes. As opposed to traditional models, notice that when using these materials there is no fixed beginning point or ending point. Students can review the materials in any order they choose. They may also choose which topics to focus on and which ones to ignore. This sounds simple enough, but traditionally students have been forced to take a more linear approach to most every learning process. By allowing students the option of studying any part in any order, we add flexibility to the learning process. A student who wants to watch a video clip again, or go back to the vocabulary list, or check the transcription, or review the translation or study Trade before Advertising is able to do so, in any order or fashion desired.

Natural Speech

One of the major hurdles for language learners is to make the transition from a theoretical understanding to a practical usage. Students generally struggle and are amazing to see how fast native speakers talk. The presentation of these materials, via technology, exposes the students to a wide variety of dialects, speeds, speaking styles and speech patterns. Another advantage of natural speech samples is that students become more receptive to these variations. Without continual exposure to natural language, students are reticent to accept language variation. For example, if prescriptive grammar says that a certain form is correct and another is grammatically wrong, student have a hard time understanding what native speakers actually say. Add to this the specialized vocabulary, idiomatic expressions, suprasegmental elements of speech, emotion and context—all are provided through the video clips and their corresponding transcripts and delivered to the student through some of the benefits of technology.

Sound and Video

The access, nonlinearity and natural style of these materials are possible because of our ability to expose students to sound and video. Clearly sound and video present greater challenges in terms of computer speed and memory load than mere text-based materials. At the same time, however, foreign language is one of the areas that can most benefit from increased exposure to sound and video. In the last few years it has become increasingly easier to record and digitize video clips, compress and store files and download or stream broadband information. Whether it be cassettes, videos, movies, songs or whatever, sound and video have to be stored and presented via some means. To date our challenge has always been how to store these data. Given the new advances, however, our challenge now is to know how to best present sound and video to the student. Again the example of society at large is our best teacher. As we look to see how society in general uses technology to transmit sound and video, we will get better ideas on how to use it in pedagogical situations.

Individual Style

By removing the linearity of presentation, we also open up the possibility for flexibility and individual learning style. For example, when preparing these materials, some learners like to have the transcript of the interviews, so that when they listen to the videos, they can also follow along with the transcript to understand them better. Others prefer to not have the written transcript and feel that it inhibits their learning and understanding of the interviews. Some learners really crave to know the translation of the comments and feel that they understand the interviews better if they already know what the interviewees are going to say. Others feel that it is detrimental to look at translations and to be in translation mode. The point is that the presentation of teaching materials through newer technologies allows for a wider range of flexibility in learner style. For example, I personally like the fact that these materials combine the naturalness of oral speech with a capacity to freeze time and analyze in great detail the speech patterns. Students who enjoy this sort of analytical review of the materials are capable of focusing in on speech patterns.

Student-Centered Learning

In general, current education tradition puts the impetus of learning on the teacher's shoulders. Our learning strategy has generally been to have the teacher tell the students what they should learn, what they should read, what part they should study and how they should be tested to prove competence. The responsibility traditionally falls on the teacher and as a result we have a tradition of passive learners. One of the byproducts of presenting materials with new technologies is that having taken away the linearity and increasing the options for access, students can also take a more active role in deciding what they want to learn and how they want to do it. The challenge now is for teachers to have enough confidence in learners that we will be willing to change our role from that of presenter to that of supporter and advisor. An educator's expertise is still an important part of the learning process, but not necessarily limited to that of presenter. Technology does not automatically improve teaching, but if nothing else, the opportunity to allow students to take on the responsibility for learning is an extremely positive step.

CONCLUSIONS

The object of this chapter was to demonstrate our materials related to Business Portuguese and to show how changes in technology have provided us with new ways of learning. We are happy to share our online materials with all who would like to use them in their own instruction. As to the use of technology, one word of caution: We need to be careful not to fall into the trap of limiting ourselves to a comparison of teaching with and without technology. I have observed that language instructors sometimes exert a lot of energy debating whether it is better to teach with or without computers. The danger of such argumentation is that we become tempted to make a traditional versus a computer version of identical curricula. A better approach, as mentioned in this brief chapter, is to look at how technology is used within society in general, and then see how these situations can be modified for language instruction. That way we are not making the traditional versus the computer version of the same materials, but we are looking at the most effective way to reach a specific objective. For example, traditionally language instructors have used poetry, short stories and other literary texts to reinforce and teach foreign language. This is a good example of how society has used language in a certain way, and how language instructors have imitated similar activities for language learning. Similarly, in a new context, if we observe how people chat online we will learn a lot about communication and learning. These sorts of communication patterns can also be modified for language teaching purposes. Consequently our challenge is not to use or not use technology, but to look for innovative ways to communicate and expose our students to language and language practice. In this context, *Falando de Negócios: Português Comercial* and *Cultural Interviews with Brazilian Executives* serve as examples to represent this sort of shift.

22

Portuguese Language without Frontiers

Regina L. P. Dell'Isola

PORTUGUESE LANGUAGE IN THE WORLD

The Portuguese language is among the 12 most spoken languages in the world when compared to the actual number of idioms in existence (approximately 3,000). Portuguese enjoys an important position, although it is not a language of significant international expression as is the case of English and Spanish. Despite the low rate of internationalization of the Portuguese language, Portuguese-speaking countries, especially Brazil and Portugal, have made efforts to promote the idiom beyond their frontiers. Among the initiatives to spread the language it is worth mentioning the Instituto Camões, in Portugal, and the Sociedade International de Português Língua Estrangeira (SIPLE, The International Society of Portuguese as a Foreign Language), in Brazil. Committed to internationally promote the Portuguese language, the Institute Camões has a network of 145 visiting professors (22 in Africa, 19 in North America, 7 in Asia, 96 in Europe, and 1 in the South Pacific). Besides this, the Institute invests in the promotion of Portuguese as a language of international communication and encourages the teaching of Portuguese as a foreign language (PFL) through activities of support to the Unidade de Avaliação do Ensino de Português Língua Estrangeira (The Unit for Evaluation of Portuguese as a Foreign Language), and to the institutions that teach free courses, and overseas Portuguese Summer courses.

Agencies such as SIPLE are also very important for the international projection of the idiom. SIPLE, created in 1992, has as its goal to contribute to international acknowledgement, diffusion and use of Portuguese; to encourage research and the teaching of Portuguese as a Foreign Language and as a Second Language (PSL); to promote knowledge and interinstitutional exchange of scientific, pedagogic and cultural production; to support the creation and updating of undergraduate and graduate programs, courses and disciplines of PFL and PSL; and to promote cooperation between graduate programs with respect to the exchange of faculty, students and others.

In Brazil the number of institutions that develop activities in the area of PFL is growing. There are now 21 Brazilian faculties working in this area: Fluminense Federal University (UFF), the Federal University of Rio de Janeiro (UFRJ) and the Catholic University (PUC-RJ) in the State of Rio de Janeiro; the Federal University of Minas Gerais (UFMG) and the Federal University of Juiz de Fora in the State of Minas Gerais; the University of Campinas (UNICAMP), the University of São Paulo (USP), the Catholic University of São Paulo (PUC-SP) and the Methodist University of the State of São Paulo (UMESP) in the State of São Paulo; the Federal University of Pernambuco (UFPE) in the State of Pernambuco; the Federal University of Rio Grande do Sul and the University of Caxias do Sul (UCS) in the State of Rio Grande do Sul; the Federal University of Santa Catarina (UFSC) in the State of Santa Catarina; the Federal University of Paraná (UFPR) in the State of Paraná; the University of Brasilia (UnB) in the Federal District; the Federal University of Bahia (UFBA) in the State of Bahia; the Federal University of Rio Grande do Norte (UFRN) in the State of Rio Grande do Norte; and the Federal University of Paraíba (UFPB) in the State of Paraíba. Besides teaching Portuguese to foreigners some universities offer graduate programs (masters and doctoral) in the area of linguistics applied to the teaching of PFL, and others are beginning to seek advisories to develop specific programs for the formation of specialists in the teaching of PFL. There are also free language courses in the country and among other options the teaching of Portuguese for foreigners.

Outside Brazil, Centers of Brazilian Studies (CEBs), Brazilian Cultural Institutes and visiting professors are responsible for promoting and valuing the Portuguese language and the Brazilian culture in different parts of America, Europe and Africa. In the countries forming Mercosur, since the beginning of the 1990s, the number of students learning Portuguese has been increasing in the Brazilian Study Centers. This is justified since the countries forming the southern cone in South America have adopted the international policy to make Portuguese and Spanish the language of diplomatic relationships. So, these languages begin to be integrated in the school syllabuses.

The growing number of people interested in learning the Portuguese language, especially Brazilian Portuguese, is evident. This fact can be easily verified by the growing number of candidates for the exam of proficiency in Portuguese language (CELPE-BRAS): the number grew from 141 registered candidates in 1998 to 1,161 in 2000. Out of this total, 210 did their exams in Brazil and 951 abroad. The candidates who were interviewed during the application of the exams demonstrated an interest in establishing specific relations with Brazil—whether professional, business, research, study or leisure-type relationships.

In the United States, Canada and Mexico, although the academic prerequisites differ according to the course or area of university studies in the cases in which a foreign language is a compulsory discipline, the student most frequently has the right to choose the language he or her wishes to study. In these countries the number of students who have sought courses of Portuguese—aiming at establishing some kind of contact with Brazil—is very large. At the University of Texas, Austin, where some 23 languages are taught, the number of students registered in the course of Portuguese has grown over the last years, and the percentage of students who have voluntarily registered is higher than that of those who are compulsory. According to Prof. Orlando Kelm, the classes of Portuguese have students who really want to learn this idiom; they are well motivated and this culminates in a dynamic learning and a very satisfactory performance at the end of the course.

INTERCULTURALISM AND THE TEACHING OF THE PORTUGUESE LANGUAGE

Generally, there is a favorable atmosphere for the teaching of Portuguese. At the moment the teaching of idioms, especially in the universities, is characterized by interculturalism and by the multiplicity of foreign language courses offered for the students to choose from. The intercultural vision generates a new orientation in the foreign language teaching scene. The teacher begins to play a double role: that of being a teacher and an intercultural agent. He or she will show the adequacy of social, cultural, ethnic and scientific values that permeate our cultures. This focus, associated with the high speed in which information travels around the world, particularly via the electronic medium, could be considered stimuli for the Federal University of Minas Gerais and the University of Texas at Austin to sign a partnership to cooperate in the form of an exchange whereby students and teachers from both institutions could experience the American or the Brazilian culture, respectively.

Right now, the teaching of a foreign language is founded on the acquisition and use of communicative functions that allow the student in any given level to have a satisfactory performance in the new language. The communicative approaches rest on the same guiding principle—that is, the language is an instrument of communication and, above all, of social interaction; hence, in the communicative methods, the focus is on meaning, on the acts of language and in the exchange of ideas in the target language. The student is seen as an agent of language and this eliminates the passive character of learning. Language teachers are responsible for creating the conditions for the students to interact in the language they are studying. Therefore, a Portuguese course in the context of immersion can offer students the opportunity to gain a more thorough knowledge of the language; that is, opportunities to perceive linguistic, semantic and discursive dimensions of the idiom.

Our summer course is centered on interactive activities that encourage communicative procedures such as greetings, making invitations, saying good-bye, accepting/refusing, demonstrating gratitude or disapproval, among other social skills as inferential procedures from a social-cultural basis. The use of authentic texts extracted from Brazilian books, magazines and newspapers is only the point of departure.

In consonance with the theoretical presuppositions of the communicative approaches, the texts selected are characterized by authenticity, for they were written for native speakers and respond to specific communicative functions. But this is not enough. Besides a good selection of texts, the advantage of the themes they deal with, their actuality and the fact that they are representative of the Brazilian scene, it is necessary to make their use very attractive to make the course really dynamic and efficient and to help the student to actually know a little more of the cultural diversity of our country.

Generally, it is important to promote opportunities for the student to understand discursive situations in which the language is not the only vehicle of communication but also everything else around it. For example, we selected a city publicity item, used by the City of Belo Horizonte, in which drivers present various excuses to the traffic officer to avoid getting a fine (these excuses are very common among Brazilians). So the student is invited to understand who is speaking, how is he or she speaking, what is the fault, the social status of the speaker, sex, approximate age and such. Following this item we begin a discussion about traffic in Brazil. There arises the possibility of establishing cultural comparisons and the debate can end in the formation and exchange of opinions. The richness of this drill is in not focusing on the language. The focus is the theme to be discussed, the language is the instrument of discussion. Thus, students express themselves in the target language, hence the recommendation for the teacher to interfere as little as possible.

The emphasis is in the production of activities that are meaningful and which encourage the student to interact using the target language, based on activities that are considered relevant. In communicative methods, it is expected that the activities proposed are of real interest or represent a need of the students in order to train them in the use of the new language in concrete communicative and interactive situations along with other users of the same language.

Index

Advanced scientific training, 83
Aerotropolis, 104
Agglomeration, 169
Agglomeration externalities, 137, 169
Aggregate increasing returns, 165
Airport-linked e-commerce, 101
Associational economy, 64
Automotive industry, 307
Average labour productivity, 165
Average national deficits, 76

Benchmarking, 26, 40, 156
Brokering, 156
Business infrastructure, 347
Business-to-business, 100
Buyer-pulled chain, 152

Capacity, 387
Capability building, 157
Capital markets, 23
Change agency, 156
Cluster support, 277
Clusters, 145
Codification, 34
Codified knowledge, 28
Cohesion gap, 185
Collective action, 2
Collective efficiency, 64, 346
Collective learning, 11, 23

Collective tacit knowledge, 31
Commodities, 66
Common markets, 89
Communication democratization, 424
Competence, 4, 7
Competence building, 375
Competence to learn, 32
Competition, 9, 127, 167
Competition policies, 125
Competitive economies, 21
Computer-based technologies, 405
Connectivity, 387
Consulting services, 276
Cooperation networks, 342
Corporate governance, 112
County productivity function, 173
Creative destruction, 4
Critical institutions, 92
Culture, 405, 423

Decentralization, 90
Decentralized responsibility, 36
Decision-making, 23
Degree of specialization, 165
Democratization of the Internet, 436
Density, 165
Deregulation, 88
Developed countries, 261
Diffusion of ideas, 168

Diffusion of information, 30
Diffusion of information technology, 11
Digital cities, 375
Digital divide, 406, 423, 435
Digital networks, 375
Dispersion of income, 8
Distributed knowledge bases, 2, 10, 393
Diversified economic base, 347
Division of labour, 27
Doctoral degrees in science and
    engineering, 83
Dynamics of innovation, 245

E-commerce, 100
Economic specialization, 109
Economics of agglomeration, 64, 166
Education and training system, 30
Education systems, 37, 144
Effective learning, 148
Elasticity, 171
Electronic language, 410
Employment density, 137
Endogenous growth theories, 9
Enterprise management, 116
Entrepreneurial scientists, 116
Entrepreneurial spirit, 227
E-tailers, 100
Ethical values, 33
EU regional policy, 138, 185
EU structural funds, 187
Experience-sharing, 156
Experiential learning, 147, 148
Expert consulting, 156

Financial system, 144
Fiscal autonomy, 90
Flexible firms, 36
Flexible organization, 317
Flexible production, 316
Flexible production systems, 99
Food industry, 363
Foreign language, 443

Gateway airports, 101
Geographical agglomeration, 165
Geographical concentration, 168
Globalization, 11, 62, 307
Governance structures, 22
Government expenditures, 75
Governmental R&D expenditures, 81
Growth theories, 3

High-speed transportation, 99
Human capital, 168, 240
Human competence, 45
Human skills, 45

Incentive mechanism, 22
Incentives, 375, 389
Inclusive development, 2
Inclusive learning, 4, 8
Increasing returns, 137, 165
Individual learning, 147
Industrial clusters, 138, 168
Industrial restructuring, 138, 307
Industrial systems, 266
Informal institutions, 110
Information and communication
    technologies (ICTs), 227
Information distribution, 435
Information technology, 60, 99
Infrastructures, 10, 74, 375, 389
Innovation agents, 154, 157
Innovation clusters, 261, 342
Innovation policy, 25
Innovation strategy, 196
Innovative capabilities, 155
Innovative clusters, 138
Institutional approaches, 138
Institutional frameworks, 13, 375
Institutional mechanisms, 5
Institutions, 2
Intangible assets, 3
Integrated competence-building
    strategy, 37
Integrated learning processes, 10
Intellectual productivity, 83
Intellectual property rights, 34
Interculturalism, 455
Intergovernmental relations, 90
Intermediaries, 154
International trade agreements, 89
Intraorganisational processes, 148

Just-in-time systems, 307

Knowledge accumulation, 3, 4, 6
Knowledge base of the firm, 34
Knowledge diffusion, 31
Knowledge economy, 58
Knowledge Era, 61
Knowledge integrated communities, 375
Knowledge management, 26, 37

Knowledge networks, 4, 102, 137
Knowledge society, 21
Knowledge spillovers, 165, 168
Knowledge-based economies, 21, 314
Knowledge-based firms, 275
Knowledge-based view of the territory, 138, 375
Knowledge-integrated communities, 395
Knowledge-intensive industries, 341

Labour market, 144
Labour market institutions, 39
Labour productivity, 137, 165
Language, 405, 423
Language engineering, 409
Language instruction, 443
Language server, 414
Learning and innovation poles, 350
Learning cycle, 148
Learning economy, 3, 25, 34
Learning networks, 146, 149, 240
Learning organizations, 33, 36, 147, 240
Learning partnership, 150
Learning skills, 26
Learning societies, 2, 8, 23, 137
Learning trajectories, 2
Learning-by-doing, 23, 27
Learning-by-interaction, 64
Learning-by-searching, 27
Learning-by-using, 27
Linear models of innovation, 8
Local productivity, 165, 170
Localised agglomeration externalities, 165
Logistic, 102

Management skills, 143
Market failures, 9
Market mechanism, 9
Market regulations, 12
Market-driven experiences, 23
Markets, 9, 88, 278
Mobilization of information and communication technologies, 138

National systems of innovation, 43, 62, 143, 202, 267,
Natural speech, 450
Neoclassical economics, 29
Neoliberal philosophies, 73
Network analysis, 360
Network building, 245

Network of suppliers, 100
Network readiness index, 380
Network society, 1, 375
Networking, 64, 149
Networks, 145
New growth theory, 27, 29
New technology-based firms (NTBFs), 141
New trade theory, 27

Organization of production, 138, 307
Organization theory, 31
Overregulation, 121

Patents, 34
Physical infrastructure, 347
Polarisation, 35
Policy makers, 121
Policy system, 144
Political systems, 74
Portuguese language, 453
Privatization, 123
Problem-solving approaches, 31
Production of knowledge, 28
Production services, 313
Production systems, 307
Productivity, 27, 165, 169
Public decision-making, 91
Public policy, 203
Public sector, 87

Quality control strategies, 138, 307

Reflection, 148
Regional density, 165
Regional economic policy, 227
Regional inequality, 357
Regional innovation policy, 201
Regional innovation strategy, 201
Regional systems of innovation, 14, 205
Regional technology policy, 185
Regulation, 88, 122
Regulatory agencies, 121, 127
Regulatory bodies, 124
Regulatory environments, 11
Regulatory mechanisms, 126
Regulatory reform, 121, 123
Regulatory structures, 88
Research networks, 138, 247
Routines, 26, 31

Scale economies, 109
Science and technology systems, 80
Shared experimentation, 147
Single European Act, 186
Size of government, 75
Skill-biased technological change
   hypothesis, 7
Skills, 5, 26, 343
Small and medium-sized enterprises
   (SMEs), 137, 141, 245
Small and medium-sized software
   firms, 342
SME Innovation, 143
Social capital, 2, 26, 35, 138
Social exclusion, 35
Software, 5, 59, 342
Spatial density, 165, 167
Specialized inputs, 276
State reform, 74
Stock of knowledge, 27
Structural connections, 158
Student-centered learning, 452
Subcontracting, 318
Supplier-pulled learning, 152
Supplier-pushed programmes, 153
Supply chain learning, 150
Supply chains, 142, 153, 314
Supply-chain transactions, 100
Systems of governance, 91

Systems of innovation, 64, 375
Systems of innovation and competence
   building, 1

Tacit knowledge, 26, 28, 34, 46, 147
Tacit skills, 25, 33
Target investors, 112
Team work, 36
Technological change, 1, 266
Technology gap, 185
Technology infrastructure, 144
Technology policy, 23
Technostructure, 10
Telecommunication regulators, 126
Telecommunications, 121, 307
Total factor productivity, 170
Total quality systems, 317
Trade unions, 38
Transfer of skills, 160
Trust, 35

Unemployment protection, 13
Universal network language, 409
Universal words, 411
Urban agglomeration economies, 169

Virtual networks, 342

Wetware, 5
Wisdom, 28

About the Contributors

Rui Miguel L. N. Baptista is Associate Professor of Strategy and Organizations at the Instituto Superior Técnico, Technical University of Lisbon, and Senior Research Fellow of the Entrepreneurship, Growth and Public Policy Group at the Max Planck Institute for Research into Economic Systems, Jena.
Contact: rui.baptista@ist.utl.pt

John Robert Bessant is Professor of Innovation Management at Cranfield University, United Kingdom.
Contact: john.bessant@cranfield.ac.uk

Isabel Bortagaray is a PhD Student at the School of Public Policy, Georgia Institute of Technology.
Contact: isabel.bortagaray@pubpolicy.gatech.edu

José Eduardo Cassiolato teaches in the Institute of Economics at the Federal University of Rio de Janeiro and coordinates RedeSist—a research network on local systems of innovation in Brazil.
Contact: cassio@ie.ufrj.br

Pedro Conceição is Assistant Professor at the Instituto Superior Técnico, Technical University of Lisbon, and a researcher at the Center for Innovation, Technology and Policy Research at IST. He is also a Deputy Director and Senior Policy Analyst at the Office of Development Studies, United Nations Development Programme (UNDP), New York.
Contact: pedroc@dem.ist.utl.pt

Fabienne Corvers works for the European Commission—DG Research, as Scientific Officer where she is involved in the development, collection and analysis of Science and Technology Indicators for the European Research Area.
Contact: fabienne.corvers@cec.eu.int

Regina L. P. Dell'Isola is Professor and Director of the Program for the Teaching of Portuguese as a Second Language, Federal University of Minas Gerais, Brazil.
Contact: isolare@horizontes.net

Marina van Geenhuizen is Associate Professor at the Faculty of Technology Policy and Management, Delft University of Technology, the Netherlands.
Contact: m.s.vangeenhuizen@tbm.tudelft.nl

Lawrence S. Graham is a former Associate Vice President for International Programs and Professor of Government at the University of Texas, Austin.
Contact: lsgraham@uts.cc.utexas.edu

Michael Guth is working as Senior Consultant for ZENIT GmbH in Germany. Michael's areas of specialisation have been regional innovation strategies, management of RTDI projects and innovative employment actions.
Contact: mg@zenit.de

Manuel V. Heitor is Full Professor at the Instituto Superior Técnico in Lisbon, and is the director of the Center for Innovation, Technology and Policy Research, IN+. He is a Senior Research Fellow of the IC² Institute, The University of Texas at Austin.
Contact: mheitor@ist.utl.pt

Alejandro Ibarra-Yunez is Professor of Economics and Public Policy at the Graduate School of Business and Leadership (EGADE), Monterrey Institute of Technology (ITESM), Mexico.
Contact: aibarra@itesm.mx

John D. Kasarda is Kenan Distinguished Professor of Management and Director of the Kenan Institute of Private Enterprise at the Kenan-Flagler Business School, University of North Carolina at Chapel Hill.
Contact: john_kasarda@unc.edu

Orlando R. Kelm is Associate Professor of Hispanic Linguistics in the Department of Spanish and Portuguese at the University of Texas at Austin. He currently serves as the Associate Director of Business Language Programs for the Center for International Business Education and Research at UT, Austin.
Contact: orkelm@mail.utexas.edu

Junmo Kim is currently a Senior Researcher at the Korea Institute of Public Administration (KIPA) where he also serves as the managing director of the Innovation Center.
Contact: junmokim@unitel.co.kr

Helena Maria Martins Lastres is a Senior Researcher and Coordinator of RedeSist at the Federal University of Rio de Janeiro, Brazil.
Contact: hlastres@ie.ufrj.br

Bengt-Åke Lundvall is Professor of Economics at the Department for Business Studies, Aalborg University, Denmark.
Contact: bal@business.auc.dk

Mambillikalathil G. K. Menon is the former Minister for Science and Technology of India, and presently member of the Board of the UNDL Foundation.
Contact: mgkmenon@ren02.nic.in

José Luiz Moutinho is a researcher at the Center for Innovation, Technology and Policy Research at the Instituto Superior Técnico, Technical University of Lisbon. He is also founder and CEO of Glassbox Management Consulting.
Contact: jmoutinho@dem.ist.utl.pt

Décio Estevão do Nascimento is Professor of Management of Technological Information at the Federal Center of Technological Education of Paraná (CEFET-PR), Brazil.
Contact: decio@ppgte.cefetpr.br

Carlos Quandt is Professor at the Graduate School of Business Administration, Pontifícia Universidade Católica do Paraná (PUC-PR), Brazil.
Contact: quandt@rla01.pucpr.br

Cristina Casanueva Reguart is a Researcher in the Institute for the Development of Education (INIDE) at the Universidad Iberoamericana, Mexico.
Contact: cristina.casanueva@uia.mx

Tarcisio Della Senta is founder and former Director of the UNU/Institute of Advanced Studies, and currently President of the newly established Universal Networking Digital Language Foundation (UNDL), Geneva.
Contact: dellasenta@undl.org

Luiz Márcio Spinosa is Professor at the Graduate School of Business Administration, Pontifícia Universidade Católica do Paraná (PUC-PR), Brazil.
Contact: spinosa@rla01.pucpr.br

Scott Tiffin is Professor at the Business School of the Universidad Adolfo Ibañez, Chile, and was Director of Babson College's Institute for Latin American Business Studies, Boston.
Contact: scott.tiffin@uai.cl

Mercedes Lynn de Uriarte is a former editor and writer with the *Los Angeles Times*, and Associate Professor of Latin American Studies and Journalism at the School of Journalism, The University of Texas at Austin.
Contact: merlyn@mail.utexas.edu

Ramiro Wahrhaftig Ramiro Wahrhaftig is at the PUC-PR. He was the Secretary of Science, Technology, and Higher Education of the state of Paraná, Brazil during the period 1998-2022.
Contact: ramiro@rla01.pucpr.br

Robert H. Wilson is the Mike Hogg Professor of Urban Policy at the Lyndon Baines Johnson School of Public Affairs, the University of Texas at Austin, where he is also Director of the Urban Issues Program and of the Brazil Center.
Contact: rwilson@mail.utexas.edu

Leslie Young is Professor of Finance and Executive Director of the Asia-Pacific Institute of Business, The Chinese University of Hong Kong.
Contact: leslieyoung@cuhk.edu.hk